Redesigning the Global Seed Commons

There is much current controversy over whether the rights to seeds or plant genetic resources should be owned by the private sector or be common property. This book addresses the legal and policy aspects of the multilateral seed management regime. First, it studies in detail the International Treaty on Plant Genetic Resources for Food and Agriculture (the Treaty) to understand and identify its dysfunctions. Second, it proposes solutions – using recent developments of the 'theory of the commons' – to improve the collective seed management system of the Treaty, a necessary condition for its Member States to reach the overall food security and sustainable agriculture goals.

Redesigning the Global Seed Commons provides a significant contribution to the current political and academic debates on agrobiodiversity law and governance, and on food security and food sovereignty, by analysing key issues under the Treaty that affect the design and implementation of regulatory instruments managing seeds as a commons. It also examines the practical, legal, political and economic problems encountered in the attempt to implement these obligations in contemporary settings. In particular, it considers how to improve the Treaty implementation by proposing ways for Contracting Parties to better reach the Treaty's objectives taking a holistic view of the human-seed ecosystem. Following the tenth anniversary of the functioning the Treaty's multilateral system of access and benefit-sharing, which is currently under review by its Contracting Parties, this book is well-timed to examine recent developments in the field and guide the current review process to design a truly Global Seed Commons.

Christine Frison is an F.W.O. post-doctoral researcher at the Law Faculty of the University of Antwerp, Belgium. She is also part of LPTransition, a scientific research platform, both pluri-disciplinary and trans-sectorial, as well as holding an associate researcher position at the Centre for Philosophy of Law of the Université catholique de Louvain, Belgium. She has worked as a Legal Research Fellow with the Centre for International Sustainable Development Law (CISDL), based at the McGill University in Montréal, Canada, where she specialized in the field of international environmental law, biodiversity, biosafety and agrobiodiversity international law. Her latest book is *The Commons, Plant Breeding and Agricultural Research* (Routledge, 2018).

Earthscan Food and Agriculture

Environmental Justice and Soy Agribusiness
Robert Hafner

Farmers' Cooperatives and Sustainable Food Systems in Europe
Raquel Ajates Gonzalez

The New Peasantries
Rural Development in Times of Globalization (Second Edition)
Jan Douwe van der Ploeg

Sustainable Intensification of Agriculture
Greening the World's Food Economy
Jules N. Pretty and Zareen Pervez Bharucha

Contested Sustainability Discourses in the Agrifood System
Edited by Douglas H. Constance, Jason Konefal and Maki Hatanaka

The Financialization of Agri-Food Systems
Contested Transformations
Edited by Hilde Bjørkhaug, André Magnan and Geoffrey Lawrence

Redesigning the Global Seed Commons
Law and Policy for Agrobiodiversity and Food Security
Christine Frison

For further details please visit the series page on the Routledge website: www.routledge.com/books/series/ECEFA/

Redesigning the Global Seed Commons

Law and Policy for Agrobiodiversity and Food Security

Christine Frison

Routledge
Taylor & Francis Group
LONDON AND NEW YORK

earthscan
from Routledge

First published 2018
by Routledge
2 Park Square, Milton Park, Abingdon, Oxon OX14 4RN

and by Routledge
52 Vanderbilt Avenue, New York, NY 10017

First issued in paperback 2020

Routledge is an imprint of the Taylor & Francis Group, an informa business

British Library Cataloguing-in-Publication Data
A catalogue record for this book is available from the British Library

Library of Congress Cataloging-in-Publication Data
Names: Frison, Christine, author.
Title: Redesigning the global seed commons law and policy for agrobiodiversity and food security / Christine Frison.
Description: New York, NY : Routledge, 2018. | Includes bibliographical references and index.
Identifiers: LCCN 2018020907 | ISBN 9781138300644 (hardback) | ISBN 9780203733257 (ebook)
Subjects: LCSH: International Treaty on Plant Genetic Resources for Food and Agriculture (2001 November 3) | Germplasm resources, Plant—Law and legislation.
Classification: LCC K3876.A42001 F75 2018 | DDC 346.04/69534—dc23
LC record available at https://lccn.loc.gov/2018020907

ISBN 13: 978-0-367-58297-5 (pbk)
ISBN 13: 978-1-138-30064-4 (hbk)

Typeset in Bembo
by Apex CoVantage, LLC

To Moeke (Josée Thiry, my grand-mother) and to Simone Desonne, the two farmer women of my life, who taught me perseverance, humility before the living, and that the power to grow is inherent in the seed.

Contents

Figures and tables

Figures

Tables

About the author

Dr Christine Frison is an F.W.O. post-doctoral researcher at the Law Faculty of the University of Antwerp, Belgium, collaborating with Prof. Esther van Zimmeren and the research group 'Government and Law' of the Law Faculty. Christine is part of LPTransition, a scientific research platform, both pluri-disciplinary and trans-sectorial. Its specific purpose is to explore paths towards ecological and social transition of our societies from a transdisciplinary perspective. She is also an associate researcher at the Centre for Philosophy of Law of the Université catholique de Louvain, where she earned a grant (Chargée de recherche) from the Fund for Scientific Research – FNRS.

Christine's area of expertise and research revolves around public international law and international environmental law in general and international biodiversity and agricultural biodiversity law and governance in particular (food security and food systems, sustainable agriculture, fair and equitable sharing of the benefits deriving from the use of biodiversity, farmers' rights, agroecological transition, intellectual property rights over biodiversity, etc.). She has published extensively on these topics, her latest book (2018) being *The Commons, Plant Breeding and Agricultural Research. Challenges for Food Security and Agrobiodiversity* (co-edited with Prof. Fabien Girard), Routledge.

She holds a PhD jointly conducted at the Centre for Philosophy of Law of the Université catholique de Louvain and at the Centre for IT & IP Law at the KU Leuven, under the co-supervision of Prof. Tom Dedeurwaerdere, Prof. Olivier De Schutter and Prof. Geertrui Van Overwalle. She received a LL.M from the Université Libre de Bruxelles (Belgium) in Public International Law, after carrying out university legal studies in France and the UK.

She has been a consultant for various international organizations and governments, such as the United Nation Environment Programme (UNEP), francophone African Countries' governments, Bioversity International (Italy), the International Development Law Organization (IDLO, Italy) and the Belgian Federal Ministry of Environment. She has worked as a legal research fellow with the Centre for International Sustainable Development Law (CISDL), based at the McGill University in Montréal (Canada), where she specialized in the field of international environmental law, biodiversity, biosafety and agrobiodiversity international law.

Preface

From the unexperienced young consultant at my first Treaty meeting back in 2004, I've grown into a reflexive research-actor, no longer scared of pointing to what I consider dominant positions. A criticism I received at my PhD defence related to the 'passionate', the 'almost-activist' position I take in my work. I accept the criticism, but I believe that it is necessary to take a wider approach to science.

From the 'whole' to the 'part' and vice-versa, from mono-disciplinarity to transdisciplinarity

Too much, not enough. One is always assessed by reference to one's own reference scale, which varies according to the reader's origin, background, discipline, political standpoints, etc. I hope that in reading this work, the reader will be open, understanding and kind. Being a lawyer has been useful to provide a new understanding of the issues at stake. However, I quickly recognized my own disciplinary limits. I attempted to mitigate them by experimenting transdisciplinarity. Therefore, I have developed my research methods 'on the spot', 'in the doing'. In 2018, transdisciplinarity is far more welcome and recognized as a valuable research trend than it was ten years ago. I hope that my sometimes-clumsy attempts to integrate real people (i.e. those I name stakeholders in my work) and other disciplines in my research process has contributed to improve our understanding of all the issues and interests at stake.

From gender discrimination to feminist research

Evolving in a male-dominant field (there were few women at my first Treaty meeting), becoming a mother during my PhD research and facing the invisible (to many eyes) yet significant discrimination women are facing in academia, I have walked my path opening my eyes on this institutional discrimination. Although I was a lucky researcher supported by both male and female feminist guides and therefore not suffering from harsh sexism, I realized that institutional gender discrimination is still a fact in everyday university life (even though I am perfectly conscious to be part of the privileged group – white, rich, educated young women – and that gender discrimination arises elsewhere too).

Both my personal and professional itinerary have transformed me into a proud feminist. This experience has led me to question my way of conducting a valuable scientific research. I have not (yet) directly implemented feminist theory in my work, but my recent readings in the field are inspiring and enriching my current thinking, and I hope to conduct my future research following that direction. I want to contribute to rendering half of the humanity visible again.

From Eurocentric domination to a humble open mind

Yet, experiencing gender discrimination acquainted me with the colonialism of our Eurocentric research approaches imposed on other populations around the world. From a strong, unquestioned Eurocentric vision, which I embodied when I first started working on the Treaty, I have opened up my eyes to the diversity of minorities' views – i.e. all other views than the dominant one. This has allowed me to question my initial assumptions, their validity and their 'unquestionability'. As a consequence, I have more and more found interest in spiritual and holistic concepts – such as Ubuntu, Pachamama and the like – to grasp an understanding of my surrounding (i.e. taking into account Haraway's 'situated knowledges') and open my research to 'non-rational' (i.e. non-Eurocentric colonial) thinking. I was eager to listen not only to what my mind can rationally understand or what my eyes can objectively see but also to what my heart can 'un-judgingly' feel and what my guts can intuitively sense.

From objectivity to subjectivity of the research-actor

This slide away from a 'rational objectivity' towards opening up all my senses in my work led me to conduct my research by recognizing my own subjectivity. This move has been sometimes interpreted as activist-research versus what some academics consider 'objective science' should be. I believe objectivity in academia is a fake. I believe no human being is objective. I prefer recognizing subjectivity in my work and starting from there, trying to implement rigorous research methods to guarantee that all sides of the coin are considered. I have received some criticism that my work favours too much one vision/one side of the coin, i.e. the 'defence of the small' (hence of farmers' communities, in particular in developing countries). Yet, for other fellow research-actors and for activists, my words are far too mild. I take responsibility for my own subjectivity in choosing to give the word rather to 'the small' and a bit less to 'the dominant' on what agriculture, breeding innovation and development is or should be. I consider this is not imbalanced or un-rigorous. It only contributes a little to render visible what remains too invisible in the arenas of 'the dominant', to rebalance the debate and to refocus academic research on a less anthropocentric view of 'development' (i.e. the neo-colonialist and paternalist approach to 'development', although I am conscious that my work can also be seen as embodying this very approach, which I sincerely try to avoid, with more or less success, I admit).

From the lonely researcher to the community of people

Finally, working as an academic researcher has not always been an easy process because of its sometimes-lonely character. I am a collaborative person, and I value exchanges with people as a source of richness and creativity. I am only able to work on my own with great difficulty. My mind becomes creative and productive out of interaction, interconnection and relation with people, whether academics or not. Therefore, I have developed a strong network of people with whom I have been, and still am, able to exchange. I thank all of them for contributing to my work and I hope to continue collaborating with them all and many new people![1]

I would be grateful to receive feedback and comments on the present book.

I wish you enjoyable reading and look forward to interacting with you!

Warm regards,
Christine Frison

Note

1 Some of the new persons I hope to engage with in the future are the authors of the following work, as 2016 and 2017 have been prolific years for agrobiodiversity law and policy in terms of PhD theses defended: Valérie Tilman (Université de Namur, Belgium); Sonya Morales (Université de Laval, Canada); Archana Patnaik (Wageningen University, the Netherlands).

Acknowledgements

I hereby acknowledge the financial support of the *Fonds Wetenschappelijk Onderzoek – Vlaanderen* (FWO), as well as the institutional support of the Law Faculty of the University of Antwerp and of the Centre for Philosophy of Law of the Université catholique de Louvain. Apologies if I forgot to mention someone in the following long series of thanking.

I am deeply grateful to my mentor, Pepe Esquinas-Alcázar, for his kindness, support, guidance and trust since we met at the FAO premises back in 2004.

I sincerely thank my PhD co-supervisors, Professor Geertrui Van Overwalle from the Centre for IT & IP Law (CiTiP) at the KU Leuven, Professor Tom Dedeurwaerdere and Professor Olivier De Schutter from the Centre for Philosophy of Law (CPDR) at the UCLouvain. Their invaluable support throughout my doctoral research enabled the long-labour birth of my PhD.

A special thank you goes to Professor Esther van Zimmeren, University of Antwerp, for her trust, constant support and dear friendship over the last ten years.

I am indebted to André Hupin for his careful reading of my thesis, as well as to Anne Liesse for her help in proofreading and formatting the present text. Both their interventions greatly facilitated the revision of my doctoral dissertation into the following book. I sincerely thank Dr Fabien Girard for his careful reading of and comments on parts of the revised manuscript. I am happy that our encounter has developed into a fruitful collaboration and a sincere friendship. I express my deepest gratitude to Anne Liesse (again) and Caroline van Schendel for their friendly technical support, as well as all the CPDR admin team for their encouragements. I thank Professor Marc Maesschalck, head of the CPDR, for his constant support throughout my research career.

I am extremely grateful to Charlotte de Callataÿ for her friendship and to her parents for hosting me during numerous short-term stays in their wonderful holiday house in the Ardennes. This beautiful and peaceful setting enabled me to finalize first the PhD thesis, and second the revision of this manuscript.

I thank Amy Johnston for her help and patience throughout the edition process and Tim Hardwick for trusting and valuing my work over the last ten years.

I thank all my friends and colleagues from Belgium and abroad, the list of which would be far too long to mention, but I trust they will recognize themselves. Interacting with them has tinted my personal life and work creativity with beautiful colours.

I thank my sisters, Séverine and Valerie, and my brother, Mathieu, as well as the members of my larger family, for their support and help in difficult moments and for sharing joy and pleasure in leisure time.

Last but not least, I owe an immense debt of gratitude to my parents, Viviane Desmedt and Emile Frison, for their constant support, help and encouragements throughout my personal life and professional path. Being a (now) single mother academic requires solid social support to meet the expectations of both a demanding family life and a challenging professional career.

I thank my dear sons, Théodore and Charlie, for their incommensurable love.

Christine Frison

Table of cases

European Patent Office

G-0002/07 Broccoli/Plant Bioscience II of 9.12.2010
G-0002/12 Tomatoes/State of Israel-Ministry of Agriculture II of 25.03.2015
G-0002/13 Broccoli/Plant Bioscience of 25.3.2015

United States

Diamond v. Chakrabarty, 447 U.S. 303 (1980)
Ex parte Hibberd, 227 U.S.P.Q. 443 (Bd. Pat. App. 1985)

Table of legislation, statutes and international conventions

European Union

European Patent Convention (EPC) of 5 October 1973

Council Regulation (EC) No 2100/94 of 27 July 1994 on community plant variety rights

Directive 98/44/EC of 6 July 1998 on the legal protection of biotechnological inventions, OJ L 213, 30.7.1998, pp. 13–21

Regulation (EC) No 1151/2012 on quality schemes for agricultural products and foodstuffs, OJ L 343, 14.12.2012, pp. 1–29

Regulation (EU) No 1257/2012 of the European Parliament and of the Council of 17 December 2012 implementing enhanced cooperation in the area of the creation of unitary patent protection, OJ L 361, 31.12.2012, pp. 1–8

Council Agreement on a Unified Patent Court of 19 February 2013

European Parliament Resolution of 10 May 2012 on the patenting of essential biological processes, 2012/2623(RSP)

European Parliament Resolution of 17 December 2015 on patents and plant breeders' rights, 2015/2981(RSP)

Directive 98/44/EC of the European Parliament and of the Council on the legal protection of biotechnological inventions (2016/C 411/03)

Decision of the Administrative Council of 29 June 2017 amending Rules 27 and 28 of the Implementing Regulations to the European Patent Convention (*CA/D 6/17*)

Costa Rica

Biodiversity Law, No. 7788, 27 May 1998

France

LOI No. 2016–1087 du 8 août 2016 pour la reconquête de la biodiversité, de la nature et des paysages, JORF n°0184 du 9 août 2016, texte n° 2

Germany

Patentgesetz, PatG in der Fassung der Bekanntmachung vom 16.12.1980 (BGBl. I 1981 S. 1), zuletzt geändert durch Gesetz vom 01.09.2017 (BGBl. I S. 3346) m.W.v. 01.03.2018

India

Protection of Plant Varieties and Farmers' Rights Act, 2001 (Act No. 53 of 2001)

International

International Convention for the Protection of New Varieties of Plants of 2 December 1961, as Revised at Geneva on 10 November 1972, on 23 October 1978, and on 19 March 1991

Vienna Convention on the Law of Treaties, 23 May 1969, 8 I.L.M. 679

Declaration of the United Nations Conference on the Human Environment, Stockholm 16 June 1972, 11 ILM 1416

International Undertaking on Plant Genetic Resources, International Undertaking on Plant Genetic Resources, adopted 23 November 1983 (FAO Resolution 8/83).

Nairobi Final Act of the Conference for the Adoption of the Agreed Text of the Convention on Biological Diversity, Resolution 3, adopted 22 May 1992 in Nairobi, Kenya, UNEP

Convention on Biological Diversity, Rio de Janeiro, 5 June 1992 [United Nations, Treaty Series, vol. 1760, I-30619]

Rio Declaration on Environment and Development, UN Report of the United Nations Conference on Environment and Development, Rio de Janeiro, 3–14 June 1992

Agreement on Trade-Related Aspects of Intellectual Property Rights, Annex 1C of the Marrakesh Agreement Establishing the World Trade Organization, signed in Marrakesh, Morocco on 15 April 1994

Global Plan of Action for the Conservation and Sustainable Utilization of Plant Genetic Resources for Food and Agriculture, adopted by the Fourth International Technical Conference, Leipzig, Germany, FAO 17–23 June 1996

Rome Declaration on World Food Security, adopted by the Fourth International Technical Conference, Leipzig, Germany, FAO 17–23 June 1996

General Comment No. 12: The Right to Adequate Food (Article 11 of the Covenant) of the United Nations Committee on Economic, Social and Cultural Rights (CESCR) 12 May 1999

African Model Legislation for the Protection of the Rights of Local Communities, Farmers and Breeders, and for the Regulation of Access to Biological Resources, Algeria, 2000

Millennium Development Goals (MDG), UN, 6–8 September 2000

International Treaty on Plant Genetic Resources for Food and Agriculture (with annexes), Rome, 3 November 2001 United Nations, Treaty Series, [vol. 2400, I-43345]

Bonn Guidelines on Access and Benefit-Sharing as Related to Genetic Resources (Conference of the Parties to the Convention on Biological Diversity) COP 6 Decision VI/24The Hague, 7–19 April 2002

The Biosafety Cartagena Protocol to the Convention on Biological Diversity, was finalized and adopted in Montreal on 29 January 2000 and entered into force on 11 September 2003

Nagoya Protocol on Access to Genetic Resources and the Fair and Equitable Sharing of Benefits Arising from their Utilization to the Convention on Biological Diversity, Nagoya, 29 October 2010, UNEP/CBD/COP/DEC/X/1, entered into force 12 October 2014

Arusha Protocol for the Protection of New Varieties of Plants, African Regional Intellectual Property Organization (ARIPO), 6 July 2015

2030 Agenda for Sustainable Development (2030 Agenda) and its 17 Sustainable Development Goals (SDGs), United Nations Sustainable Development Summit, New York, 25–27 September 2015

UN Declaration on the Rights of Peasants and Other People Working in Rural Areas, draft version of the 5th OEIWG session (pursuant to paragraph 1 of Human *Rights* Council Resolution 36/22), Geneva, 9–13 April 2018

Malaysia

Protection of New Plant Varieties Act 2004 (Act 634)

Thailand

Plant Varieties Protection Act B.E. 2542 (1999)

The Philippines

Executive Order 247, Prescribing Guidelines and Establishing a Regulatory Framework for the Prospecting of Biological and Genetic Resources, Their By-Products and Derivatives, for Scientific and Commercial Purposes, and for Other Purposes, 18 May 1995

United States

Plant Patent Act of 1930, Pub. L. No. 71-245, 46 Stat. 376, enacted on 17 June 1930, codified as title 35 United States Code

Plant Variety Protection Act of 1970 (PVPA), 7 U.S.C. §§ 2321–2583

Acronyms and abbreviations

3PB	Third Party Beneficiary (Plant Treaty)
ABS	Access and Benefit-sharing
ACFS	Ad Hoc Advisory Committee on the Funding Strategy (Plant Treaty)
ACSU	Ad Hoc Technical Committee on Sustainable Use of Plant Genetic Resources for Food and Agriculture (Plant Treaty)
AEGIS	A European Genebank Integrated System
BSF	Benefit-sharing Fund (Plant Treaty)
CBD	Convention on Biological Diversity
CAPRi	System Wide Programme on Property Rights and Collective Action (CGIAR)
CePaCT	Centre for Pacific Crops and Trees
CFS	Committee on World Food Security (UN, FAO)
CG	Centres or CGIAR
CGIAR	Consultative Group for International Agricultural Research
CGRFA	Commission on Genetic Resources for Food and Agriculture (FAO)
CIMMYT	International Maize and Wheat Improvement Centre (CGIAR)
CIP	Centre International de la Pomme de Terre (CGIAR)
COP	Conference of the Parties (CBD)
CPR	Common-Pool Resource (theory of the commons)
CPGR	Commission on Plant Genetic Resources (then CGRFA, FAO)
DHS	Distinction, Homogeneity and Stability
DNA	Desoxyribonucleic Acid
DOI	Digital Object Identifier
DSI	Digital Sequence Information
DUS	Distinct Uniform and Stable criteria (UPOV)
EMBRAPA	Brazilian Agricultural Research Corporation, Ministry of Agriculture, Livestock and Food Supply

ENB	Earth Negotiation Bulletin of International Institute for Sustainable Development
EPC	European Patent Convention
EPO	European Patent Office
ESA	European Seed Association
ETC	Group Action Group on Erosion, Technology and Concentration (formerly RAFI)
EU	European Union
FRs	Farmers' Right (Plant Treaty)
FAO	Food and Agriculture Organization (UN)
FWO	Fonds Wetenschappelijk Onderzoek-Vlaanderen (Research Foundation Flanders)
G-77	Group 77 – Loose coalition of developing nations in UN fora
GATT	General Agreement on Tariffs and Trade (WTO)
GB	Governing Body (ITPGRFA)
GCDT	Global Crop Diversity Trust
GEF	Division of Global Environment Facility Coordination – UNEP
GFAR	Global Forum on Agricultural Research
GIs	Geographical Indications
GLIS	Global Information System (ITPGRFA)
GM Crops	Genetically Modified crops
GMO	Genetically Modified Organism
GPA	Global Plan of Action (Voluntary instrument in CGRFA, FAO)
GPG	Global Public Good
GRULAC	Latin American and Caribbean Group (FAO)
GURT	Genetic Use Restriction Technology
IAARD	Indonesian Agency for Agricultural Research and Development
IAASTD	International Assessment of Agricultural Science and Technology for Development (under sponsorship of the UN and the World Bank)
IAD	Institutional Analysis and Development
IARC	International Agricultural Research Centres (supported by CGIAR)
IBPGR	International Board for Plant genetic Resources (then IPGRI, then Bioversity International)
IBP	International Biological Programme
ICARDA	International Centre for Agricultural Research in the Dry Areas (CGIAR)
IDLO	International Development Law Organization
IFAD	International Fund for Agricultural Development (UN agency)

IGC	Intergovernmental Committee on Intellectual Property, Genetic Resources, Traditional Knowledge and Folklore (WIPO)
INGO	International Non-Governmental Organizations
INRA	Institut National de Recherche Agronomique (France)
IP	Intellectual Property
IPC	International Planning Committee for Food Sovereignty (alliance of small-scale producers)
IPES-Food	International Panel of Experts on Sustainable Food
IPGRI	International Plant Genetic Research Institute (the Bioversity International, CGIAR)
IPR	Intellectual Property Right
IR	International Relations
IRD	French Institute for Development Research
IRRI	International Rice Research Institute (CGIAR)
ISF	International Seed Federation (formerly ASSINSEL)
ITPGRFA or the Treaty	International Treaty on Plant Genetic Resources for Food and Agriculture (or the Plant Treaty)
IU	International Undertaking on Plant Genetic Resources (FAO)
IUCN	International Union for Conservation of Nature
MDGs	Millennium Development Goals
MFO	Maison Française d'Oxford
MLS	Multilateral System of access and benefit-sharing (Plant Treaty)
MTA	Material Transfer Agreement
Nagoya Protocol	Nagoya Protocol on Access to Genetic Resources and the Fair and Equitable Sharing of Benefits Arising from their Utilization to the CBD
NGO	Non-Governmental Organization
NSSL	National Seed Storage Laboratory (Genebanks)
OECD	Organization for Economic Cooperation and Development
OSSI	Open Source Seed Initiative
PBR	Plant Breeders' Right
PGR	Plant Genetic Resource
PGRFA	Plant Genetic Resources for Food and Agriculture
PIC	Prior Informed Consent
PPB	Participative Plant Breeding
PUD	Product Under Development (Plant Treaty)
PVP (A)	Plant Variety Protection (Act)
R&D	Research & Development
RdS	Red de Semillas (Resembrando e Intercambiando)
RSP	Réseau Semences Paysannes

SDGs	Sustainable Development Goals
SEARICE	Southeast Asia Regional Initiatives for Community Empowerment
SkE	Simalikalatone E
SMTA	Standard Material Transfer Agreement (Plant Treaty)
SPC	Pacific Community (scientific and technical organization in the Pacific region 26 countries and territories)
SPC Community	South Pacific Community
TK	Traditional Knowledge
TRIPS	Trade-related Aspects of Intellectual Property Rights (Agreement, WTO)
TWN	Third World Network
UN	United Nations
UNCTAD	United Nations Conference on Trade and Development
UNDP	United Nations Development Programme
UNEP	United Nations Environment Programme
UPOV	International Union for the Protection of New Varieties of Plant
USDA	United States Department of Agriculture
VCU	Value for Cultivation and Use
WG-MLS	Ad Hoc Open-ended Working Group on the Multilateral System
WIPO	World Intellectual Property Organization
WP-SU	Work Programme on Sustainable Use
WTO	World Trade Organization

1 Introduction

Harvesting the benefits of the commons to grow a food secure world

> The first person that after having built a fence around a piece of land, declared: This is mine, and found people simple-minded enough to believe him, was the real founder of society. How many crimes, wars, murders, misery and horrors would have been spared to humankind if someone would have pulled out the fence posts or filled up the ditch, and warned his fellows: Beware of listening to this impostor; You are lost if you forget that the fruits belong to everyone and the earth to no one.
>
> – Jean-Jacques Rousseau (1755: 68)

For those lucky enough to live in rich countries, most of us are able to eat quality food three times a day. With almost 8 billion people to be fed on earth, food production is undeniably an important and highly profitable market. It is also a complex matter, involving many different aspects, such as agricultural knowledge and practices of various farming systems, land and water management policies (whether national or international), technological innovation (chemical inputs, seeds and breeds innovation, mechanization, etc.), geopolitical relationship between major producer countries and main import countries, climate changes hazards, global market regulation and policies, etc. War, climate change, poverty, land grabbing, water scarcity, etc., are so many factors to be taken into account and tackled when addressing food insecurity crises worldwide. Similarly, health issues (whether related to malnutrition or to junk food overconsumption – IPES-Food, 2017a) or the cultural role of food and cooking in our societies are key to tackling endemic world food insecurity. While it is undeniable that all these aspects are important and merit research interest, this book will focus on the international management and regulation of plant genetic resources for food and agriculture (PGRFA),[1] i.e. what I will call for short the 'international governance of seeds'. By doing so, I do not claim that addressing the problems relating to crop variety conservation and sustainable use is the best or only way to respond to food insecurity, which is rising again (SOFI, 2017) after several decades of diminishing (SOFI, 2014). However, the governance of seeds, as an essential resource (Pistor and De Schutter, 2015) that is indispensable to produce our food, should be addressed in an urgent, serious and collective manner.

The diversity of plant genetic resources for food and agriculture (in short: seeds)[2] is an essential condition for our survival (Altieri, 1999; FAO, 2015). Given the current context of climate change, access to the diversity of seed varieties is even more vital to produce the food we need (IPES-Food, 2016; Beddington et al., 2012). However, at least two main reasons threaten access to seeds. First, the ongoing massive erosion of genetic diversity constitutes an urgent and serious risk to our food production, in particular in the context of climate change (Saab, 2015; Wheeler and von Braun, 2013). Biodiversity erosion is caused by a multitude of drivers, such as climate change, land use changes and the introduction of invasive species (FAO, 1998, 2010; Esquinas-Alcazar, 2005; Butchart et al., 2010), but also micro-level farming decisions and macro-level policies (Pascual and Perrings, 2007, 2012) including agricultural intensification, shifting cultivation and indiscriminate crossbreeding. Second, hyperownership over genetic resources (Safrin, 2004; Chiarolla, 2006; Aoki, 2008, 2010), understood as 'exclusive ownership and restrictions on the sharing of genetic material' (Safrin, 2004: 641), impedes an easy access to seeds whether being 'traditional varieties' or 'biotechnologically improved'.

Both elements limit the use and sharing of seeds among food and agriculture stakeholders and constrain international seed management within an appropriation paradigm. This appropriation paradigm stands at odds with what is seen by an important part of the literature as the dynamic flow of germplasm inherent to agricultural practices since the first forms of plant domestication (Pistorius, 1997). This narrative of the unfettered access to seeds is contested by recent academics, who show that '[a]lthough such networks are open systems, this does not mean that seeds are free or that seed flow goes unimpeded among communities, farmers, families, ethnic groups or polities. Seeds bear social costs and meanings' (Coomes et al., 2015: 47). One might argue that considering seeds as having always been available in an open access type of system prepares people's minds to accept that seeds are available for appropriation. This appropriation paradigm considers resources as commercial objects to be appropriated and sold on the global market as a commodity, reducing resources to merely an economic value and erasing all the other (cultural, social, religious or spiritual, etc.) values of the resources (Bavikatte, 2014: 232; Posey, 1999). This commodification process eludes any other contributions and values related to resources and confines seeds into a vicious circle of appropriation and exclusion, which contradicts the intrinsic interdependence of such resources and their exchange systems within ecosystems, as well as the necessary and evolutionary diversity of seed varieties. Indeed, market mechanisms – regulated by an international regime complex (Raustiala and Victor, 2004; Gerstetter et al., 2007; Andersen, 2008) – control the access to genetic resources and have gradually reinforced market control over increasingly homogenized varieties (Howard, 2009; Bonneuil and Thomas, 2009). This has gradually impeded the use and exchange of seed diversity (Fowler and Hodgkin, 2004; Fowler and Mooney, 1990; Kloppenburg, 2004; Pautasso et al., 2013). Acknowledging that feeding the world's population (2015 Sustainable Development Goal n°2 'Zero Hunger'; see UN, 2015a) requires changing our agricultural practices (IPES-Food, 2016, 2017b), there is an urgent need to unlock access to these essential resources (Pistor and De Schutter, 2015) by stepping out of the dominant appropriation paradigm. This means facilitating the use, conservation and exchange of the diversity of local and traditional seeds and protecting it from (mis)appropriation and homogenization. Indeed, one may question the usefulness of biotechnologically improved seeds – and their related proprietary tools (whether legal or technological) – as a solution to hunger (Saab, 2015). It is undeniable that post-WWII developments in the agricultural sector have allowed people to produce more food and to feed a growing population. However, already in the 1980s Sen had identified the roots of hunger elsewhere than in the limited quantity of food production (Sen, 1981). Moreover, the challenges we face today in the Anthropocene are different and require different and necessarily sustainable solutions. Producing sufficient and quality food for all, in a growing world population remains a challenge, in particular if we are to transform the current industrial farming system destroying the planet into sustainable practices, i.e. what the IPES-Food panel of experts promotes as the necessary paradigm shift switching from uniformity of industrial agriculture to diversity of agroecological systems (IPES-Food, 2016). We do not necessarily need to produce more, but better and to share the production equitably.

Involving all seeds stakeholders in the 'PGRFA regime complex' (Aoki and Luvai, 2007; Jungcurt, 2007), in particular farmers who are producing our food, is an unescapable choice, if we are to face world food insecurity in an equitable manner. Moreover, protecting the diversity of local and traditional seeds goes hand in hand with the protection of traditional and innovative farming systems, particularly well adapted to local needs, practices, culture and social behaviours. Engaging in a virtuous circle – englobing seed conservation and use in a holistic system where social, cultural and spiritual needs are taken into account besides the food, nutrition and economic aspects – promises to enable local populations to produce sufficient and nutritious food for all in a resilient manner (Altieri and Merrick, 1987; IPES-Food, 2016).

I hypothesize that this challenge can be met through collective governance of crop diversity addressed in a holistic manner inspired from Capra and Mattei's views (2015). By holistic, I mean that regulation of seed management (variety improvement, conservation and use) should be addressed in a systemic way as opposed to an analytic approach where the improvement of crop varieties would be tackled independently from its economic or political context, its environment, its users' culture or needs and its ecosystems. A holistic approach acknowledges the role of seed diversity in their ecosystems, of all seed stakeholders in the governing system and of the diversity of farming systems, needs and practices. It acknowledges the relationship between the various elements of an ecosystem, as well as their interdependency and mutuality. As Rival puts it (2018: 155)

> Capra and Mattei take a more spiritual and historical approach to ecology to show that individual property rights, far from being natural, have been crafted over centuries as part of a modernist worldview. In its blindness to

> nature's relational webs and patterns, this worldview has marginalized or even erased possibilities for future human development. Reasserting the principles of ecology and co-evolution today thus amounts to an acknowledgment of this loss, while offering the tools to re-harmonize human laws with the laws of nature.
>
> (Capra and Mattei, 2015: 152, 177)

Indeed Capra and Mattei argue for a radical change in the legal order because human institutions must be attuned to ecological facts and principles. According to them

> [. . .] law is always a process of '*commoning*' a long term collective action in which communities sharing a common purpose and culture, institutionalize their collective will to maintain order and stability in the pursuit of social reproduction. Thus the commons – an open network of relationships – rather than the individual, is the building block of the ecology of law and what we call an 'ecolegal' order.
>
> (2015: 14–15, emphasis original)

I believe this holistic approach to seed governance, which is key to transition our food and agriculture global system, can be reached through 'commoning' (Bollier, 2014). Commoning is defined by Commons scholars as assembling three dimensions: a resource, a community and a set of rules to manage the resource collectively and sustainably (Bollier, 2014). Commoning recognizes the relationship between humans and their environments. It recognizes the interdependence existing between different elements and living beings of an ecosystem. It emphasizes the importance of collective management. Over the last 15 years, authors have (re)discovered common ownership regimes as a potential solution against the limits of individual property rights over seeds (Halewood and Nnadozie, 2008; Girard, 2015; Frison, 2016). Based on the extensive literature on the commons, which has studied examples of local common-pool resources, Ostrom's design principles have been used to analyse the governing mechanisms of seeds in the International Treaty on Plant Genetic Resources for Food and Agriculture (Halewood, 2013). However, various limitations (see Chapter 4) render the parallel insufficient to fully understand why the Treaty is not reaching its objectives of food security and sustainable agriculture (Frison, 2016). To overcome these limitations, I believe that we need to switch our perception of seeds as extractable resources to understanding seeds as living elements of our environment/ecosystem with which a tight interdependency is correlated. Hence, a holistic approach to the global seed commons is promoted, one that transforms ownership from being 'extractive' of profit to making it 'generative' of values. Rival (2018: 156) confirms this analysis of Capra and Mattei by adding that

> [t]his requires that the law as a system of knowledge and jurisprudence fully recognizes the 'community' as a sovereign principle. Such a recognition

> implies in turn a radical revision of the concept of private property as well as of its relationship with state sovereignty, given that the boundaries between private and common property will always be negotiated through the state. From an ecological perspective, however, the state remains a legitimate institution only so long as it is capable of protecting the commons.
>
> (Capra and Mattei, 2015: 139, 141)

Recent literature has explored the idea that managing seeds as a commons would unlock access to seed diversity and promote their efficient conservation and sustainable use (Helfer, 2005; Halewood and Nnadozie, 2008; Dedeurwaerdere, 2012, 2013; Frison, 2016). However, one must be careful when promoting an unfettered access to seeds not to spread the idea that seeds belong to all and nobody at the same time, and therefore that they are de facto available for commodification. I argue that seeds should be governed as commons by the global seed community, encompassing all local farming communities, in order to ensure both their food security and sovereignty, as well as to support the transition towards a more sustainable agriculture. The theory of the commons sheds light on the efficiency of local collective management systems over natural resources, where private (market) or public (State) controlling failed to sustainably manage specific resources (Ostrom, 1990, 2010b). However, Ostrom's work remains within the boundaries of the above-mentioned appropriation paradigm by considering resources as economic objects to be governed for the sustainability of the community's living (Schlager and Ostrom, 1992). Her work remains within the economic efficiency framework, i.e. the best allocation of resources. It fails to explore the governance of resources outside of the anthropocentric influence, in a more holistic manner (Capra and Mattei, 2015), i.e. by focusing on the relationship between culture and nature rather than on the institutional arrangements designed mainly to manage an object considered exclusively as a 'resource/good'. As commons, I argue that seed exchange schemes are considered networked knowledge-goods with non-exclusive access and use conditions, which are governed, produced and consumed by communities (Girard and Frison, 2018a).

Appropriation of plant varieties and their collective management: a challenging equilibrium between the promotion of agricultural innovation and Farmers' Rights

On 25 March 2015, the Enlarged Board of Appeal of the European Patent Office (EPO) widened the extent of patent claims over plants with the *Broccoli II* (G 0002/13) and *Tomatoes II* (G 0002/12) cases. This 'decision leaves considerable leeway for patenting novel and inventive plants and products thereof, which have been produced by "conventional" methods including breeding steps' (Minssen and Nordberg, 2015: 81). Plant products such as fruits, seeds and parts of plants are now considered patentable even if they are obtained through 'essentially biological breeding methods' involving crossing and selection. This decision goes counter to a European Parliament Resolution (2012/2623(RSP), which is not binding) adopted on 10 May 2012 on the patenting of essential biological processes. This loose interpretation of the European Patent Convention (EPC) Article 53(b)[3] allows to patent native genes and traits,[4] which can be found naturally in any variety of a specific species. This means that a patent owner may impede breeders from using any variety that happens to contain this patented native trait, even if he/she was already using this variety previously (Girard, 2015: 4, 9–10). In Europe, variety protection normally happens under Plant Breeders' Rights (PBRs), where a breeder's exemption allows other breeders to use the protected variety when breeding a new variety.[5] In the US,[6] besides plant variety protection, utility patents may cover seeds, plants, cuttings and plant varieties since the *Ex parte Hibberd* case law in 1985 (227 USPQ 443) (Llewelyn and Adcock, 2006: 88). This protection prohibits a breeder to use the initial material for further breeding and the farmer to sow the material without paying royalties (Janis, 2001: 105–117).

These EPO decisions extend further the appropriation and enclosure of plants and seeds, which accelerated at the end of the twentieth century (Helfer, 2004) following a general trend in the intellectual property field (Cohen, 2004), and which Sabrina Safrin called 'hyperownership' (Safrin, 2004). It shrinks even more the rights of farmers to save, grow and sell their seeds. Farmers' rights had already been reduced to nothingness with the revision of the UPOV Convention in its 1991 Act, where Articles 14(1), 14(5), 15(1)(iii), and 15(2) define the scope and exceptions of Breeders' Rights. Previously, under the 1978 Act, the UPOV Convention left flexibility for States to recognize farmers' right to replant seeds from the crop produced by protected seeds for his own subsequent use (save seeds); to exchange seeds with other farmers without paying additional royalties to the breeder; and to use a protected variety to create new varieties without prior authorization of the original breeder. The 1991 Act reduced the right to freely exchange seeds and imposed limitations on their replanting (against royalty payment). As for the right to use seeds for further breeding, the 1991 Act limits it to new varieties that are not 'essentially derived' from protected varieties.[7] The overall result of the amendment has narrowed the exemption and expanded the rights of first-generation breeders (Helfer, 2004).

As a reaction, recent regulatory developments occurred in Europe to reaffirm the unpatentability of plants derived from biological processes. France adopted a law on biodiversity protection excluding plants derived from biological processes from patentability.[8] Germany,[9] Italy and the Netherlands have similar legislations. Under the pressure from the European Parliament and the European Commission,[10] the EPO modified its implementing regulations.[11] Under the new Article 28(2) 'European patents shall not be granted in respect of plants or animals exclusively obtained by means of an essentially biological process'. However, it should be noted that this ruling does not prevail on the EPC and that if a conflict arises, the latter will prevail (Article 164(2), EPC).

Notwithstanding this latter development in Europe, the appropriation trend is reinforced by the current 'dematerialization' of crop varieties (i.e. detach the genetic sequence information of a variety from the physical seed out of which the code has been extracted). Dematerialization supports the consideration that seeds are to be tackled as commercial goods (i.e. resources out of which profit can be extracted). Avoiding envisaging all the other dimensions of the seeds (cultural, religious, ethical, biological, societal, etc.) and forcing to decouple the information relating to the variety from its physical shell is a hypocritical nonsense serving the only commercial dimension of the seed for a limited number of seed stakeholders (i.e. the seed industry). The digital/genetic sequenced information (DSI) of a variety does not come from nowhere. It is extracted from the crop. Even if one may then send it and use it independently from the plant, its link with the physical variety cannot be ignored. Indeed, it is not information that will then be sowed back in the field, but a new variety, improved thanks to the use of the DSI. When talking about food production, one cannot use the information as such. One cannot eat information. DSI is used to recreate new varieties to be put on the global market, and hence the link and interdependence between the DSI and the physical shell of the seed cannot be passed over in silence. This superficial separation of elements and relationships goes against the holistic approach to agrobiodiversity governance as a sustainable commons. It also goes counter to elementary principles of legal interpretation (Vienna Convention on the Law of Treaties 1980), as demonstrated in Chapter 4.

This first (physical and informational) appropriation trend strengthens the increasing domination of food and agriculture markets by a few corporate multinational companies (De Schutter, 2009b, 2010c; Altieri and Nicholls, 2012; Foresight, 2011; Clapp and Fuchs, 2009).

As a reaction, farmers (e.g. *La Via Campesina*, see Claeys, 2015), researchers (such as the *Open Source Seed Initiative*, promoted by Prof. Jack Kloppenburg at the University Wisconsin-Madison, see Deibel, 2018), breeders (such as the *Garden Organic* UK-based association and its *Heritage Seed Library*) and citizens (e.g. *Graines de Troc*) are acting collectively worldwide to promote sustainable conservation, use and exchange patterns (e.g. the French *Réseau Semences Paysannes*, see Demeulenaere, 2018) for so-called 'non-industrial varieties'.[12] This second trend represents an alternative path to produce local, diverse, sustainable and healthy food (Altieri and Nicholls, 2012; IPES-Food, 2016), taking into account the interdependent and holistic approach mentioned above.

Concurrently with these two trends emerges the global challenge of feeding a growing world population in the face of increasing social, economic and environmental vulnerabilities (Sumner and Tiwari, 2011; SDSN, 2013) together with the more specific issue of conservation and access to seeds for food security and sustainable agriculture (Cullet, 2004, 2005). Indeed, on 25 September 2015, the United Nations Sustainable Development Summit adopted 17 Sustainable

Development Goals (SDGs), of which 'Goal 2 Zero Hunger', provides in target 2.5 that

> [b]y 2020, [States should] maintain the genetic diversity of seeds, cultivated plants and farmed and domesticated animals and their related wild species, including through soundly managed and diversified seed and plant banks at the national, regional and international levels, and promote access to and fair and equitable sharing of benefits arising from the utilization of genetic resources and associated traditional knowledge, as internationally agreed.

Since the middle of the twentieth century, policies (through the green revolution, Evenson and Gollin, 2003; Tangley, 1987; Brush, 2001) have promoted the large-scale production of uniform, high yielding monocultures of a few staple crops, in highly artificial environments, as the solution to feed a growing population (Evenson and Gollin, 2003). The focus was on increasing yields through the development of new breeding technologies, thereby quickly replacing local and diverse varieties with uniform crops worldwide (FAO, 1998) and shifting the economic and legal classification of seeds from public goods to highly privatized goods (Helfer, 2004; Chiarolla, 2006). Today, numerous studies show that this approach has had various serious consequences. First, a rapid diversity loss resulting from the widespread intensive monocultures (FAO, 1998, 2010; Esquinas-Alcazar, 2005). Second, the domination of a few multinational corporations over the entire agriculture input sector (De Schutter, 2009b; Howard, 2009). Third, the hyperownership and enclosure of seeds through legal and technological means (Safrin, 2004), leading to the increasing brittleness of traditional informal seed systems (Louwaars, 2008; De Schutter, 2009b; Wattnem, 2016). Fourth, the continuing reduction in numbers of small-scale farms on which most of developing countries' population rely for their food production (Altieri and Nicholls, 2012), etc.

While agrochemical companies have systematically used the argument of reducing hunger and malnutrition to promote policies (Magretta, 1997; Simanis and Hart, 2001) that strengthen their dominant position worldwide[13] and expand the commodification process (Chiarolla, 2006), it is undeniable that these strategies have not reached the 'official objective' of eliminating hunger and malnutrition (SDSN, 2013; De Schutter and Vanloqueren, 2011; Dorward et al., 2004), despite the significant yield increases (Morley et al., 2014). Indeed, although the number of hungry people had diminished,[14] the first Millennium Development Goal (MDG) to eradicate poverty and hunger and the 'Zero Hunger' 2015 Sustainable Development Goal (SDG) are far from being achieved (FAO, 2014b; De Schutter, 2008),[15] in part because hunger crises are not only due to food production problems but rather related to general policy directions (Sen, 1981) including the (mis)appropriation paradigm.

The problem is therefore not so much about our capacity of producing enough food (indeed studies have shown that within our limited world resources, we are able to feed the world population; Altieri and Nicholls, 2012;

De Schutter, 2009b), but rather about managing and facilitating the access to food and to the seeds (and other agricultural inputs) needed for its production in a fair and equitable manner (Lappé et al., 1998; De Schutter, 2008; Pautasso et al., 2013; Sen, 1981).

A growing number of studies show that a different type of agriculture could better address the above-mentioned needs (IPES-Food, 2016; SDSN, 2013; Rodale Institute, 2011), taking into account the social, economic and environmental hazards.[16] In December 2010, the Special rapporteur on the Right to Food, Olivier De Schutter, pointed out that 'States can and must achieve a reorientation of their agricultural systems towards modes of production that are highly productive, highly sustainable and that contribute to the progressive realization of the human right to adequate food' (De Schutter, 2010a, De Schutter and Vanloqueren, 2011; Pautasso et al., 2013). Drawing on an extensive review of the scientific literature on the topic, the Special Rapporteur identified agroecology[17] as a mode of agricultural development to be promoted (Altieri and Faminow, 1996; Sánchez Carpio and Becheva, 2014). Agroecology demonstrates that yields can be doubled within ten years' time and that favouring diversity increases productivity while facing environmental challenges (Altieri et al., 2012b; Altieri and Nicholls, 2012; Rodale Institute, 2011). In an interview in the *New Zealand Herald*, Steve Wratten confirms these observations,[18] saying we 'have the protocols or recipes' to do this, 'but getting governments to adopt it has a major barrier: international corporations' (Reider, 2015). Wratten points here to a crucial issue: the necessity for political will to cooperate and promote collectively a fair and equitable access regime to food and seeds. Besides, other crucial aspects should also be tackled, such as access to land and the issue of land grabbing, or access to water.

This observation highlights the imperative need for all stakeholders in the world food chain – and especially states – to cooperate to operate a transition towards a sustainable agriculture and food system. As mentioned above, access to seeds for producing food is of vital importance for reaching food security worldwide (Frison, Cherfas and Hodgkin, 2011). Indeed, states are highly interdependent with regard to the provision of food and agriculture plant varieties (Khoury et al., 2015; Fowler and Hodgkin, 2004; FAO, 1998). Countries' interdependence justifies a 'compulsory' cooperation between states in establishing and protecting a fair and equitable access to seeds. This international cooperation challenge can be addressed by setting up (global) institutional arrangements (Keohane and Ostrom, 1995; Jungcurt, 2007). This is precisely why the International Treaty on Plant Genetic Resources for Food and Agriculture[19] (hereafter the Treaty) was shaped and adopted in 2001. Steered by sustainable development principles, the Treaty designs several tools to help countries reach their food security and sustainable agriculture overall goals (Frison, 2006). Two major provisions – the Multilateral System of access and benefit-sharing (MLS; Treaty Articles 11–13) and the recognition of Farmers' Rights (FRs; Treaty Article 9) – are designed as incentives for Contracting Parties to provide a facilitated access to seeds to all food and agriculture stakeholders, including smallholder farmers. The MLS is viewed as a global commons system (Halewood, 2013; Santilli, 2012), where stakeholders manage together the access to seeds, their conservation and sustainable use. Both tools – the Multilateral System and Farmers' Rights – aim at proposing an alternative path to the current food and agriculture system blocked in the middle of a private/public good dilemma. However, little thorough research has been conducted on analysing whether these tools adequately respond to the need for reaching food security and sustainable agriculture through the collective management of plant genetic resources for food and agriculture (PGRFA, or seeds).[20]

Aim and structure of this book

The present research explores the consideration of seeds and the MLS as a global commons system to facilitate the conservation and provision of seeds worldwide for food security and sustainable agriculture. The purpose of the thesis is to bring a theoretical insight to the Treaty, using the theory of the commons, in order to understand how the Treaty is (dys-)functioning and to make normative proposals so as to improve its implementation. The work is designed around three steps.

First, the aim is to set the contextual field in which the Treaty has its origins (Chapter 2) and identify the general challenges related to PGRFA management (Chapter 3). The historical evolution of PGRFA management has shifted the consideration that seeds were considered public goods available to all to the consideration that seeds are overly privatized goods, accessible to few following strict (legal, economic and technical) access conditions. This evolution has crystallized an imbalance of rights pertaining to seeds and contributed to further limit access to and exchanges of seeds between all stakeholders, thereby endangering seed conservation and sustainable use. This descriptive first part highlights major tensions resulting from the above-mentioned developments: i.e. the international regime complex (Jungcurt, 2007; Gerstetter et al., 2007) for PGRFA and the hyperownership of seeds (Safrin, 2004). These tensions express an imbalance of recognition in the rights pertaining to seeds: private individual hyperownership of seeds (through legal and technological tools) overpower collective rights over seeds (e.g. through (in-)effective Farmer's Rights). This historical narrative explains that the international community needed to design a new international convention to overcome these tensions: the International Treaty on Plant Genetic Resources for Food and Agriculture.

The second step elucidates why seed exchanges remain problematic notwithstanding the negotiation and implementation of the Plant Treaty. By creating the MLS, Contracting Parties have attempted to strike an equitable balance between public and private interests in access to seeds, but countries face difficulties in implementing the Treaty. The de facto imbalance of rights pertaining to seeds needs to be re-balanced to implement efficiently the MLS and allow stakeholders to reach the Treaty's objectives. The objective of this second research step is to draw a precise portrait of the Plant Treaty's functioning, of the constraints in its text and of the difficulties in its implementation, in order to understand why the Treaty does not reach its objectives. First, a classical legal analysis of the Treaty is carried out (Chapter 4), to explain if and how the Treaty attempts to overcome the public/private paradigm.[21] However, this legal study provides insufficient appreciation to understand fully the slow implementation of the Treaty and the difficulties in fulfilling its objectives. Therefore, as a complementary investigation, a stakeholder analysis is carried out (Chapter 5), where actors have identified limitations and constraints they face in their experience with the Treaty negotiation and implementation (Frison et al., 2011a).[22] This second step therefore focuses on a technical level, by identifying constraints in the implementation of the Treaty, to which recommendations will be formulated for Treaty stakeholders.

Third, Chapter 6 proposes a way to overcome the deficit of Contracting Parties' obligations in reaching their food security and sustainable agriculture overall goals. Following the technical analysis of the Treaty' dysfunction, eight thematic group of conceptual constraints are highlighted as problematic in the current design of the common management of seeds. These conceptual constraints all seem to point to a lack of holistic vision to the problems we are faced with, including the interconnected character of many of these problems. According to Capra and Mattei, 'recogniz[ing] the interconnectedness of our global problems [would] enable us to find appropriate, mutually supportive solutions that [. . .] would mirror the interdependence of the problems they address' (Capra and Mattei, 2015: 159; see also Capra, 1996; Moore, 2015). Therefore, they call for our legal institutions to be attuned to ecological facts and principles and to community practices. By elaborating on the *Theory of the Commons*[23] (in particular regarding States' recognition of FRs, their conservation and sustainable use responsibilities, as well as their access and benefit-sharing obligations) and using the concept of 'ecology of law' developed by Capra and Mattei, six invariable principles are identified for an effective *Global Seed Commons.* These invariable principles are: 1) sustainability, 2) interdependence, 3) anticommons dilemma, 4) physical and informational components inextricably bound to the use of seeds, 5) diversity, heterogeneity and complexity, and 6) community. Implementing Capra and Mattei's view on a new ecolegal order, our legal instruments should respect the invariable principles of the global seeds commons. Enhancing the MLS as a global seed commons, i.e. a legal instrument 'in tune with nature and community', would contribute to better reaching the Treaty's goals of food security and sustainable agriculture. It would constitute an alternative way to overcome the dichotomy that appeared in the Treaty analysis between seeds defined exclusively as private goods (individual property rights) and seeds characterized as public goods (collective rights to use and conserve). Eight recommendations are made to enhance the functioning of the global seed commons, presented as an alternative to overcome the limits of the current seed regulatory setting resulting from the public/private good dichotomy. One cross-cutting aspect that appears all along the analysis is the lack of recognition of the role and rights of smallholder farmers. Recognition of Farmers' Rights at the international level could overcome the imbalance of rights pertaining to seeds and contribute to reach the food security and sustainable agriculture overall goals of the Treaty. Chapter 6 makes normative proposals as to what can be done in the implementation of the Treaty for it to reach its objectives. Thereby, the author hopes to contribute clarifying challenging issues at stake during the Treaty's review process and guiding the redesign of an effective global seed commons for reaching food security and sustainable agriculture.

Theoretical framework: the theory of the commons

The international management for the conservation, sustainable use and access to seeds is a global challenge that requires multilateral and multi-stakeholder cooperation. Globalization has significantly increased this fundamental interdependence between States and between stakeholders. Analysing the Plant Treaty from an exclusively legal perspective would miss out much of the issues at stake. Indeed, policy is deeply intertwined with the international law-making process. To integrate this multilateral and multi-stakeholder cooperation dimension in support to the legal analysis of the Treaty, this research is framed by a major theory pertaining to political sciences: the theory of the commons. Indeed, it provides a governance dimension necessary to understand the role played by stakeholders in the creation and implementation of the Treaty as a set of international, formal and binding norms. Such a governance approach allows examining the legal shortcomings of the Treaty and understanding the interplay between stakeholders in the negotiation and implementation of the Treaty.

In this book, it is argued that the wide international cooperation (Vollan and Ostrom, 2010) between all stakeholders for the provision of PGRFA has resulted in the creation of a seed commons-type mechanism through the design of the Treaty's MLS (Halewood et al., 2013). Indeed, the Treaty is the result of global cooperation based on commons principles (Holder and Flessas, 2008; Falcon and Fowler, 2002; Halewood, 2013); and the mechanism that the Treaty puts in place is evidence of a 'new multilateralism', echoing what former UN Secretary General Ban Ki-moon called for at the Fifth Summit of the Americas in 2009:

> We need a new vision, a new paradigm, a new multilateralism. A multilateralism that is organized around delivering a set of global goods. A multilateralism that harnesses both power and principle. A multilateralism that recognizes the interconnected nature of global challenges.
>
> (Ki-Moon, 2009)

The legal and stakeholder analyses carried out in Chapters 4 and 5 of this work highlight this multilateralism in managing global challenges, but they also point to constraints in the Treaty implementation that need to be overcome for an efficient provision of the Treaty's objectives. Analysing these limitations through the lens of governance may contribute to clarify why the system is not functioning well and propose actions and directions for all stakeholders to improve the implementation of the Treaty. The theory of the commons is proposed as the governance approach to study the Treaty, similarly to what has been done to analyse the international governance of microbial genetic resources (Dedeurwaerdere et al., 2009; Halewood, 2010). The concept of commons is not very well defined and has even grown fuzzier with globalization and the complexification of wider resource governing systems. In 1968, Garrett Hardin published an (over-exploited) allegory named the 'Tragedy of the Commons', where he analysed the problems related to over-exploitation of finite resources under unlimited and free access conditions to all (Hardin, 1968). He took the example of grazing and posed the precondition that rational people would always try to get the maximum and immediate profit from the use of a 'common resource'. In game theory, this has been modelled under the prisoner's dilemma (Rapoport and Chammah, 1965). Hardin concludes that following this 'rational' leads to overgrazing and the destruction of the common pasture. Hardin states that

> [e]ach man is locked into a system that compels him to increase his herd without limit – in a world that is limited. Ruin is the destination toward which all men rush, each pursuing his own best interest in a society that believes in the freedom of the commons. Freedom of the commons brings the ruin to all.
>
> (Hardin, 1968: 1244)

This view is supported by Mancur Olson in his work on the logic of collective action, who states that 'unless the number of individuals is quite small, or unless there is coercion or some other special device to make individuals act in their common interest, rational, self-interested individuals will not act to achieve their common or group interest' (Olson, 1965: 2, emphasis in original text). Although Olson was much more precautionary than Hardin in the proposed solution to the 'tragedy', leaving the question of common management open. Hardin proposes three solutions to his tragedy: reducing world population to avoid overconsumption; or establishing an external institution to regulate the use of the resource, whether through public management (State) or through the market (i.e. enclose the commons). Yet, according to Berge and Laerhoven, his 'explanation for the need to enclose the commons confounded the resource with its governance regime' (Berge and Laerhoven, 2011: 161), mixing the governance under a commons with open access. Other criticism can be formulated against Hardin's views, including the fact that in real life, people communicate and are rarely put in a situation where a common resource is used by different person who do not talk to each other and discuss how to manage the resource commonly (Ostrom, 1990: 7). Another criticism relates to the rational character or the human being. According to Sen, who worked on welfare economics, peoples' values and commitments will also influence economic policies in terms of their effects on the well-being of the community. Therefore, ethical aspects are also important and may counterbalance the 'rational part' of human's behaviour (Sen, 2003: 15, 40; Sen, 1970).

Later on, as a reaction to the supremacy of property rights (whether state or private) as the 'best' system to manage resources, Elinor Ostrom[24] studied the management of common resources using a 'bundle of rights' approach, where she distinguishes between operational-level property rights and collective-choice property rights. Schlager and Ostrom define operational activities as

> constrained and made predictable by operational-level rules regardless of the source of these rules. By the term "rules" we refer to generally agreed upon and enforced prescriptions that require, forbid, or permit specific actions for more than a single individual. [. . .] Operational rules are changed by collective-choice actions. Such actions are undertaken within a set of collective-choice rules that specify who may participate in changing operational rules and the level of agreement required for their change. With regard to common-pool resources, the most relevant operational-level property rights are "access" and "withdrawal" rights. In regard to common-pool resources, collective-choice property rights include management, exclusion, and alienation.
>
> (Schlager and Ostrom, 1992: 250–251)

Indeed, according to Schlager and Ostrom, '[a]ssigning full ownership rights does not guarantee an avoidance of resource degradation and over-investment' (Schlager and Ostrom, 1992: 259). To get a deeper comprehension of the conditions for sustainable resource use and governance regimes, she analysed Common-Pool Resource (CPR) institutional arrangements based on extensive field studies (Ostrom, 1990).[25] In her famous book *Governing the Commons*, Ostrom focused on case studies in agricultural production systems, e.g. irrigation, forestry or fishery management systems. In her understanding, a commons is 'any natural or manmade resource that is or could be held and used in common' (Berge and Laerhoven, 2011: 161). Ostrom showed that stakeholders[26] can effectively set up rules together (i.e. self-organization) to manage resources established in a local common-pool for their own use, and outside the market or governmental intervention (i.e. self-governance). Thanks to these data and to her observations, she designed eight principles useful to govern an efficient CPR system (Ostrom, 1990: 90, see table 3.1):

1 Clearly defined boundaries (i.e. effective exclusion of external unentitled parties);
2 Congruence between appropriation and provision rules and local conditions;
3 Collective-choice arrangements (i.e. allow most resource appropriators to participate in and modify the operational rules);
4 Effective monitoring (by monitors who are part of or accountable to the appropriators);
5 Graduated sanctions (scale of sanctions for appropriators violating community rules);

6 Conflict-resolution mechanisms (cheap and of easy access);
7 Minimal recognition of rights to self-organize (the self-determination of the community is recognized by higher-level/governmental authorities);

Plus, for CPRs that are parts of larger systems

8 Nested enterprises (organization in the form of multiple layers of nested enterprises, with small local CPRs at the base level).

These design principles are helpful (but not compulsory) to identify whether other resource management systems can qualify as CPR or not (Ostrom and Delville, 2009; Quinn et al., 2007). Indeed, Ostrom leaves much space for heterogeneity and diversity in systems and places, insisting on the fact that the institutional arrangement should always be adapted to local needs and conditions in order to be efficient (which implies that other design principles may be better adapted to different situations) (Ostrom, 1990, 2005, 2009a; Cox et al., 2010).

In 2008, there was little legal scientific literature talking about the Plant Treaty (Cooper, 2002; Batta-Bjørnstad, 2004; Ruiz-Muller, 2006; Frison, 2006; Gerstetter et al., 2007), and hence very little on the relationship between the Treaty and the commons theory. Since then, some authors, including non-academics, have assimilated the Treaty's MLS to a commons-type management regime (Dedeurwaerdere, 2010). The MLS, as a virtual pool management mechanism for selected plants, has been qualified as 'global commons' or 'global crop commons' (Halewood et al., 2013), 'PGRFA Commons' (Halewood and Nnadozie, 2008), 'communal seed treasury' (Raustiala and Victor, 2004; Helfer, 2004), 'common seed pool' (Raustiala and Victor, 2004), 'global genetic commons' (Safrin, 2004; Falcon and Fowler, 2002; Helfer, 2005), 'positive commons' (Van Overwalle, 2008), etc. These terms are often used loosely in relation to the Treaty, without a precise definition or application[27] of existing solid scientific literature on the commons theory. Building on this trend, Ostrom's theory is used as a framework to screen the MLS through the lens of the commons. The aim is to analyse whether the MLS, as it has been created by Contracting Parties to the Treaty, can be understood as a CPR, according to Ostrom's definition. My analysis leads to the conclusion that it is not fully the case. One of the difficulties relates to the global dimension of the MLS (Dedeurwaerdere, 2013; Henry and Dietz, 2011; Stern, 2011). Another problem lies in the fact that Contracting Parties have designed the institutional arrangement (even if it is based on prior existing practices by specific PGRFA stakeholders), and are managing it, with no formal space for all stakeholders to participate in the management of the MLS, and with little trust emanating from stakeholders' collaboration in the Governing Body forum (Six et al., 2015). Recent developments in the theory of the commons have expanded its frontier within the economic field (Coriat, 2013, 2015) or to other disciplines such as law (Rochfeld, 2009; Mattei, 2011a; Lucarelli, 2011b; Dani, 2014), or philosophy (Dardot and Laval, 2010, 2014; Gutwirth and

Stengers, 2016). These expansions have allowed for reconceptualizing Ostrom's institutional analysis into envisaging the commons as a collective political construct (Dardot and Laval, 2014), where a transformation of our legal system(s) is necessary to be sustainably compatible with nature and communities (Capra and Mattei, 2015). These new 'commons narratives' are explored and used to make normative proposals to mitigate the identified conceptual constraints in the Treaty functioning.

Building on a thorough transdisciplinary study of the MLS, six invariable principles are identified as core elements for an efficient global seed commons: 1) sustainability; 2) interdependence; 3) the anticommons dilemma; 4) the physical and informational components inextricably bound to the use of seeds; 5); diversity, heterogeneity and complexity and 6) the global seed community. It is argued that these invariable principles should be better articulated in the tools, rules and procedures of the global seed commons in order to realize the Treaty's overall goals. The six invariable principles are not design principles which have to be adapted to a specific resources commons regime. In my view, these invariable principles express the very essence of seeds evolution: they constitute core principles enabling the existence of seed diversity with human communities. Humans cannot continue to avoid respecting them when designing the commons regime to govern the global seed commons.

Scope of the book

The present research has a legal, a material and a temporal scope.

The legal scope

Regarding the legal scope, the research focuses on the analysis of the Treaty. Related international instruments such as the Convention on Biological Diversity (CBD) and its Nagoya Protocol, the Trade-related Aspects of Intellectual Property Rights (TRIPS Agreement), and the International Union for the Protection of New Varieties of Plant (UPOV) will be touched upon, but only to describe the context and serve the arguments made on the Treaty. A human rights approach will similarly not be addressed, although it is contended that it is an important component (De Schutter, 2008; Cullet, 2007; Van Overwalle, 2010), which requires further research, in particular since the opening of negotiations on a 'UN Declaration on the rights of peasants and other people working in rural areas'.

The Convention on Biological Diversity and the Nagoya Protocol

Due to its 'fall-back-regulatory-instrument' position, the CBD and its Nagoya Protocol (Morgera et al., 2014) will partly be addressed in Chapter 2, but only to explain the access and benefit-sharing concept and mechanism (for an assessment of the linkages between the Treaty and the CBD see Garforth and Frison, 2007).

TRIPS, UPOV and intellectual property rights issues

The TRIPS agreement and UPOV will be mentioned when talking about intellectual property rights (IPRs) issues related to plants (Cullet, 1999, 2001; Cullet and Kolluru, 2003; Cullet and Raja, 2004). Although intellectual property is an important aspect of seed governance, it is clearly stated that this research is not an IPR piece of work (see Batur, 2014; Dutfield, 2004). Future negotiation outcomes in the World Intellectual Property Organization (WIPO) fora[28] could be relevant to this research when dealing with the IP questions for genetic resources and traditional knowledge (Van Overwalle, 2005, 2009), but as no legally binding instruments have been adopted yet and as the negotiations are on hold, these negotiations will not be addressed.[29] Consistently with this choice, the theoretical framework of this work is not the theory of property, which Brahy has addressed in studying the management system for genetic resources and traditional knowledge using law and economics as a background (Brahy, 2008). Analysing further international seed management through the lens of property has also been carried out, in an interdisciplinary perspective, through the 'Common Plant – ANR project'[30] between 2015 and 2017. In this research project, the assumption is that the concept of property can be rethought against the background of the 'right to include', so as to endow others with a common 'right to access' genetic resources (Girard and Frison, 2018b).[31]

Besides, the author is currently conducting a research on the dematerialization of genetic resources, i.e. genetic sequenced information and their link with the Treaty. As this topic is very recent and complex, it will not be addressed in the present publication. Outcomes of the research project will be published in separate articles by the author.

Biosafety and GMOs

Biosafety international regulations[32] could also be relevant as the PGRFA covered by the Treaty are potentially genetically modified organisms (GMOs). However, the Cartagena Protocol on Biosafety regulates the safe transfers and the commercialization aspects of GMO products (Cordonier Segger et al., 2013; Frison and Joie, 2006a, 2006b) and does not touch upon the common management for the conservation, sustainable use and access and benefit-sharing of PGRFA. Therefore, this instrument is considered outside the limits of this book.

International law versus national legislations on biodiversity or seed management

This work remains exclusively at the international level and will not dig into the implementation of the Treaty at national levels; it only refers to it in a few cases as examples. Seed legislations are largely national and will therefore not be dealt with (even though they have a strong impact on seed exchange; see Anvar, 2008; Louwaars, 2008; Girard and Noiville, 2014), except when mentioned as illustrations. Indeed, attention is centred on the international level as the aim is to understand governing mechanisms set by stakeholders in the Treaty at the global level, using global theories.

Human rights

Due to time, resources and scope restrictions, the human rights' perspective on access to seeds and the right to food have not been deeply explored (Van Overwalle, 2010). According to the UN Human Rights office of the high commissioner, the right to food can be defined as

> the right to have regular, permanent and unrestricted access, either directly or by means of financial purchases, to quantitatively and qualitatively adequate and sufficient food corresponding to the cultural traditions of the people to which the consumer belongs, and which ensure a physical and mental, individual and collective, fulfilling and dignified life free of fear.[33]

The former United Nations Commission on Human Rights (now United Nations Human Rights Council) has established a specific mandate on the right to food[34] since the year 2000 by appointing a special rapporteur on the right to food[35] to promote the full realization of the right to food *inter alia* through the adoption of measures at the national, regional and international levels (Frison and Claeys, 2014). The special rapporteur on the right to food pleads to improve the global governance of food security. He stresses that '[i]n times of crisis, more than ever, only by strengthening multilateralism can we hope to effectively realize the right to food'.[36] Although the right to food has been enjoying a growing recognition for the last decade (Sage, 2014; Cullet, 2005; Lambek et al., 2014; De Schutter, 2009c), it suffers serious implementation and enforcement problems (Niada, 2006; Claeys, 2015). In 2015, several studies were published on the between human rights, seeds laws and Farmers' Rights (Christinck and Walloe Tvedt, 2015; La Via Campesina and GRAIN, 2015; Braunschweig et al., 2014; Helfer, 2015), paving the way for further research on the intersection between access to seeds and the right to food.

International law and international relations

Finally, state cooperation within international negotiating fora holds a key place and impact in this research. Notwithstanding the fact that the analysis shows that cooperation between states constitutes a very important aspect in understanding the seed regime complex (Rosendal, 2001), due to the legal focus of the present research, I do not claim to conduct research following IR methods. Again, further research in this field would benefit the resolution of the issues at stake, in continuation of Stefan Jungcurt's work (Jungcurt, 2007).

The plant genetic resources for food and agriculture material scope

As for the material scope, it is limited to plant genetic resources for food and agriculture covered by the Treaty (Article 3). The Treaty defines PGRFA as 'any genetic material of plant origin of actual or potential value for food and agriculture' (Treaty Article 2). The Treaty defines 'genetic material' as 'any material of plant origin, including reproductive and vegetative propagating material, containing functional units of heredity'. However, the MLS creates a more restrictive sub-category of seeds, listed in Annex I to the Treaty. Article 11.2 stipulates that the MLS covers only the 'PGRFA listed in Annex I that are under the management and control of the Contracting Parties and in the public domain'. This distinction might fade away in the future as discussions are ongoing in the governing body forum of the Treaty on expanding the MLS to all PGRFA, thereby erasing the Annex I distinction.

Article 12.3(a) further specifies that 'access shall be provided solely for the purpose of utilization and conservation for *research, breeding and training for food and agriculture*, provided that such purpose does not include chemical, pharmaceutical and/or other non-food/feed industrial uses' (emphasis added). This means that PGRFA that are used for another purpose, such as the production of biofuels, cosmetics or pharmaceuticals, are not considered as PGRFA under the MLS. This distinction is important as the exchange mechanism and applicable law will differ when the subject matter is PGRFA or other plant genetic resources (i.e. plant genetic resources used for biofuels, cosmetics, pharmaceuticals, etc.). Indeed, for the latter, it is most likely that the CBD and its access and benefit-sharing (ABS) obligations under the Nagoya Protocol are applicable.

The temporal scope

Last, regarding the temporal scope, the historical analysis of seed management (Chapter 2) is limited to a period covering the second half of the twentieth century (1960–2000), and the legal analysis of the implementation of the Treaty (Chapter 4) is concentrated on the last decade (2004–2017). This temporal scope covers the important historical facts that have influenced the design of the Treaty, i.e. the development of IPRs and plant breeders' rights; the rise of Farmers' Rights and the recognition of States' sovereign rights over their genetic resources; and the emergence of a common governing tool to manage PGRFA.

Governing seeds and humans in a holistic manner: towards a sustainable global seed commons

This work shows that one needs to move away from the regulation of property rights over seeds (individual ownership of objects) and shift towards regulating the relationship governing the diverse elements – humans and seeds – of a global ecosystem – nature (i.e. embracing collective rights with a systemic approach). To do so, implementing the six invariable principles can help designing a sustainable global seed commons, in that these principles respect the human-seed relationship within its ecosystem and encompasses the diversity of values humans devote to seeds. Shifting from exclusive individual rights to inclusive collective rights will allow embracing the necessary holistic approach to seed governance, which I believe constitutes one way to govern seeds in a collective, equitable and sustainable manner.

Notes

1 In the present work, the words 'seed', 'plant', 'PGRFA', 'material' or 'genetic resource' are used interchangeably to talk about the 'plant genetic resources for food and agriculture' (PGRFA) as defined under Article 2 of the Plant Treaty. In simple terms, PGRFA are crops and forages (physical and informational components) used as nutriments for humans and animals; see also note 12 below.

2 For a detailed definition, see below Chapter 3.

3 Article 53(b) 'Exceptions to patentability' of the European Patent Convention states that 'European patents shall not be granted in respect of: (b) plant or animal varieties or essentially biological processes for the production of plants or animals'.

4 According to the French High Council of Biotechnologies (HCB, 2014: 177), a native trait means 'a trait of any kind, whether physical or chemical, the expression of which is likely to be observed in any plant, wild ecotype or cultivar of a plant species or of a sexually-compatible species (including through embryo rescue), and can be recombined by sexual crossing (with or without the use of DNA markers). A trait that results or may result from random mutation(s) of genes responsible for the expression of a native trait remains a native trait'. As for native gene (allele), it may be defined as 'a non-scientific term which refers to any gene (allele) part of the gene pool of the target species or species interbreeding with (including through embryo rescue), and including wild ecotypes and cultivars, which may be a product from another native gene (allele) as a result of random mutation events, and can be introduced into a variety of the species by sexual crossing (with or without the use of DNA markers)'.

5 International Convention for the Protection of New Varieties of Plants (UPOV Convention) of December 2, 1961, as revised in Geneva on November 10, 1972, on October 23, 1978, and on March 19, 1991. Text available at www.upov.int/upovlex/en/conventions/1991/content.html/; see also Council Regulation 2100/94/CE on Community Plant Variety Rights. For an extensive description of these matters Chiarolla (2006: 31–42). For a comparison of European and American approaches to patent protection of plants before the widening of protection scope, see Van Overwalle (1998).

6 In the USA, the Plant Patent Act of 1930 (enacted on 17 June 1930, codified as title 35 United States Code) Section 161 states that: '[w]hoever invents or discovers and asexually reproduces any distinct and new variety of plant, including cultivated sports, mutants, hybrids and newly found seedlings, other than a tuber propagated plant or a plant found in an uncultivated state, may obtain a patent therefor, subject to the conditions and requirements of title (Amended September 3, 1954, 68 Stat. 1190)'.

7 A further important aspect that will not be addressed here relates to the laws on certification and commercialization of seeds, which also significantly impact the flexibility breeders and farmers may have in using and exchanging seeds (Anvar, 2008; Wattnem, 2016).

8 The law introduces a new paragraph I. 3° bis in Articles L. 611–191 CPI stating that essentially biological processes are not patentable 'L. n° 2016–1087 du 8 août 2016 pour la reconquête de la biodiversité, de la nature et des paysages' (I. – Ne sont pas brevetables: [. . .] 3 bis Les produits exclusivement obtenus par des procédés essentiellement biologiques définis au 3°, y compris les éléments qui constituent ces produits et les informations génétiques qu'ils contiennent).

9 German law 'Patentgesetz, PatG', Section 2a introduces the non patentability of plant varieties, animal breeds and essentially biological processes.

10 Commission Notice on certain articles of Directive 98/44/EC of the European Parliament and of the Council on the legal protection of biotechnological inventions (2016/C 411/03), https://eur-lex.europa.eu/legal-content/EN/TXT/?uri=uriserv:OJ.C_.2016.411.01.0003.01.ENG&toc=OJ:C:2016:411:TOC

11 Decision of the Administrative Council of 29 June 2017 amending Rules 27 and 28 of the Implementing Regulations to the European Patent Convention (*CA/D 6/17*). www.epo.org/law-practice/legal-texts/official-journal/2017/07/a56.html

12 I call 'non-industrial seeds' seeds that are not registered in official plant variety catalogues, thereby seeds that do not fulfil one or several of the criteria for certification of seed i.e. distinctness; uniformity, stability and value for cultivation and use – for agricultural crops. This notion covers 'non-conventional seeds, old / ancient / forgotten varieties', etc. (Hecquet, 2015 unpublished; see also SEED, 2017).

13 See for example Monsanto's website www.monsanto.com/pages/default.aspx, Syngenta's website www.syngenta.com/global/corporate/en/Pages/home.aspx, or Bayer CropScience's website, available at www.cropscience.bayer.com/.

14 See the Millennium Development Goals (MDG) and in particular Target 1.C to halve, between 1990 and 2015, the proportion of people who suffer from hunger. See www.un.org/millenniumgoals/. The 2015 MDG report states that 'current estimates suggest that about 795 million people are undernourished globally. This means that nearly one in nine individuals do not have enough to eat. The vast majority of them (780 million people) live in the developing regions. However, projections indicate a drop of almost half in the proportion of undernourished people in the developing regions, from 23.3 per cent in 1990–1992 to 12.9 per cent in 2014–2016. This is very close to the MDG hunger target. Rapid progress during the 1990s was followed by a slower decline in hunger in the first five years of the new millennium and then a rebound starting around 2008. The projections for the most recent period mark a new phase of slower progress' (UN, 2015a: 20).

15 MDG Goal 1 Eradicate extreme poverty and hunger. 'Although the MDG targets of halving the proportion of people living in extreme poverty and hunger have been met or almost met, the world is still far from reaching the MDG goal of eradicating extreme poverty and hunger. In 2015, an estimated 825 million people still live in extreme poverty and 800 million still suffer from hunger'. Eradicating poverty and hunger remains at the core of the post-2015 development agenda.

16 Numerous studies exist for different plants. Here are few examples. For wheat: Jones (2004). For fungi: Stamets (2005) who obtained a patent for his invention 'application Ser. No. 09/678,141 for MYCOPESTICIDES, filed Oct. 3, 2000, now US Pat. No. 6,660,290'. For a general overview: Caplat (2014). For an example in El Salvador: (Laplace, 2015).

17 Altieri defines agroecology as an application of ecological science to the study, design and management of sustainable agro-eco systems. This is applied at the farm-level, but also across the global network of food production, distribution and consumption (i.e. including food production systems, processing and marketing, the role of the consumer and the policy level). Agroecology uses knowledge from many disciplines, *inter alia* agricultural and ecological science and traditional knowledge systems. It questions conventional approaches, which are centred on the use of science to promote economic growth. De Schutter (2010b) further specifies that agroecology seeks ways to enhance farming systems by mimicking natural processes, using biological interactions and synergies to support production.

18 Although a recent study from the Metaforum thinktank of the KU Leuven University doubts that agroecology can really feed the world, pointing to the vague definition of the concept, to the fact that agroecology cannot replace conventional agriculture and questioning whether it is judicious to replace a performant system with an agricultural system, which objectives and producing techniques are not sufficiently clear (Metaforum

KU Leuven, 2015: 30–33). However, this report is easily contestable on these points when looking at the very limited number of studies and references referred to on agroecology and when keeping in mind that 'the funding available for organic research is again negligible, remaining at about 2% or the total investment into agricultural research in Flanders' (showing that conventional agriculture strongly remains the dominant position) (Baret et al., 2015: 8).

19 International Treaty on Plant Genetic Resources for Food and Agriculture, FAO Res. 3/2001, 3 November 2001 (entered into force 29 June 2004); 2400 UNTS 379. Throughout the present research, the words 'Treaty', 'Plant Treaty', and 'ITPGRFA' are used interchangeably.

20 The terms 'seed' is used in lay term to designate PGRFA. This use is not in conformity with the actual definition of PGRFA or seeds. It is done so for simplicity of writing.

21 This chapter is a summary of the full legal analysis to be published in a separate book foreseen for 2019–2020.

22 This chapter derives from a stakeholder study published in 2011 by Routledge.

23 The theory of the commons developed following Hardin's paper on the 'Tragedy of the Commons'. It was then widely addressed by Elinor Ostrom, whose seminal book 'Governing the Commons' revolutionized the field. The theory is explained below under the theoretical framework and is detailed in chapter 5.

24 The theory of the Commons gained much visibility in 2009 when Elinor Ostrom received the Nobel Prize in Economic Sciences.

25 Ostrom conducted wide meta-analysis of existing common-pool resources case studies.

26 Ostrom takes stakeholders as a point of departure for her research (whether empirical or theoretical); this approach is close to the research method I have implemented in this book.

27 With the notable exception of Michael Halewood's article (2013).

28 Such as the Intergovernmental Committee (IGC) on Intellectual Property, Genetic Resources and Traditional Knowledge and Folklore, or within the context of the negotiations on Substantive Patent Law Treaty.

29 The negotiations at the IGC are currently on hold, due to political holdbacks and blockages from certain countries. See 'US Proposes Suspension of WIPO TK Committee; Switzerland and Others Counter' (11/09/2015) by Catherine Seas for 'IP Watch' available at www.ip-watch.org/2015/09/11/us-proposes-suspension-of-wipo-tk-committee-switzerland-and-others-counter/ (accessed on September 10, 2015).

30 French National Research Agency (ANR) project 'ANR-15-CE21–0004', *Common Plant*, conducted by Fabien Girard.

31 The contributions to this volume were first presented at an international workshop titled: 'The Commons, Plant Breeding and Agricultural Research. How to face the challenges of an increasing world population ant the preservation of agrobiodiversity', co-organized by the Maison Française d'Oxford (MFO) and the Centre de Recherches Juridiques (CRJ) of Grenoble on the 30 September and 1 October 2016.

32 The Biosafety Cartagena Protocol to the CBD. Signed in Montreal on 29 January 2000 and entered into force on 11 September 2003. 171 Member States on 5 April 2018.

33 See www.ohchr.org/EN/issues/food/Pages/FoodIndex.aspx This definition is in line with the core elements of the right to food as defined by General Comment No. 12 of the United Nations Committee on Economic, Social and Cultural Rights (the body in charge of monitoring the implementation of the International Covenant on Economic, Social and Cultural Rights in those States, which are party to it). The Committee declared that 'the right to adequate food is realized when every man, woman and child, alone or in community with others, has physical and economic access at all times to adequate food or means for its procurement. The right to adequate food shall therefore not be interpreted in a narrow or restrictive sense, which equates it with a minimum package of calories, proteins and other specific nutrients. The right to adequate food will have to be realized progressively. However, States have a core obligation to take the necessary action to mitigate and alleviate hunger even in times of natural or other disasters'. Available at www.ohchr.org/EN/issues/food/Pages/FoodIndex.aspx

34 The Office of the High Commissioner for Human Rights defines the special rapporteur 'as an independent expert appointed by the Human Rights Council to examine and report back on a country situation or a specific human rights theme. This position is honorary and the expert is not a staff of the United Nations nor paid for his/her work. Since 1979, special mechanisms have been created by the United Nations to examine specific country situations or themes from a human rights perspective. The United Nations Commission on Human Rights, replaced by the Human Rights Council in June 2006, has mandated experts to study particular human rights issues. These experts constitute what are known as the United Nations human rights mechanisms or mandates, or the system of special procedures'. Available at www.ohchr.org/EN/issues/food/Pages/FoodIndex.aspx

35 The first special rapporteur on the Right to Food was Jean Ziegler. He performed two mandates from 2000 to 2004 and then to 2008. Olivier De Schutter succeeded with two mandates from 2008 to 2014. The current rapporteur is Hilal Elver.

36 Report 'Crisis into opportunity: reinforcing multilateralism' presented to the Human Rights Council, Follow-up session on the Global Food Crisis at the 12th session, 17 September 2009, at pp. 1 and 22–25, available at www.srfood.org/images/stories/pdf/officialreports/20090917_a-hrc-12-31_en.pdf. See also *Report of Special Rapporteur on the Right to Food*, UNITED NATIONS GENERAL ASSEMBLY (22 August 2010), available at www.righttofood.org/new/PDF/A62289.pdf.

2 A history of the seed international regulatory setting

Alike Harlan's vision, it is important to understand the relational character of agricultural evolution, defined as 'the activities of man that have shaped the evolution of crops and [. . .] the influences of crops in shaping the evolution of human societies' (Harlan, 1975: 3). Alongside the evolution of agricultural crops, regulatory developments through history around the exchange of seeds bring an important insight in the way stakeholders and institutions have managed these exchanges. Understanding the historical timeline of PGRFA exchanges, including the wider picture of all international instruments relevant to seed exchanges (Bragdon, 2004: 12), contributes to identifying challenges in the functioning of the Plant Treaty and its Multilateral System of access and benefit-sharing. The aim of this chapter is to describe the intertwined forces and rules that have grown fast during the second half of the twentieth century in the Global North, and which form the basis on which the current legal setting of the Treaty is constructed (Aoki, 2010). To reach this objective, two questions are addressed. What has been the evolution in plant genetic resources for food and agriculture management since the birth of agriculture? What regime has the relevant international instruments created in the international seed management?

To answer these questions, a descriptive research method is used to 'systematically analyze a legal phenomenon in all its components to present it in an accurate, significant and neatly arranged way' (Kestemont, 2015: 5). This systematic analysis leads to divide this chapter into six sections. Section 1 briefly describes the birth of agriculture. Section 2 focuses on the loss of biological diversity and traces back to early collection missions and ex situ conservation programmes. Section 3 moves on to the modern biotechnology era and explains how these technological advancements and their correlated opening up of IPRs regulations led to the increase of the economic value of genetic resources and to the further commodification of PGRFA. Section 4 explains how the International Undertaking on Plant Genetic Resources has failed to keep PGRFA in the public domain, while section 5 relates to the rise of states' sovereign rights over genetic resources and its concretization through the Convention on Biological Diversity and its Nagoya Protocol. Finally, section 6 expands on the reinforcement of seed appropriation through the International Union for the Protection of New Varieties of Plants (UPOV) 1991 and the 1994 Trade-related aspects of Intellectual Property Rights Agreement (TRIPS Agreement) of the World Trade Organization. Each section of this chapter ends with a concise 'history box' highlighting the important events that took place over that period of time.

The birth of agriculture and its developments

Ten thousand years ago, domestication of crops began, and human movements initiated the wide geographical spread of crops. Humans started their transition from nomad hunters-gatherers to sedentary farmers (Cochran and Harpending, 2009: 65–84), which has allowed for the development of agriculture and of agricultural biodiversity over the last millennia. For example, Sumerians and Egyptians actively collected PGRFA (Esquinas-Alcazar, 2005). Cultural contacts and interactions have resulted in extensive crop diffusion and global transfer of PGRFA (Gepts, 2004). The discovery of the Americas further boosted the intercontinental exchanges.

> For millennia, common heritage has been implicitly used as the principle governing the diffusion of crop and animal genetic resources from centres of domestication, their exchange among farmers, and their introduction into new continents, in particular between the Old and the New Worlds after 1492.
>
> (ibid: 1295)

The birth and expansion of agriculture[1] was made possible thanks to easy exchanges of seeds between farmers, resulting in domestication and diversification of cultivated crops (Harlan, 1995), although this does not mean that seed exchanges and systems happened under no rules (Santilli, 2012: 266; Coomes et al., 2015). This trend still constitutes the core pattern of the present PGRFA management system. However, the historical descriptions below will not focus so much on these early periods of time but rather on the nineteenth century onwards. One should recall, however, that throughout agricultural history, many conflicts have occurred and revolved around the access to and use of genetic resources (Kloppenburg, 2004; Aoki, 2008). As Fowler and Mooney remark:

> [f]rom the earliest times, ownership and control of plants and their diversity have been much more than a merely scientific or technical concern. They have been and will continue to be profoundly political. The strength of nations has risen and fallen; great fortunes have been made and lost; and people have enjoyed plenty or suffered hunger at least in part because of who owned, controlled, used and benefited from genetic diversity, and who did not.
>
> (Fowler and Mooney, 1990: 200)

Box 2.1 The birth of agriculture

- 10,000 years ago: Domestication and geographic spread of crops
- c. 8000 BC: Humans turn from nomad hunters-gatherers to sedentary farmers
- c. 3000 BC: Sumerians and Egyptians actively collect PGRFA
- Last millennia: Development of agriculture and agricultural biodiversity
- 1492: The discovery of America boosts intercontinental exchange

The loss of biological diversity: wide collection and international ex situ conservation programmes as a response

Setting up an international collection and conservation agenda

In the nineteenth century, discoveries by Charles Darwin[2] and Gregor Mendel[3] proved the importance of genetic diversity for biological evolution and adaptation. Scientists started to pay attention to the importance and potential value of genetic diversity (Biber-Klemm and Cottier, 2006: 7–10), and governments of developed countries began to subsidize major collecting campaigns (Aoki, 2008: 12–16). European nations set up a global network of botanical gardens (Aoki, 2008: 11, citing Brockway, 1979; Kloppenburg, 2004: 156–157), and the US (which is fairly poor in genetic diversity) played a key role in the collection, transfer and exploitation of PGRFA (Shands, 1995; Pistorius, 1997: 5–7). Aoki provides several examples of American-funded seed collection projects during the nineteenth century, explaining that only a state power could bear the weight and costs of such collecting and breeding activities that constituted the basis of a stable American agricultural foundation (Aoki, 2008: 13–14). Later, in the twentieth century, Nikolai Vavilov,[4] a famous Russian botanist and geneticist, identified eight centres of origin of cultivated plants (Brush, 2004: 25–28; Andersen, 2008: 17–19) and showed that cultivated plants originated in primary and secondary centres of origin and diversity.[5] He spent his lifetime collecting, studying and improving wheat, corn and other cereal staple crops (Vavilov and Dorofeev, 1992) and created in St. Petersburg one of the biggest gene banks in the world: the N.I. Vavilov Institute of Plant Industry.

After World War II, conscious about the value of genetic diversity and worried by their continuing loss (Dirzo and Raven, 2003: 158–160),[6] Member States of the newly born United Nations' Food and Agriculture Organization (FAO) decided to put the issue on their agenda. In 1958, the US established the first national genebank of long-term storage (the National Seed Storage Laboratory (NSSL) in Fort Collins, Colorado) (Brush, 2004: 195–196). In 1948, FAO created a clearinghouse for plant exploration, recording living collections and removing artificial barriers in the exchange of plants (Pistorius, 1997: 10–14). Following this initiative, three international technical conferences on plant genetic resources were held by FAO in 1967, 1973 and 1981 (Esquinas-Alcazar, 1993). These conferences contributed to raise awareness on the importance of PGRFA conservation. Between 1964 and 1974, the objective of FAO and the International Biological Programme (IBP)[7] was to make a wide survey on genetic resources in the fields to identify where priority exploration and conservation programmes were needed; to develop long-term conservation of seeds (which mainly took the form of ex situ gene banks); to develop a cooperative network of seed storage laboratories and to document in a more systematic manner all genetic information through the creation of a Global Ex Situ Conservation Network (Frankel and Hawkes, 1975: chapters 1 and 37; Pistorius, 1997: Chapter 5). The overall objectives were to

> assemble, and in many instances to salvage what is left of the crop genetic resources, to see that it is preserved against loss and deterioration, to make it generally available to those who can evaluate and use it, and to process and publish all available evaluation records for the benefit of all users.
>
> (Frankel and Hawkes, 1975: 473)

In the early 1970s, Prof. Harlan, a famous botanist and geneticist teaching at the University of Illinois Urbana-Champaign, warned that the increase of the world's population would lead to major hunger crises and that although there would be industrial solutions to food problems, in reality 'there seems to be no way out' (Harlan, 1975: 267). Furthermore, Frankel and Hawkes (1975: chapter 37; see also Christensen, 1987: 281) argued that the amplification of pollution and of seed loss made it urgent for stakeholders to organize themselves through a global exchange network to 'save what can be saved'. In 1972, the UN Conference on the Human Environment held in Stockholm strengthened the international agenda for collecting PGRFA and building long-term storage banks (Brush, 2004: 196). 'Strongly worded recommendations were carried urging governments and UN agencies to save and preserve irreplaceable genetic resources for the good of present and future generations' (Frankel and Hawkes, 1975: 4–5). The Governing Council of the United Nations Environment Programme (UNEP) allocated substantial funds to this aim. Besides, the Consultative Group on International Agricultural Research (CGIAR)[8] created the International Board for Plant genetic Resources (IBPGR) in 1974 which, together with other CGIAR-supported International Agricultural Research Centres (IARC), set up wide collection missions in response to the rapid erosion of PGRFA (Plucknett et al., 1983; Moore and Frison, 2011) resulting *inter alia* from the 'Green Revolution'.[9] Andersen notes that 'the erosion of PGRFA had been increasing at an unprecedented rate due to the "green revolution", and the efforts of the IBPGR were vital to saving plant varieties in danger of extinction' (Andersen, 2008: 89; see also Bragdon, 2004: 13–14). The financial administration of these collection and conservation activities was assigned to the newly created IBPGR[10] (Christensen, 1987: 292–295), while the FAO Expert Panel constituted in 1972 at a conference in Beltsville, US, acted as its technical committee (Pistorius, 1997: Chapters 1 to 4). The die was cast for 20 years of conservation activities (Johnston, 1993) mainly focused on ex situ collections, at the expense of in situ conservation and of crop diversity on farms (Fowler, 1994: 184–188; see also Brush, 2004: chapter 9).

Importance of crop diversity and the continuing loss of PGRFA

In spite of their vital importance for human survival, PGRFA are still being lost at an alarming rate, both in situ and ex situ (Starr and Hardy, 1993: 86–87). Hundreds of thousands of farmers' heterogeneous plant varieties and landraces, which have been developed for generations in farmers' fields until the beginning of the twentieth century, have been substituted by a very small number of modern and highly uniform commercial varieties (FAO, 1996a: 165).[11] In the US alone, more than 90 per cent of the fruit trees and vegetables that were grown in farmers' fields at the beginning of the twentieth century can no longer be found. Today only a few of them are maintained in gene banks. In Mexico, only 20 per cent of the maize varieties described in 1930 are still known. In China, in 1949 nearly 10,000 wheat varieties were known and used. By the 1970s, only about 1,000 remained in use. A similar picture is reported for melon varieties in Spain. In 1970, over 350 local varieties of melons were collected and documented; today no more than 5 per cent of them can still be found in the field. The picture is much the same throughout the world (FAO, 1996b). The FAO's first report on the state of the world's PGRFA (FAO, 1996a) estimated that some 7000 species had been used by humankind to satisfy human basic needs, while today no more than 30 cultivated species provide 90 per cent of human calorific food supplied by plants. Furthermore, 12 plant species alone provide more than 70 per cent of all human calorific food and a mere four plant species (potatoes, rice, maize and wheat) provide more than half of all human calorific food.

The genetic vulnerability[12] caused by the loss of agricultural biological diversity (Christensen, 1987) has not only affected small farmers' livelihoods but has also drastically reduced the capability of present and future generations to adapt to changing conditions. In addition, many neglected crops and many wild relatives of crops are expected to play a critical role in food, medicine and energy production in the near future. Efforts for the conservation of crop diversity and their sustainable use therefore need to be continued and increased (Fowler and Hodgkin, 2004). In this spirit, already in 1990, Fowler and Mooney had identified five 'laws' of genetic diversity conservation, which are still valid even 25 years later:

> (1) Agricultural diversity can only be safeguarded through the use of diverse strategies; (2) What agriculture diversity is saved depends on who is consulted. How much is saved depends on how many people are involved; (3) Agricultural diversity will not be saved unless it is used; (4) Agricultural

> diversity cannot be saved without saving the farm community. Conversely, the farm community cannot be saved without saving diversity; (5) The need for diversity is never-ending. Therefore, our efforts to preserve this diversity can never cease.
>
> (Fowler and Mooney, 1990: 2018)

Box 2.2 The loss of biological diversity

- 1859 and 1866: Discoveries by Charles Darwin and Gregor Mendel
- End nineteenth century: Plant expeditions and first national gene banks
- Beginning twentieth century: Colonialism opens new markets for PGRFA
- 1930s: Nikolai Vavilov identifies 8 centres of origin of crop diversity
- Post World War II: Creation of international institutions dealing with plant genetic resources (FAO, CGIAR)
- 1948: FAO starts technical work on PGRFA collection and conservation
- 1960s: Green Revolution R&D and technology transfer initiatives
- 1974: Creation of the International Board for Plant Genetic resources (now Bioversity International) to support collection and conservation

The rise of the breeding industry, modern biotechnology and IPRs: genetic resources gain economic value

At the beginning of the twentieth century, hundreds of small companies producing seeds, mostly adapted to local conditions, constituted the seed industry in developed countries. Universities and government agencies were the main creators of new varieties through publicly funded plant breeding programmes (Herdt, 1999: 6), which were then provided to seed companies for further improvements at little cost or for free (ibid: 4). At that time, farmers could reproduce seeds of new varieties easily and

> [t]he incentive for private companies to invest in developing new plant varieties was limited because open-pollinated seeds are like a public good – once they exist it is difficult to prevent any farmer from using them (non-excludable), and because they self-reproduce, their use by one farmer does not compete with their use by another (non-rival).
>
> (ibid: 5)

However, the agricultural field and the seed industry landscapes changed dramatically, switching drivers from a general public interest to more private ones (Fowler, 1994: 137–151; Jaffe and Srivastava, 1994). Agricultural science had previously been largely supported and financed by public institutions, such as the United States Department of Agriculture (USDA), or the Institut National de Recherche Agronomique (INRA) in France. This trend had reversed by the end of the twentieth century (Evenson, 2005: 194; Schimmelpfenning and Heisey, 2010). The progressive commodification[13] of PGRFA and the increase of their economic value (Christensen, 1987: 289)[14] were made possible *inter alia* thanks to two trends: the development of modern biotechnologies with high agricultural research and development (R&D) costs and the political decision to widen IPRs over plants (Safrin, 2004: 641; see also Fowler, 1994; Aoki, 2010: 83–105).

On the scientific side, the development of hybrid seeds, and later of DNA-based techniques (i.e. modern biotechnology) and their application to plants, allowed for the production of seeds of special varieties (i.e. genetic modification), for which companies sought IP protection. Indeed, Evenson (2005: 194) says that the 'gene revolution' had depended on the development of IPRs. Agricultural economist Robert Herdt states that, contrary to open-pollinated seeds, '[h]ybrid seeds are not public goods because it is possible for companies to exclude farmers from [re]producing their seed'.[15] DNA techniques allow for the development and improvement of a wide range of new plant traits that increase the economic value of plants (Sedjo, 1992: 210), which opened new possibilities for financial benefits (Bragdon, 2004: 16). This ability to commodify what was once easily taken and used without any control has opened the door to major landscape changes (for a negative assessment of this commodification process Kloppenburg, 2004). What was once a myriad of family-size seed companies producing seeds of hundreds of locally adapted crops, merged to create major multinational agrochemical and seed companies.[16] These mega agro-companies promote uniform high-yielding varieties and control important market shares.[17] Bragdon (2004: 16; FAO, 1996a: 165) specifies that by 1996, the world's top ten agrochemical corporations accounted for 82 per cent of global agrochemical sales and that the top ten seed corporations accounted for 40 per cent of the commercial seed market. These companies have little incentive to produce crops that are important to the poor and disadvantaged farmers who primarily save their own seeds (Herdt, 1999: 8–9). To protect the interests of these companies, developed countries promoted the expansion of the scope of application of IPRs.[18]

On the regulatory side, the development of IP legislations was justified by the argument that R&D in the field of agriculture was very long and costly (Evenson, 2005: 203). Breeders and developers therefore needed to be able to get returns on their investment through the exclusive right to exploit their invention for a limited period of time.[19] In the US since 1930, varieties of vegetatively propagated plants have been patentable under the 1930 Plant Patent Act,[20] while sexually propagated plants[21] have been protected under the Plant Variety Protection Act (PVP Act) since 1970.[22] However, Chiarolla notes that the Plant Patent Act voluntarily limits the scope of patent application for food security reasons. According to him, 'the 1930 Plant Patent Act was specifically designed for the protection of vegetatively propagated (mainly ornamental) plants only, with the stated intention to exclude staple foods, such as grain, from its scope of application' (Chiarolla, 2006: 42). At the international level, the Union for the Protection of New Plant Varieties (UPOV)[23] adopted in 1961 recognized sui generis property rights to breeders over improved plant varieties, although still allowing an access policy for research and further breeding.

With the rapid progress achieved by science in the 1980s and 1990s, in particular DNA-based techniques, new regulatory structures were needed. In 1980, the US Supreme Court decided in the *Diamond vs. Chakrabarty* case to allow patenting of biological organisms, traits and genes. The court held that under Title 35 USC. 101, a live, human-made micro-organism is a patentable subject matter. The 'respondent's micro-organism constitutes a "manufacture" or "composition of matter" within that statute'.[24] Along with the 1991 modification of the UPOV Convention,[25] the adoption of the trade-related Aspects of Intellectual Property (TRIPS) Agreement of 1994[26] led to the culmination of a drive towards internationally recognized protection of IPRs (Chiarolla, 2011: 76) affecting genetic resources generally.[27,28]

What is important to recall from this brief historical background is the shift of perception in the control and ownership over PGRFA. Before World War II, seeds were exchanged easily, for commercial, social and cultural purposes. However, this trend was hampered by several facts, including: 1) the realization of the loss of genetic diversity and its subsequent massive collection campaigns by developed countries; 2) the development of new technologies leading to the various revolutions (industrial revolution in the eighteenth century, green revolution in the 1960s and biotech revolution in the 1990s); 3) the consecutive increase in financial value of specific PGRFA, and the strengthening of IPRs over these PGRFA. These facts have created an atmosphere where poor people, small farmers, and developing countries felt robbed by the multinational companies and developed countries who expanded their control over PGRFA (and exclusion of use) through technological and legal means. Scholars have named this conflict the 'Seed Wars' (Aoki, 2008; Christensen, 1987: 299–301; Mooney, 2011). In the 1980s, Christensen (1987: 280) argued that if 'the rift between the poor and the rich countries continues to widen, access to germplasm may be lost'. This tension pushed for the development of international regulation recognizing the nature of 'common heritage of mankind' to PGRFA in order to secure a free access to these resources for the benefit of present and future generations (Mgbeoji, 2003; Baslar, 1998).

Box 2.3 Modern biotechnology and intellectual property rights

- 1920s: Development of the first (Filial 1) F1 hybrid crop varieties
- 1953: The double-helix structure of DNA is discovered
- 1960s: Green Revolution

- 1961: Creation of the International Union for the Protection of New Varieties of Plants (UPOV) then revised in 1972, 1978 and 1991.
- National legislations restrict access to PGRFA (including through IPRs)
- 1980s: *Diamond v Chakrabarty* U.S. Supreme Court case
- 1995: First genetically engineered crop commercialized (GMOs)

The International Undertaking on Plant Genetic Resources: a failed attempt to keep resources in the public domain

In the 1980s and 1990s, concern arose that genetic diversity from developing countries was being used for profit in the North with no return to the countries of origin (Dutfield, 2004: 52–59). Initial worry focused on plant breeding and the development of new plant varieties in the North through the use and appropriation of the genetic diversity from the South.[29] Developing countries wanted unrestricted access to these new varieties in the same way that developed countries had access to resources in the South. They also wanted to clarify the access and property regime for PGRFA (Fowler, 1994: 159). To this end, the FAO members designed a new instrument: the International Undertaking on Plant Genetic Resources (IU).[30]

The political discussion and negotiating process began in the FAO Conference[31] in November 1979,[32] when the Spanish delegation, later supported by numerous countries, proposed the development of an international agreement on PGRFA and the establishment of a germplasm bank under the jurisdiction of the United Nations. During the 1981 FAO Conference, this proposal became a draft resolution[33] written by Mexico and presented by the Latin American and Caribbean Group (GRULAC)[34] region on behalf of the G-77.[35] As a result, the next FAO Conference (November 1983)[36] approved the first intergovernmental agreement on this subject – with the reservation of eight countries.[37] The same conference established an intergovernmental body – the FAO Commission on Plant Genetic Resources (CPGR)[38] (today the Commission on Genetic Resources for Food and Agriculture (CGRFA), which includes 167 member countries and the European Community)[39] to monitor its implementation (Mekouar, 2002: 2).

The Undertaking is a non-binding agreement based on the principle that plant genetic resources are a heritage of humankind that should be available without restriction (Fowler, 1994: 159–167). FAO Conference Resolution 8/83 states indeed that:

> Recognizing that: (a) plant genetic resources are a heritage of mankind to be preserved and to be freely available for use, for the benefit of present and future generations.

Article 1 (Objective) of the Undertaking specifies that:

> 1. The objective of this Undertaking is to ensure that plant genetic resources of economic and/or social interest, particularly for agriculture, will be explored, preserved, evaluated and made available for plant breeding and scientific purposes. This Undertaking is based on the universally accepted principle that plant genetic resources are a heritage of mankind and consequently should be available without restriction.

It was argued that in this way, unrestricted access to PGRFA would continue to be ensured (Tsioumani, 2005: 121). However, developed countries had precisely used that principle to promote their free access to resources in gene-rich countries (Aoki, 2008: 71). For their parts, developing countries believed that recognizing the heritage of humankind principle in an international agreement would safeguard their interests (Tilman, 2016: 273–297). Developing countries thought that this principle would allow them to access the improved varieties developed using PGRFA originating from the South but improved in laboratories situated in developed countries where financial and technical means were available (Bordwin, 1985: 1069). The IU recognized plant genetic resources to be the heritage of humankind,[40] but this principle was subject to the scope of application of the IU, which had a very broad definition of PGRFA (Aoki, 2008: 71). This meant that PGRFA including land races, wild and weedy species but also breeders' lines, commercial varieties and other products of biotechnologies should be available to everybody.[41] Unsurprisingly, a number of developed countries did not support the IU as they felt that this clause ran counter to their economic interests.[42] This particular issue explains why the IU was approved with eight reservations. Commenting upon these reservations by developed countries, Brush recognizes that

> this position soon leads to the argument that common heritage and intellectual property are incompatible [and that] the intellectual and political divide between these views overshadows discussions of the future of crop resources. Despite the recognition of the importance of international public goods, the subtleties that characterize public goods and the public domain were lost in the political rhetoric of the international debate over plant resources. The contest between these two views takes on political significance because of the increased value of genetic resources, the need to secure a firm financial base to support conservation, and the need to provide for continued access and movement of genetic resources between countries.
>
> (Brush, 2004: 232)

Stoll further argues that it

> cannot be overlooked that the different claims made in this case clearly represent the conflicting interests involved. The recognition of plant breeders' rights and the proprietary character of breeding lines were of comfort to the

> breeding industry – which in those days was mainly situated in the North. The so-called 'farmers' rights' and the concept of a sovereign right on [genetic resources] can be roughly considered a counterclaim of the South. In sum, the example amply shows that plant [genetic resources], which can be considered a public good, have become the subject of claims of different stakeholders. This is likely to cause conflicts in demand, intensive negotiation and result in inefficiencies.
>
> (Stoll, 2009: 7)

In an effort to bring the reluctant developed countries on board and resolve this conflict, three Agreed Interpretations of the IU were negotiated in the FAO Commission between 1983 and 1991.[43] International non-governmental organizations (INGOs) played an essential role in this part of the process (Mooney, 2011). One particularly important initiative was the Keystone International Dialogue Series on Plant Genetic Resources (Frison et al., 2011a: 8). These neutral and non-governmental dialogue series were informal in nature and convened between 1988 and 1991. The process was chaired by Dr. M. S. Swaminathan, who brought together key individuals from government, the private sector, the research community, civil society, international organizations, and others in their individual capacity, to systematically discuss and seek consensual solutions to a range of critical issues. This initiative was very useful in paving the road for the formal intergovernmental negotiations in the Commission.[44] Several points of consensus were identified in these series of informal meetings (Frison et al., 2011a: 8).[45] These included language that IPRs, in particular plant breeders' rights under the UPOV Convention, were not in conflict with the IU.[46] It was also stated that free access does not mean 'free of charge'.[47] The concepts of plant breeders' rights and farmers' rights[48] were simultaneously recognized, while the expression 'heritage of humankind' was recognized as subject to 'the sovereignty of the states over their plant genetic resources'[49] and new concepts such as 'global concern' and 'fair and equitable sharing of benefits' were introduced (Ten Kate and Lasen Diaz, 1997).

The agreed interpretations thus largely led to the abandonment of the common heritage approach (Esquinas-Alcázar and Hilmi, 2008),[50] and set the rationale for compensating traditional farmers as conservationists and providers of PGRFA (Fowler, 1994: 190–192). Brush (2004: 233) regrets that 'conceptualizing crop genetic resources as common heritage leaned towards rules derived for rivalry goods rather than towards the rule regulating public goods in the public domain'. He argues that common heritage intrinsically includes the concept of 'property over any and all biological material'. He concludes that 'the motifs of tangible property and theft dominated this discourse and thwarted consensus about protection of the public domain while promoting social utility'. Furthermore, in my opinion, the recognition of sates' sovereign rights over their PGRFA as a reaction to 'biopiracy' slogans[51] and to the 'Seed Wars', led to a further appropriation of these resources by those very actors (states from the South) condemning the appropriation of their resources through IPRs by seed companies. By affirming sovereign rights over their genetic resources, States reinforced the 'hyperownership' logic. Furthering this trend, in parallel to these developments, countries were negotiating the Convention on Biological Diversity. In response, in 1993, the FAO Conference adopted Resolution 7/93 at its 27th Session, requesting the FAO Director-General to provide a forum for negotiations among governments, for (a) the Revision of the IU, in harmony with the CBD; (b) consideration of the issue of access on mutually agreed terms to plant genetic resources, including ex situ collections not addressed by the Convention; and (c) the issue of the realization of Farmers' Rights.

Besides, within FAO related fora, FAO Member States were building a 'Global System for the Conservation and Sustainable Utilization of Plant Genetic Resources for Food and Agriculture' (the Global System) through the adoption of an International Code of Conduct for Plant Germplasm Collecting and Transfer, Genebank Standards, a Global Plan of Action on Plant Genetic Resources for Food and Agriculture (GPA)[52] and the first report on the State of the World's Plant Genetic Resources for Food and Agriculture (Frison et al., 2011b: 281–291). By implementing these tools, the Global System had the objective to

> ensure the safe conservation and promote the availability and sustainable utilization of plant genetic resources, for present and future generations, by providing a flexible framework for sharing the benefits and burdens. The System covers both the conservation of plant genetic resources (ex situ and in situ, including on-farm) and their sustainable utilization.
>
> (CGRFA, 1997: § 3)

For their part, the CGIAR (also called the CG Centres) noted a serious decrease in access to varieties for their gene banks, thereby justifying the need for a clearer legal access system at the international level. Moore and Frison explain that

> [t]he need to find a more appropriate system of access to PGRFA [. . .] coupled with a lack of clarity over the legal status of the ex situ collections acquired before the entry into force of the CBD, led directly to the conclusion of the In Trust agreements between the CG Centres and FAO.
>
> (Moore and Frison, 2011: 154)

Twelve centres of the CGIAR, and subsequently other institutions, signed in 1994 agreements with FAO to place most of their collections (some 500,000 accessions) in the realm of the IU under the auspices of FAO. Through these agreements, the Centres agreed to hold the designated germplasm 'in trust for the benefit of the international community' and could neither claim legal ownership over the material nor seek IPRs (Gotor and Caracciolo, 2009). According to Brahy (2008: 225), '[t]he heart of the Agreement is the safeguarding of shared access and reciprocity social norms. The Agreement insists on the continuity with past policies and social norms'. Indeed, The Centres also had to make samples of these resources available to users for the purpose of scientific research, plant breeding and genetic resources conservation without restriction.[53] This was carried out through the use of a Standard Material Transfer Agreement (SMTA), identical for all the CGIAR material. Brahy (ibid: 227) notes that '[b]eyond its content, this agreement is an interesting attempt to "formalize social norms", i.e. insert their provision into contracts'. It should be noted that the text of this SMTA was later considered and amended by the Governing Body of the Treaty as the basis for the current SMTA (Moore and Frison, 2011: 157). The agreements provided an interim solution, until the revision of the IU was completed.

Box 2.4 A failed attempt to keep resources in the public domain with the IU on Plant Genetic Resources

- 1979: Beginning of FAO policy discussions on PGRFA
- 1983: Creation of the Commission on Plant Genetic Resources and adoption of the IU on plant genetic resources
- 1989 and 1991: International Keystone Dialogue Series on PGRFA
- 1993: 27th session FAO Conference adoption of Resolution 7/93 to open negotiation for the Plant Treaty
- 1994: 12 CG Centres sign the 'In Trust Agreements' with FAO
- 1995: Broadening of the Commission on Plant Genetic Resources with a mandate to cover all components of biodiversity for food and agriculture
- 1996: Publication of the first report on the State of the World's PGRFA

The CBDs contractual approach to access genetic resources: the rise of States' sovereign rights

Genetic resources as common heritage was further weakened by the CBD,[54] which granted states sovereignty over the genetic resources found within their borders (CBD Articles 3 and 15(1)).[55] The CBD, which aimed at becoming the first legally binding international agreement covering all biological diversity, was negotiated from 1988 to 1992 by the UNEP (Johnston, 1997). It was presented for signature at the Río De Janeiro Earth Summit in June 1992. Unlike the IU, the CBD entered into force as a binding international instrument, thus officially marking the end of PGRFA as common resources as provided for in the IU (Garforth and Frison, 2007: 17).

In 1993, the American National Research Council (1993: 14) had forecast that with the CBD, 'the era of free and open exchange of agricultural germplasm will soon be over'. Indeed, the CBD, which also covers agricultural biodiversity,[56] did not sufficiently take into account the latter uniqueness and the specific needs of the agricultural sector, partly because the CBD was negotiated primarily by ministries of environment and agricultural experts were barely present during the negotiation process (Frison et al., 2011a: 8).[57] Negotiators adopted the CBD to take effective steps to stem the loss and degradation of biodiversity. The major breakthrough of this agreement was the legally binding recognition of states' sovereign rights over their biological resources,[58] which allowed developing countries to nationally control and regulate the access to the resources situated in their territories.

The rationale behind placing genetic resources within national jurisdiction can be further explained by reading the third objective of the Convention (Article 1). This objective is

> the fair and equitable sharing of the benefits arising out of the utilization of genetic resources, including by appropriate access to genetic resources and by appropriate transfer of relevant technologies, taking into account all rights over those resources and to technologies, and by appropriate funding.

The CBD was thus based on both a premise that developing countries have a right over their own genetic resources and a promise that these can be used to engender funding for development and conservation objectives (Garforth and Frison, 2007: 17–18).

However, regarding PGRFA, Wilkes (1987: 216) poses an interesting question when he notes that '[n]ational sovereignty over resources is a recognized right, but can germplasm that left a region long ago still be claimed by the current government? Very few landraces of our modern crops evolved under current governments'. Rosell (1997: 280) or Glowka (1997) point out that states' sovereign right over their genetic resources does not grant states property rights over these resources preserved ex situ (i.e. under this interpretation, property rights regimes are left to national legislation). Stoll (2009: 5) asserts that it 'is the very essence of State sovereignty over genetic resources that States can freely dispose upon such resources for their own uses and to provide them to third parties upon terms and conditions they may deem appropriate'.[59] Brahy goes further by saying that

> [n]ational sovereignty is a way for those countries to negotiate access to their genetic resources, and in so doing, appropriate some of the benefits of their conservation efforts. It is, therefore, an attempt to solve a public goods problem through the creation of property rights.
>
> (Brahy, 2008: 192)

Indeed, under Articles 3 and 15 of the CBD, countries may now set the terms for access to their resources, thus allowing them to profit from their biodiversity, further encouraging conservation. Article 15 of the Convention is entitled 'Access to Genetic Resources'. It reiterates the sovereign right of states over their natural resources and declares national governments to have the authority to determine access to genetic resources. Access to these resources is premised on the negotiation of access contracts. Although there was no obligation within the CBD to limit these relationships to a bilateral approach, most countries which implemented the CBD have developed an administrative procedure where access contracts are negotiated on a bilateral basis (IUCN, 2004; Garforth et al., 2005).

Article 15 states the general principles on which access is to be granted, namely 'mutually agreed terms' and 'prior informed consent', but the specific bargain between access to the resources and the sharing of benefits was left open for negotiation by the parties concerned, thereby leading to a market-based bargaining approach. The article also requires the Parties to the Convention to take measures for sharing the benefits from the use of genetic resources with the Party providing such resources.

However, 12 years after the entry into force, not much progress has been made in implementing the CBD by its 193 contracting parties. Stoll notes that

> [i]n the case of ABS, the environmental objective would clearly be the conservation of biological diversity. If the ABS system were to be judged upon this criterion, the result would be disappointing. There is no indication that, in the 15 years since the adoption of the Convention, the ABS system has had any significant impact on conservation, be that benefits being used to undertake certain conservation measures or halting of the human degradation of biodiversity.
>
> (Stoll, 2009: 4)

Some developing countries created national systems[60] for controlling access to genetic resources and requiring benefit-sharing. Most of these regulatory mechanisms have clearly opted for a purely commercial approach whereby access to national genetic resources is exchanged for benefit-sharing (whether financial or non-monetary, or both) on a bilateral basis. This approach follows a 'coasian-type' of reasoning (Coase, 1960; Merges, 1994), where it is believed that the free market is the best mechanism for valuing and exchanging resources, which are treated as pure commodities (Bertacchini, 2008). The international regulation of contractual access to genetic resources was aimed at promoting such market-based exchanges, especially in the pharmaceutical and cosmetic industries, by generating legal certainty, which in turn is supposed to lower transaction costs. Transaction costs include the cost of communication among parties, search costs, negotiation costs, enforcement costs and the cost of excluding others from sharing the benefits exchanged by parties to the contract (Narasimhan and Robinson, 2008: 16; Brush, 2005: 73).

However, these national systems and the contractual negotiations they require appeared to be expensive to operate and called for highly skilled human resources from a number of disciplines (Dutfield, 2002). Moreover, the link between conservation of genetic resources and the financial arrangement made for their access was not necessarily obvious, and the valuation of the resource was highly controversial (e.g. how can one be sure that a resource with little economic value today would not have a very high economic, social or environmental value in 50 years?) (Aoki, 2008: 92).

Furthermore, developing countries felt that the national implementation of access and benefit-sharing (ABS) was being hindered by a lack of cooperation from developed countries where most of the users of genetic resources are located (Dutfield, 2002). After more than ten years, there was no sign that access to genetic resources had been facilitated through the CBD, or that the conservation of these resources had made any progress (IUCN, 2004: 275). On the contrary, in the field of biological control, van Lenteren et al. (2011) argue that recent applications of access and benefit-sharing principles have made it difficult or impossible to collect and export natural enemies for biological control research in several countries. If such an approach is widely applied, it would impede this very successful and environmentally safe pest management method based on the use of biological diversity. Some experts warned that exchanges through bilateral contracts might limit the access to (and therefore the conservation of) genetic resources (Stoll, 2009: 4; Fowler, 1994: 119–122), in particular within the field of food and agriculture, by rendering the whole system extremely complex, costly and with heavy administrative burdens.[61] Furthermore, a high-level international Panel of Experts on ABS recognized that 'there is a risk that access legislation under consideration in a number of countries might foreclose or restrict the option of multilateral approaches that those same countries may be pursuing in international forums'.[62]

In 2001, an international study (FAO, 2001) prepared at the request of the Global Forum on Agricultural Research (GFAR)[63] showed that access through bilateral agreements resulted in a significant increase in global transaction costs (i.e. negotiation costs, pre-distribution tracking costs and post-distribution tracking costs for both users and providers), which could possibly result in an impasse. This study estimates that a fully bilateral system would cost on a worldwide basis from US$22 to US$78 million annually, while a limited multilateral system would cost from US$8 to US$19 million and a fully open system only from US$1.2 to US$1.9 million (FAO, 2001: 16).[64]

In an attempt to elaborate the ABS requirements of the CBD and help countries develop measures to implement these requirements, the Parties agreed to the Bonn Guidelines on Access to Genetic Resources and Fair and Equitable Sharing of the Benefits Arising out of their Utilization at the sixth meeting of the Conference of the Parties (COP) to the CBD in 2002. However, the Bonn Guidelines are voluntary and therefore do not place obligations on users or user countries. This led developing countries, and a group of mega-diverse countries in particular, to call for the negotiation of an 'international regime' on ABS. Language to this effect was included in the Johannesburg Plan of Implementation adopted at the 2002 World Summit on Sustainable Development and translated into an agreement to launch such negotiations at the seventh COP in 2004. At COP 8, in 2006, two permanent co-chairs of the Ad Hoc Open-Ended Working Group on ABS were named to lead the negotiations. The Working Group completed its work in 2010 when the Nagoya Protocol on Access to Genetic Resources and the Fair and Equitable Sharing of Benefits Arising from their Utilization to the Convention on Biological Diversity (hereafter the Nagoya

Protocol) was adopted.[65] Under both agreements, access to genetic resources is subject to prior informed consent from the Contracting Party providing the resources (CBD, Article 15 § 5) and when appropriate, to the consent of indigenous and local communities (Nagoya Protocol, Article 6 § 2; Morgera et al., 2014: 145). Furthermore, fair and equitable sharing of benefits that arise from the use of genetic resources as well as subsequent applications and commercialization are guaranteed (CBD, Article 15 §§ 3 and 7; Nagoya Protocol, Article 5 § 1) (Coolsaet et al., 2015; Morgera et al., 2012, 2014). An important point to make is that the CBD and in particular the Nagoya Protocol constitute the applicable law for PGRFA that are not used for food/feed purposes, thereby functioning as a fall back general regulation for PGRFA used for biofuels or pharmaceutical purposes for example.

Box 2.5 The rise of States' sovereign rights through the Convention on Biological Diversity

- 1992: Adoption of the Convention on Biological Diversity
- 1993: The CGRFA agrees to renegotiate the IU
- 2002: The CBD COP adopts the Bonn Guidelines on Access to Genetic Resources and Fair and Equitable Sharing of the Benefits Arising out of their Utilization
- 2004: The CBD COP opens new negotiations for an international regime on access and benefit-sharing
- 2010: The Nagoya Protocol on Access to Genetic Resources and the Fair and Equitable Sharing of Benefits Arising from their Utilization to the CBD is adopted (entered into force in 2014)

UPOV 1991 and the TRIPS Agreement: reinforcing PGRFA appropriation

The commodification of crop diversity is defined by Chiarolla (2011: 1) as being 'the adoption, harmonization and implementation of laws and international law instruments, which determine the allocation of legal entitlements to manage and control plant genetic resources, their derivatives and the benefits thereof'. The two major international legal instruments that have contributed to the commodification of crop diversity are the UPOV Convention and the TRIPS Agreement (Chiarolla, 2011); they are briefly described below.

The UPOV Convention[66] provides a sui generis form of IP protection for plant varieties specifically, by setting forth standards for the granting of 'breeders' rights' (Helfer, 2004: 20–32). They are known under the term Plant Breeders' Rights (PBRs). A plant variety must meet several requirements to be eligible for protection. The variety must be (commercially) novel (1991 UPOV Convention, Article 6.1), distinct (ibid Article 7), uniform (ibid Article 8) and stable (ibid Article 9) (the so-called DUS criteria); and the variety must have a denomination (ibid Article 5.2). Under UPOV 1978, it was admitted that a farmer could save seeds from his harvested material and use it for sowing on his own land (the so-called 'farmer's privilege').[67] Similarly, there was a mandatory exception for breeders, who could use protected material for their breeding and research activities. With the 1991 Act (1) the rights of holders (plant breeders) are extended to the harvested material (ibid Article 14(2));[68] (2) a breeder may seek simultaneously PBRs and patent protection; (3) the protection is extended to all plant genera and species (ibid Article 3); and the exceptions are permitted rather than mandatory. The protection is granted for 20 or 25 years, depending on the plant. During that period, the owner of the PBR may exclusively exploit and commercialize the product (ibid Article 19).

As indicated above, under UPOV 1991, exceptions for the use of protected varieties for specific purposes, i.e. seed saving by farmers[69] and the breeder's use for further breeding,[70] are permitted (but no longer mandatory), thereby recognizing the importance of free access to research material in the breeding process (Fowler, 1994: 166). According to Blakeney (2011: 73–75), however, both these exceptions, and in particular the farmer's privilege,[71] have been fairly limited since the 1991 act, raising concerns that these restrictions hamper the informal exchanges practices of farmers and breeders (Cullet and Kolluru, 2003; Cullet, 2005; Shashikant and Meienberg, 2015: 7–9). Indeed, examining the UPOV Guidance documents made available to help member countries establish their national legislation according to UPOV obligations, it is noted that UPOV is promoting a limited interpretation of the farmers' exception (UPOV, 2013: 65). This is further confirmed by the examination of national legislations showing that countries adopt and implement a narrow interpretation of these exceptions in their legislations.[72] Academics (Morten Haugen, 2015; Oguamanam, 2015) have also examined the impact of seed laws with regard to other international obligations (such as conservation, sustainable use and ABS) and promote a more coherent and complementary implementation of the various instruments at stake (Louwaars, 2008; Anvar, 2008; Santilli, 2012; Morten Haugen, 2014).[73]

The Trade-related Aspects of Intellectual Property (TRIPS) Agreement[74] was signed at the WTO ministerial meeting held in Marrakesh in April 1994.[75] The TRIPS Agreement is the first comprehensive international agreement on IP law at the interface with international trade law and establishes minimum standards for different forms of IP legislation.[76] Its wide scope of application has led to a multilateral protection of IPRs (Chiarolla, 2011: 76) affecting genetic resources generally (Helfer, 2004: 33). Article 27 of TRIPS specifies that patents are granted[77] in all fields of technology, although exceptions are provided to protect 'order public and morality, including to protect human, animal or plant life or health or to avoid serious prejudice to the environment'.[78] Protecting plant varieties is compulsory, either through a patent system or a sui generis protection, or a combination of both. According to Blakeney (2011), most countries implement the UPOV protection although the TRIPS Agreement does not name the UPOV Convention in its text as the *sui generis* protection of rights system to be adopted. Although the sui generis option (Dhar, 2002: 7–16) could be used as a more flexible option to provide protection for plant varieties (De Jonge et al., 2015; Koonan, 2014: 1–5; Correa, 2015), the narrow implementation of the TRIPS Agreement (and UPOV) contributes to the idea that international regulations concretize the 'enclosure of the intangible commons of the mind' (Boyle, 2003b: 37).[79]

According to Helfer (2004: 2), granting IPRs to plant varieties – whether patents or PBR – results from developed countries' policy objectives to enrich society's welfare in the field of agricultural innovation. But this can only work for developed countries. Studies have shown that this system is designed for developed economies, and that it is not adapted to agricultural development in developing economies where most seeds are accessed through informal seed systems (Tripp et al., 2007; Chiarolla, 2006; Anvar, 2008; Bertacchini, 2007). Adding to this controversy, experts fear that the promotion of wide IPRs over plants leads to some impediments in research activities. For example, Van der Kooij (2010) promotes the establishment of a breeders' exemption for patent law, to mitigate the reduced accessed to breeding material due to wide patents in the field and promote breeding innovation. Difficulties in identifying proprietary rights (for many different owners, on different technologies and products situated in different jurisdictions) have contributed to tracking issues. Bragdon (2004: 75) gives the anecdotal example of the Golden Rice, where several years of searches and compromise with the rights' holders were necessary for the researchers to conduct their work lawfully, and in the end not even be able to commercialize their product. Furthermore, fears also arise as to blocking-position patents or defensive patents (ibid: 76–78), where right owners claim patents to impede their competitors from accessing the necessary material and technology.[80]

Box 2.6 Reinforcing PGRFA appropriation through UPOV 1991 and the TRIPS Agreement

- 1991: Revision of the UPOV Convention
- 1992: Adoption of the Convention on Biological Diversity
- 1993: The CGRFA agrees to renegotiate the IU
- 1994: Adoption of the Marrakech Agreement on Trade-related Aspects of Intellectual Property Rights (TRIPS)

Conclusion

Scholars have carried out research on the regime constellation for PGRFA – what Oberthür and Stokke (2011) or Jungcurt (2007) call 'regime complex' – which essentially includes five legally binding international agreements: the UPOV Convention, the CBD, the TRIPS Agreement, the Treaty and the newly adopted Nagoya Protocol. Gerstetter et al. confirm that

> wherever there are several legally binding international agreements dealing with at least partially overlapping issues, a conflict between them may arise. Such conflicts narrow the leeway that parties to more than one of the treaties have for implementation and may thus also reduce the treaties' potential for reaching their objectives.
>
> (Gerstetter et al., 2007: 259)

These international legally binding agreements aggregated in a regime complex influence the design and implementation of other regulatory instruments (Fowler, 1994: 170–184), whether national, regional or international. Indeed, regarding PGRFA, the design and implementation of the CBD, UPOV and TRIPS have influenced the negotiations on the Plant Treaty and have contributed to creating a multilateral system where IPRs are explicitly recognized and integrated in the ABS mechanism put in place. Andersen has looked in depth into the regime overlap and interaction and the resulting regime constellation for the management of PGRFA. The effects of this constellation were quite negative regarding the access to PGRFA and the development and recognition of the concept of FRs, and the regime developed has not resolved the 'tragedy of the commons' issue for PGRFA (Andersen, 2008: 352; Brush, 2005: 108–109). While it is outside the scope of this thesis to study this regime constellation in detail, it is interesting to note that Andersen finds that

> [f]rom the history of regime formation and interaction regarding PGRFA management, it appears that developments have been dominated by two main fronts, with some bridge-builders and issue-specific factions in-between. On the one side, there were a few powerful industrialized countries, [which pushed for intellectual property protection regimes through UPOV and TRIPS] [. . .]; on the other, a large majority of developing countries [which imposed sovereign rights over national genetic resources to control their access].
>
> (Anderson, 2008: 173)

It is clear, therefore, that the question of access to genetic resources has strong political and economic forces, superseding the environmental and social objectives of the CBD, the IU or the Treaty (Jungcurt, 2007: 2). The tensions resulting from this situation are examined in Chapter 3.

Notes

1 The study of the origins of agriculture and the historical development of farming has led to the development of several theories. The analysis of these theories fall outside the scope of this research, but further information can be found *inter alia* in the following work (Sauer, 1952; Reed, 1977; Rindos, 1984).

2 Charles Darwin is recognized to be the first scientist addressing the origin of species (Darwin, 1859). However, regarding plants evolution Alphonse de Candolle first discovered the geographic origin of cultivated plants. Alphonse de Candolle (1806–93) was a French-Swiss botanist who was an important figure in the study of the origins of plants and the reasons for their geographic distribution. He also created the first Code of Botanical Nomenclature (Candolle, 1883). This work is his most famous and influential book, tracing the geographic origins of plants known to have been cultivated by humans. It is one of the earliest studies of the history of crop domestication, and an important contribution to phytogeography.

3 Gregor Johann Mendel (1822–1884) is known as the 'father of modern genetics', because he demonstrated that the inheritance patterns of certain traits in pea plants. His contributions to the new science of genetics are now referred to as the laws of Mendelian inheritance.

4 Nikolai Ivanovich Vavilov (1887–1943). While developing his theory on the centres of origin of cultivated plants, Vavilov organized a series of botanical-agronomic expeditions and collected seeds all over the world.

5 A 'centre of origin' can be defined as the geographic region in which a crop first arose. The notion of centre of origin has been subject to debates and is commonly replaced by the expression 'centre of diversity', which can be defined as the geographic region in which the greatest variability of a crop occurs. A 'primary centre of diversity' is the region of true origin of a species; while 'secondary centres' are regions of subsequent spread of a crop.

6 Dirzo and Raven warn that '[f]or the past 300 years, recorded extinctions for a few groups of organisms reveal rates of extinction at least several hundred times the rate expected on the basis of the geological record. The loss of biodiversity is the only truly irreversible global environmental change the Earth faces today'.

7 The International Biological Programme was an effort between 1964 and 1974 to coordinate large-scale ecological and environmental studies (Pistorius, 1997: chapter 2; Frankel and Hawkes, 1975: chapter 1).

8 The CGIAR grew out of the international response to widespread concern in the 1950s, '60s and early '70s that many developing countries would succumb to hunger. Experts predicted widespread and devastating famine between 1970 and 1985, with hundreds of millions starving to death. The roots of the CGIAR go back almost 3 decades before its formal inauguration, beginning with a collaborative programme between Mexico and the Rockefeller Foundation. High-yielding semidwarf varieties of wheat developed in Mexico in the 1950s and of rice developed in the Philippines in the 1960s demonstrated the potential of publicly funded international agricultural research to unlock the productivity of smallholder farms in the developing world. A series of senior consultations – known as Bellagio conferences – invited the World Bank to set up a consultative group for international agricultural research, similar to other groups that it had created to coordinate and support development in individual countries. The World Bank accepted the challenge and led the effort to create the CGIAR in 1971. FAO and UNEP worked with the World Bank as cosponsors, subsequently joined by the International Fund for Agricultural Development. For more information see www.cgiar.org/who/history/index.html.

9 The Green Revolution refers to R&D and technology transfer initiatives, particularly in the developing world, that increased agricultural production worldwide, beginning most markedly in the late 1960s. This movement was led by Norman Borlaug, the 'Father of the Green Revolution', who won the Nobel Peace Prize in 1970. It was credited with

saving over a billion people from starvation, involved the development of high-yielding varieties of cereal grains, expansion of irrigation infrastructure, modernization of management techniques and distribution of hybridized seeds, synthetic fertilizers, and pesticides to farmers. However, today the Green Revolution is widely criticized by some authors for several reasons, including the erosion of a wide diversity of local and traditional varieties of crops. Lappé et al. (1979) condemn the social and economic consequences of the Green Revolution because the boost of food production in some developing countries replaced valuable traditional varieties with high-yielding new varieties of rice and wheat. For an assessment of the movement see Tangley (1987).

10 After 20 years of existence within FAO, the IBPGR became an independent intergovernmental organization, named IPGRI (International Plant Genetic Research Institute), and its scope of research was widened from the initial emergency collection of plant genetic resources to the larger promotion of conservation of plant genetic resources through use. Reflecting the fact that the mandate of the organization now covers all forms of biodiversity (i.e. plants, animals, microbials, etc.), it integrated one of the other IARC – the International Network for the Improvement of Banana and Plantain, INIBAP – and changed its name to Bioversity International since 2006, although the legal name remains unchanged.

11 A recent study on the genetic diversity of commercial varieties of vegetables and apples states that the number of apple and vegetable commercial varieties between 1903 and 2004 has been sustained (Heald and Chapman, 2012). However, maintaining numbers of commercial varieties available on the market does not mean that genetic diversity in situ is not declining, nor that the genetic diversity of the commercial varieties on the market is maintained or increased. Indeed, other studies on wheat varieties, show that the genetic diversity of varieties commercialized has diminished, even though more varieties might be available on the market (Goffaux et al., 2011; Perronne et al., 2016).

12 Genetic vulnerability may be described as the poor deployment of genetic diversity in agricultural production systems.

13 The Concise Oxford English Dictionary defines 'to commodify' as 'to turn into or treat as a mere commodity'. The *Collins English Dictionary* goes further saying that 'commodification' is 'the inappropriate treatment of something as if it can be acquired or marketed like other commodities'. Radin (1996: xi) defines commodification as 'the social process by which something comes to be apprehended as a commodity, as well as to the state of affairs once the process has taken place'.

14 Smolders (no date: 6) and Bordwin (1985: 1057) both recognize that it is hard to value PGRFA. However, Ten Kate and Laird (2000) have attempted to put numbers on the financial value of the commercialization of agricultural products

15 Herdt (1999: 5) goes on saying that '[t]hese techniques permit one to routinely distinguish any individual living organism from any other, regardless of how similar they may look. This makes it possible to identify the ancestors of any plant variety and hence the developers of a variety with near certainty, just as in the animal world a sample of DNA can be matched to any individual with near certainty. That certainty provides a way to exercise property rights over seed varieties because now one can prove who created the original variety'.

16 Hope says that the merger-mania was driven primarily by the need to avoid high transaction costs associated with clearing multiple IPR (cited in Aoki, 2008: 113).

17 A World Bank Report stressed that the way IPRs are implemented affects the structure and concentration of the seed industry and raised concerns for policymakers in the developing world (Louwaars et al., 2005).

18 According to Herdt (1999: 8–9), '[t]he business plans of the mega-seed companies seem straightforward: control everything from genetic engineering of seeds to the selling of seeds to farmers, to marketing plant-grown drugs, modified foods, and industrial products. They aggressively employ patents to claim intellectual property and defend those claims equally aggressively'.

19 IPR are said to enable the owner to capture the full value of his individual investments. This aims at encouraging everyone to put time and labour into the development of such resources. The same evaluation is made for other fields than agriculture where IPRs have developed significantly. An example with the biomedical sector can be found in van Zimmeren (2011: in particular part II). However, Chiarolla (2006: 43) doubts that the positive correlation between stronger IPR protection and a higher level of private capital investment in agricultural R&D leads *per se* to the creation of innovation incentives for private behaviour and therefore enables the increase of welfare for society as a whole. Chiarolla refers to a study by Eaton and van Tongeren demonstrate that increasing IP protection of plant varieties with either PBRs or patents had potential negative effects.

20 Plant Patent Act of 1930, Pub. L. No. 71-245, 46 Stat. 376 (1930) (Evenson, 2005: 192).

21 Sexually propagated plants are those reproduced through ordinary seeds.

22 US Plant Variety Protection Act, 1970 (PVPA), 7 USC. §§ 2321–2582.

23 UPOV was modified several times, resulting in different Acts, with Member States contracting either to the 1961, 1961 amended in 1972, 1978 or 1991 Act.

24 *Diamond v Chakrabarty*, 447 US 303, 206 USPQ 193 (1980); see also the subsequent case *Ex parte Hibbard*, 227 USPQ. 443 (1985).

25 The 1991 Act however adds restrictions on research and breeding activities.

26 The TRIPS Agreement constitutes Annex 1C of the Marrakesh Agreement Establishing the World Trade Organization (WTO) and was adopted in 1994. It strengthens the recognition of property rights over genetic resources (Cullet, 1999).

27 According to Bosselmann (1996: 128), 'the expansion of an international [property right] regime has not only encouraged the technology responsible for potential problems with introduced species, but arguably has created a monopoly situation where the bulk of agricultural seeds are owned by a few, large multinational firms with the resources and technology available to be competitive in this area. Such a monopoly rights system encourages and seeks to solidify an agricultural system that is environmentally damaging and incompatible with the concepts of sustainable development'.

28 Other regulations such as those controlling plant health inspection, pest control in plants and environmental protection also impact the management of PGRFA (Evenson, 2005: 204); however, they fall outside the scope of the present study.

29 For an opposite view, see Harlan (1995: 248), saying that '[fo]r one thing, nothing is stolen. No germplasm is removed from a country or farmer's field that does not also remain. The small samples sent back to the collector's homeland may not even be very representative of the germplasm available and are no loss to either the grower or the nation. Furthermore, little germplasm moves from "south" to "north". [. . .] The "north" invests the money for conservation, not the "south". [. . .] The major movement of germplasm is east to west and west to east'. While this statement might have been true at that time, in 2017, I believe it is not so anymore. Indeed, the development of modern biotechnologies and of IPRs now allows companies to hinder access to a variety providing from a farmer's field. Even though the seed remains in the field, the fact that the farmer may not use it anymore without paying royalties to the company detaining a patent on a gene of that variety equates to a 'removal'.

30 International Undertaking on Plant Genetic Resources, FAO Conference Resolution 8/83, adopted 23 November 1983.

31 The FAO Conference is the sovereign governing body of the organization. It determines the policy, approve the budget of the Organization and exercises the other powers conferred upon it by the Constitution, *inter alia* to make recommendations to Member Nations and Associate Members concerning questions relating to food and agriculture, in order for them to be reviewed and implemented through national action; to make recommendations to any international organization regarding any matter pertaining to the purposes of the Organization. It comprises all Members and Associate Members.

32 C 79/REP – Report of the Conference of FAO, Twentieth Session, Rome, 10–28 November 1979.
33 C 81/REP – Report of the Conference of FAO, Twenty-first Session, Rome, 7–25 November 1981, see FAO Conference Resolution 6/8.
34 The GRULAC is one of the five unofficial geopolitical regional groupings of the UN, with 33 Member States.
35 The G-77 is a loose coalition of developing nations within the UN fora, which nowadays include a total of 132 countries.
36 C 83/REP, at §§ 275–285 – Report of the Conference of FAO, Twenty-second Session, Rome, 5–23 November 1983.
37 The delegations from Canada, France, Germany (the Federal Republic of Germany) Japan, Switzerland, the United Kingdom and the United States of America made reservations with respect to Resolution 8/83 (the International Undertaking on Plant Genetic Resources) adopted in the 22nd Conference of FAO in Rome, November 1983. New Zealand expressed reservations regarding the IU text since it did not take into consideration breeders' rights. The same seven countries and the Netherlands also expressed reservations concerning Resolution 9/83 on the establishment of a Plant Genetic Resources Commission, also adopted in the 22nd Conference of FAO. A note should be made on the fact that the undertaking was not a legally binding international agreement, and that therefore, the various reservations, interpretations and understandings, expressed either verbally or in writing by member nations, are not to be considered 'reservations' within the meaning given to that term under international law and are not binding on the other Member Nations.
38 FAO Conference Resolution 9/83, adopted 23 November 1983.
39 The scope of application of the CGRFA was first limited to plants, but it was broadened to all components of agrobiodiversity in 1995 during the 28th session of the FAO Conference. See Conference Resolution 3/95, in C 95/REP, §§ 65–69 and 10th session of FAO Council CL 110/REP, §§ 13–14.
40 According to Stoll (2009: 6), '[t]he common heritage principle, developed through UN negotiations concerning the uses of the deep seabed and its resources, contains little more than an idea of free access and an air of distributional justice. The same holds true for the Undertaking: it had little to say in view of the fact that such plant [genetic resources] are not a wealth to be distributed, but require investments into the conservation of centres of origin and landraces as well as into improvement of breeds, which is mainly done by private breeders. Today, it looks quite strange that the Undertaking proclaimed such a principle to be applicable to virtually any germplasm with relevance for food and agriculture, including wild species, landraces and highly developed commercial varieties. Indeed, the undertaking was soon modified'.
41 IU, Article 2.1. This principle was reiterated several times in later FAO Conferences: see C1985, § 294: on the importance of 'unrestricted exchange of PGR'; C1989 § 105: 'The draft resolutions, preserving the *principle of unrestricted availability of germplasm*, recognized the rights of both donors of technologies and donors of germplasm to be compensated for their contributions through *the simultaneous recognition of plant breeders' and farmers' rights*. The Conference recognized that both resolutions were intended to lay the basis for an equitable and lasting global system for sharing the costs and benefits of the world's plant genetic resources for present and future generations' (emphasis added).
42 Helfer (2004: 218–219) notes that '[a]lthough the Undertaking was merely a nonbinding statement of principles, it was opposed by the United States and some European governments who argued that the document conflicted with a multilateral treaty – the [UPOV] – but also with their national patent laws, which grant intellectual property rights in isolated and purified genes'.
43 Agreed Interpretation of the International Undertaking, Res. 4/89, UN FAO, 25th Sess., UN Doc. C/89/24 (1989) being Annex I to the International Undertaking; Farmers'

Rights, FAO Res. 5/89, being Annex II to the International Undertaking; FAO Res. 3/91 being Annex III to the International Undertaking.

44 For more details on the role and influence of NGOs in the negotiation process, see Chapter 4 below.

45 In the access and benefit-sharing process, Garforth and Medaglia (2006: 144–146) note that NGO actors were much involved.

46 FAO Conference Resolution 4/89.

47 Ibid., last §.

48 FAO Conference Resolution 5/89. The concept of Farmers' Rights is explained further in chapter 3.

49 FAO Conference Resolution 3/91. Although Bordwin (1985: 1063) points that PGRFA were already subject to the sovereignty of the State in which they are located.

50 See also Stoll (2009: 7), who argues that '[t]wo years later, the FAO conference again modified the system of the Undertaking. At the FAO conference it was decided that "breeders' lines and farmers' breeding material should only be available at the discretion of their developers during the period of development", and thereby acknowledged the proprietary character of such lines. However, in turn, the Conference decided "nations have sovereign rights over their plant genetic resources". Thus, within just a few years, the former "common heritage" has been divided up into various proprietary claims'. See also Mgbeoji (2003).

51 See chapter 2 for some information on biopiracy.

52 The objective of the GPA was to identify the technical and financial needs for ensuring conservation and promoting sustainable use of plant genetic resources (Frison et al., 2011: 281–291).

53 The debate around the ownership of the material conserved in genebanks is a very interesting issue and would require deeper analysis; however, this question falls outside the scope of this study and will therefore not be dealt with in depth.

54 Convention on Biological Diversity, 5 June 1992, 31 I.L.M. 818 (entered into force 29 December 1993).

55 Stoll (2009: 4) reminds that 'the long-standing sovereign right of states over their natural resources, [. . .] is based on the international law principle of territorial sovereignty and has been further developed by UN bodies, and [is] reiterated once more in Article 3 CBD and – in view of biological resources – in the preamble of paragraph 4'.

56 Resolution 3 from the Nairobi Final Act (the relationship between the Convention on Biological Diversity and the promotion of sustainable agriculture) was adopted 22 May 1992 in Nairobi.

57 See also Garforth and Cabrera Medaglia (2006: 143–144) arguing that the lack of intragovernmental cooperation hinders legal reform for ABS.

58 Principle 21 of the Stockholm declaration provides that 'States have, in accordance with the Charter of the United Nations and the principles of international law, the sovereign right to exploit their own resources pursuant to their own environmental policies, and the responsibility to ensure that activities within their jurisdiction or control do not cause damage to the environment of other States or of areas beyond the limits of national jurisdiction'. Declaration of the United Nations Conference on the Human Environment, Stockholm 16 June 1972, 11 ILM 1416 (1972).

59 However, one may question how realistic this assertion is if the resources have left the country for example. Furthermore, how is such an approach to be enforced?

60 Two of the best-known examples are Executive Order 247, Prescribing Guidelines and Establishing a Regulatory Framework for the Prospecting of Biological and Genetic Resources, Their By-Products and Derivatives, for Scientific and Commercial Purposes, and for Other Purposes, 18 May 1995, from the Philippines; and Biodiversity Law, No. 7788, 27 May 1998 from Costa Rica.

61 In a very detailed study (IPGRI, 1996), Bioversity International showed that there is a wide spectrum of options for genetic resources exchange systems, ranging from a strictly

bilateral approach at one extreme, to an unstructured informal multilateral approach at the other. Between these two extremes lie a host of additional options which draw on elements from both ends of the spectrum.

62 CBD Panel of Experts on Access and Benefit-sharing, Costa Rica, 4–8 October 1999, document UNEP/CBD/COP/5/8 at §104.

63 The Global Forum on Agricultural Research (GFAR) is an inclusive global mechanism enabling all those concerned with the future of agriculture and its role in development around the world to come together and address key global needs. GFAR provides an open forum for stakeholders across the agricultural spectrum – from researchers and organizations to farmers – to participate in collaborative discussion and action around the current and future state of agriculture. Established in 1996, GFAR was formed as a project for resource sharing – a commitment that remains the essential purpose of the Forum today. GFAR facilitates collaboration, partnerships and sharing of objectives along the complex pathways from research through to development outcomes. Its headquarters is in Rome, Italy, where it is hosted by FAO. Available at www.gfar.net/.

64 This study was well-received by Plant Treaty negotiators. It appeared useful in convincing parties for the necessity of operating a multilateral ABS system.

65 The protocol entered into force on 12 October 2014. So far, there are 104 contracting parties.

66 The International Convention for the Protection of New Varieties of Plants (UPOV) was first adopted on 2 December 1961. This Act was amended several times, first by the Additional Act of 10 November 1972. UPOV 1978 is the act of 23 October 1978. UPOV 1991 was signed on 19 March 1991 and entered into force on 24 April 1998. As of 13 October 2017, there were 75 contracting parties to this instrument (there were 70 in 2011). The evolution of these modifications reflects the general trend strengthening the rights granted. Bragdon (2004: 64) states that the 'growth of biotechnology and the possibility of formal patent coverage created pressure leading to the 1991 revision of UPOV'.

67 Article 5(1) of the UPOV 1978 Act was interpreted as allowing implicitly farmers to re-sow or exchange protected seeds. Article 5(1) states that: 'The effect of the right granted to the breeder is that his prior authorization shall be required for

- the production for purposes of commercial marketing
- the offering for sale
- the marketing of the reproductive or vegetative propagating material, as such, of the variety.

Vegetative propagating material shall be deemed to include whole plants. The right of the breeder shall extend to ornamental plants or parts thereof normally marketed for purposes other than propagation when they are used commercially as propagating material in the production of ornamental plants or cut flowers'.

68 In comparison, patents cover also the technical processes for the production of protected varieties.

69 1991 UPOV Convention, Article 15(2) provides as an optional exception that 'each Contracting Party may, within reasonable limits and subject to the safeguarding of the legitimate interests of the breeder, restrict the breeder's right in relation to any variety in order to permit farmers to use for propagating purposes, on their own holdings, the product of the harvest which they have obtained by planting, on their own holdings, the protected variety'.

70 1991 UPOV Convention, Article 15.1(i) and (ii); and Article 17.1, which provides a sort of compulsory licensing obligation when it states that 'no Contracting Party may restrict the free exercise of a breeder's right for reasons other than of public interest'.

71 In the 1991 Act, the optional exception is restricted to the following permission: 'farmers to use for propagating purposes, on their own holdings, the product of the harvest which they have obtained by planting, on their own holdings, the protected variety or a variety

covered by Article 14(5)(a)(i) or (ii)'. The wording of the Convention clarifies that the optional exception relates to the use of the product of the harvest by the farmer on his own holding. Thus, for example, the optional exception does not extend to propagating material which was produced on the holding of another farmer (UPOV, 2009:11). For an activists' view on the question see Grain (2007).

72 A more detailed analysis of this topic goes beyond the scope of the present research. However, further information can be found in the following studies from governments (Christinck and Walloe Tvedt, 2015) and from NGOs (La Via Campesina and GRAIN, 2015; Shashikant andMeienberg, 2015). Seed legislations implementing UPOV can be found on the UPOV website at www.upov.int/upovlex/en/.

73 A more detailed analysis of this topic goes beyond the scope of the present research.

74 The Trade-Related Aspects of Intellectual Property Rights (TRIPS Agreement) constitutes Annex 1C of the Marrakesh Agreement Establishing the World Trade Organization (WTO) and was adopted as a result of the 1986–94 Uruguay Round negotiations. It strengthens the recognition of intellectual property rights over genetic resources.

75 Ratification of TRIPS is a compulsory requirement of World Trade Organization membership, which counts 164 members on 29 July 2016. It entered into force on 1 January 1995.

76 TRIPS Agreement, Part II, Articles 1 to 8 cover copyright and related rights, trademarks, geographical Indications, industrial designs, patents, layout-designs (topographies) of integrated circuits, protection of undisclosed information, and control of anti-competitive practices in contractual licences.

77 Article 33 provides that '[t]he term of protection available shall not end before the expiration of a period of twenty years counted from the filing date'.

78 TRIPS Agreement, Part II, Article 27 2. Full text states that 'Members may exclude from patentability inventions, the prevention within their territory of the commercial exploitation of which is necessary to protect ordre public or morality, including to protect human, animal or plant life or health or to avoid serious prejudice to the environment, provided that such exclusion is not made merely because the exploitation is prohibited by their law'.

3 Members may also exclude from patentability:

(a) diagnostic, therapeutic and surgical methods for the treatment of humans or animals;
(b) plants and animals other than micro-organisms, and essentially biological processes for the production of plants or animals other than non-biological and microbiological processes. However, Members shall provide for the protection of plant varieties either by patents or by an effective sui generis system or by any combination thereof. The provisions of this subparagraph shall be reviewed four years after the date of entry into force of the WTO Agreement'.

79 Notwithstanding this argument, one can question whether it is the property rights regime or the way they are implemented in national seed legislations, that most impede access to seeds and technology.

80 For a contradictory position, see Koo et al. (2004).

3 Challenges in the exchange of PGRFA to reward the custodians of agrobiodiversity and promote innovation

Demba is a small farmer in Mali who grows different varieties of millet, sorghum, cowpea and peanuts (i.e. plant genetic resources for food and agriculture) on his 0.35ha field. The varieties he grows originate from exchanging seeds with his family, his neighbours and seeds bought on markets nearby. Demba does not manage to feed his 9 persons family with his production. Droughts, diseases and pests often ruin his work so he would like to increase and diversify his production. Accessing more diverse varieties better adapted to climate hazards as well as improved varieties with higher yields would help Demba. But few are available on the market. Improved seeds sold by multinationals such as Monsanto are too expensive, require 'modern' techniques for growing them (i.e. use of purchased inputs and inaccessible technical machinery), and do not necessarily answer Demba's needs and modes of production (Temudo, 2011). Moreover, patents and 'sterilized' genetically modified varieties hinder Demba from accessing similar varieties by other means. The remaining option for Demba is to access improved varieties at national or international research centres, which develop and conserve such varieties. Hence, research centres also need to have access to the necessary genetic resources to develop these new varieties. However, the present international regime complex renders access to seeds far more difficult than at the beginning of the twentieth century. What tensions have emerged from this regime complex?

Chapter 3 analyses the tensions arising from the multifaceted international seed regime complex, as a necessary preliminary analysis to allow for the evaluation of the present international seed regime conducted in Chapters 4 and 5. Since the dawn of settled agriculture, farmers have developed, conserved and exchanged crop and forage varieties to respond to their needs (Reed, 1977; Pistorius, 1997: xxvii; Smith, 1998). Fowler and Hodgkin explain indeed that:

> [w]hen it comes to food security, all countries share something important and fundamental. All depend on crops domesticated in distant lands during the Neolithic era. As crops, the maize grown in Africa, the wheat that blankets the Canadian prairies, and the potatoes cultivated on more than 10 million acres in China are botanical immigrants, and old ones at that. None are native to those lands. Directly or indirectly, therefore, the world's six billion people depend on crops and, thus, on genetic resources that would not normally be found in and are not part of the indigenous flora of their

> country. The questions of farmer and breeder access to and of availability of genetic resources – seeds, plants and plant parts useful in crop breeding, research, or conservation for their genetic attributes – are of tremendous importance.
>
> (Fowler and Hodgkin, 2004: 144)

The generally open patterns of exchange and use, established by early farmers were continued by public researchers and plant breeders until the middle of the twentieth century (Klose, 1950). This trend perpetuated and amplified a situation that originated in the fifteenth century, where countries all over the world are reliant on PGRFA located within each other's borders (Palacios, 1997; Khoury et al., 2015). Today, there is not a single country that does not need crops originating from other countries or continents to feed its population. From the second half of last century however, access to PGRFA diversity has become far more difficult. This explains why, for the last three decades, besides the promotion of conservation and sustainable use of seeds, facilitating access to plant genetic resources for food and agriculture has been a priority for FAO and other international organizations.

What are the underlying tensions rendering the international seed management system so complex? Indeed, these tensions need to be resolved to effectively facilitate access to PGRFA and reach the Treaty's overall goals of food security and sustainable agriculture. In this chapter, four tensions are described, depicting a global picture around the access problem. These tensions all arise at a horizontal level in terms of land space (between developed and developing countries mainly) and at a vertical level in time (from the 1900s up to nowadays), which further complicates their analysis.

The method used to answer this question follows the same explanatory analysis as in Chapter 2. To grasp a thorough understanding of the regime complex, a wide literature review on the PGRFA management was undertaken at the international level – from the mid-twentieth century to nowadays – both from scientific legal and non-legal literature. This reading enabled to grasp a broad understanding of the international PGRFA management system. Additionally, as mentioned earlier, law in books and law in practice being two different things, my concomitant experience as negotiator and observer in Treaty meetings provided other useful knowledge, used as a support tool to clarify the context and understand the law accordingly. The description of the tensions arising from the international seed regime complex between 1950 and 2004 constitute the basis for the evaluation of the current International Treaty regulatory setting, covered in Chapters 4 and 5 of this book.

Below, four major challenges encountered by stakeholders in the exchanges of and access to PGRFA are exposed. These are: the existing tension between public seeds and intellectual property rights; the tension between advancements in biotechnology by mega agrochemical companies and small-scale farmers; the tension between farmers' seed systems for the exchanges of PGRFA and national or international over-regulation on access to seeds; and the North/South divide.

The tension between 'public seeds' and IPRs: ownership as a factor of rights imbalance

The purpose of this section is to highlight the tension that arose between stakeholders in the agrobiodiversity field following the shift in considering seeds as public goods to considering seeds almost exclusively as private goods. This shift has directly affected and limited the rights of access to seeds by many stakeholders.

The historical timeline for PGRFA exchange and management shows that there is a conceptual debate on the ownership and control of PGRFA. Footer (2000: 49–50) mentions a 'gradual paradigm shift in the attitude towards PGRFA', while Stoll (2009: 8) refers to the development of different entitlements relevant to the use of genetic resources. Some stakeholders would rather treat PGRFA strictly as a private resource, i.e. a commodity subject to market and private property rules. Others would rather see PGRFA as purely public goods to be supplied, financed and conserved by public authorities. Yet others understand PGRFA as containing both a private and a public dimension, calling for a *sui generis* management mechanism. The 'privateness' or 'publicness' of a good is not an intrinsic characteristic of the good. Legal constructs have created these concepts (Boyle, 2003a, 2003b; Heller and Dagant, 2001). Moreover, the dichotomy between the public and private qualification of a good is not so straightforward and comes from a neo-liberal ideological claim. Aoki (1998: 13–15) shows that both categories regularly mix and that intellectual property may be fairly characterized as a jointly social product. Most of the time, a good can easily be moved from a private to a public nature (Heller, 1998).[1] Regardless of how literature names what is called public seeds,[2] the fact highlighted here is the impact on access to seeds for various types of 'ownership' over genetic resources (Chang, 2009), which result from the dual nature of the resource, i.e. seeds as containing genetic information or data potentially protected by IPRs (Drahos and Braithwaite, 2002) and as physical material held by a farmer. Common property, as a third category complementing the public and private ones (Heller, 1998), is sought as an alternative option to overcome the enclosure of seeds. Digging into the property regimes over PGRFA would certainly be very helpful to understand this shift. However, theories of property are very complex and wide. They can be looked at through legal (Honoré, 1961; Demsetz, 1967; Alchion and Demsetz, 1973; Merges, 2001; Demsetz, 2002; Boyle, 2003a, 2003b; Santilli, 2012), economic (Coase, 1960; Merges, 1994; Barzel, 1997; Bertacchini, 2008) or philosophical (Waldron, 1988, 1994; Drahos, 1996; not to cite earlier philosophers such as Locke, Humes or Kant) lenses. From the legal perspective, an important literature led by Douglas Melamed and Guido Calabresi (1972) has developed around the notion of entitlement to property rules, or liability rules. More recently, James Boyle (2003a: 30–31; see also Aoki, 2008: 103) provides an interesting view when he states that the axis of variation is not the 'owned' versus the 'free' but rather the 'individual' versus 'collective' control. Applying an economic lens over property rights theory as Harold Demsetz (1967) and the transaction costs theory of Ronald Coase (1960; also see Merges, 1994; Bertacchini, 2008) serves as a background and allows to understand the PGRFA appropriation process that occurred since the 1960s and the rise of an economic value attached to genetic resources (i.e. the 'green gold').

The aim of this section is not to repeat the work of said eminent academics (Drahos, 2006), nor to apply these theories to biodiversity and traditional knowledge in general (Brahy, 2008) or to the agricultural innovation chain in particular (Batur, 2014), as it has already been done exhaustively. Rather, the purpose of this section is to highlight that the evolution of the property regimes applied to PGRFA – leading to the hyperownership of seeds (Safrin, 2004) – has excessively enclosed seeds and contributed dangerously to creating an 'anticommons' dilemma (Heller, 1998), where seeds are under-utilized. This under-utilization is dangerous at various levels (erosion of agrobiodiversity, lack of subsistence for smallholder farmers, impediments in public research, threats to global food security, etc.) and must be countered by facilitating access to seeds.

Until fairly recently, seeds were considered as '*free* goods' (Kloppenburg, 2004: 15) – goods in 'the public domain' – that is to say in lay terms seeds 'available to all' for further breeding and research (Pistorius, 1997). At the time of the green revolution (Tangley, 1987), improved seeds used in developing countries had the characteristic of public goods: they were easily reproduced, so many farmers could sow them (non-rivalry),[3] and there was no technical means for the seed industry to exclude farmers in developing countries from sowing protected improved varieties (non-excludability). Indeed, before modern biotechnology, open-pollinated seeds (whether improved or not) were widely used by farmers because they self-reproduce easily (Herdt, 1999: 4–5). Today, there are two main means of appropriating seeds (see Figure 3.1 below): i.e. the legal means (e.g. through IPRs) and the technological means (e.g. Genetic Use Restriction Technology (GURT) seeds).

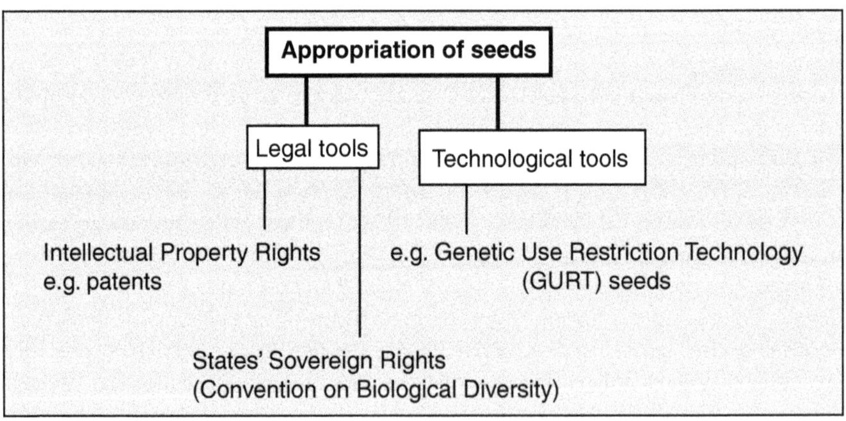

Figure 3.1 Appropriation of seeds

The relatively recent entry of private companies into the plant breeding business, from the late nineteenth century onwards, has been facilitated by the development of IPRs for plants and plant varieties (Aoki, 2008; Louwaars, 2008: 51–112). The development of intellectual property rights over PGRFA reinforced by biotechnology advancements has shifted this 'public' feature to a private one. Previously 'uncontrollable' seeds can now be identified, traced and effectively protected through the enforcement of breeders' rights or IPRs. Tansey states that

> [c]ompanies naturally want to stop others from copying – or buyers reproducing – new products if they can. This can be done in two ways. One is by legal means, through IPRs where such rights can be enforced. The other is through attempting to develop technologies that will stop seeds germinating or specific traits being activated without a purchased input – these are projected genetic use restriction technologies (GURTs) also dubbed 'terminator' and 'traitor' technologies.
>
> (Tansey, 2002: 579)

The knowledge embedded in the seed can now 'legitimately' be managed as a private good (Roa-Rodríguez and Van Dooren, 2008: 187). While these IPRs were intended to create incentives for investment in the private plant breeding sector,[4] they also represented a break in the earlier tradition of unfettered access for a majority of stakeholders (i.e. farmers and small breeders around the world). In particular, it has been argued that IPRs' power to interrupt the open flow and use of plant genetic resources threatens food security and poverty alleviation in developing countries in particular, by reducing their access to essential PGRFA (Runge and Defrancesco, 2006; Bonadio, 2007). As portrayed by the former UN Special Rapporteur on the Right to Food (UN, 2004), the current PGRFA management system seems to experience: 'a marked paradigm shift from a system seeking to foster food security on the basis of the free exchange of knowledge to a system seeking to achieve the same goal on the basis of private appropriation of knowledge'. Furthermore, as Boyle puts it

> [d]o we know that property rights in this sphere will yield the same surge of productive energy that is claimed for the enclosure of arable land? There, I think the answer is a resounding 'No'. We rush to enclose ever-larger stretches of the commons of the mind without convincing economic evidence that it will help our processes of innovation and with very good reason to believe it will actually hurt them.
>
> (Boyle, 2003b: 49–50, and footnote 70)

This position is confirmed by Heller (2010) who contends that private ownership usually creates wealth, but that too much ownership has the opposite effect in that it creates gridlocks.

To grasp the existing legal framework, as well as its strengths and weaknesses, it is important to consider the effect of two trends (Safrin, 2004). One may envision the TRIPS Agreement as the culmination of a drive for multilateral protection of intellectual property rights affecting genetic resources generally (i.e. strong recognition of property rights over genetic resources). On the other hand, one may also view the CBD as the outcome of a pushback by States – especially developing countries – to re-establish control over their basic resources (Çoban, 2004). Indeed, these resources would serve as inputs into the burgeoning new sectors where genetic resources are now being commercially exploited, but states would participate in the benefits of their uses by imposing a benefit-sharing obligation on the user of the resources. However, the implementation of the CBD at the national level has led to the opposite result, i.e. the limitation in the exchanges of genetic resources instead of their expanded access (IUCN, 2004; Garforth et al., 2005). In addition, it has acted as a complementary tool to the appropriation and commodification of biological diversity, reinforcing the TRIPS/UPOV trend. Analysing the interrelationships between the two international regimes of sovereignty and property rights over biodiversity, namely the CBD and the TRIPS agreement, Çoban confirms indeed that

> [w]e have also seen that both property and sovereignty regimes are manifestations of exclusivity and power relations; that both work together in the commodification of life forms; that the creation and realization of IPRs entails the exercise of state-sovereignty rights; and that the structural relationship between the economic/private actors/property rights and the political/state authorities/sovereignty rights manifests a 'separation-in-unity'. In contrast to views that see these two regimes of property and sovereignty rights as contradictory, the paper has thus shown that they are complementary in the process of capitalist accumulation.
>
> (Çoban, 2004: 755)

Consequently, access and availability are subject to (costly) property rights. Tansey explains clearly how the current IPR regime was fitted into the GATT negotiations,[5] and how it was created by a small number of actors representing powerful corporate interests and professionals to fit their specific commercial objectives into global public policies. Burris et al. (2005: 31) confirm that it is important to grasp the ins and outs of the establishment of such regime as '[u]nderstanding how power is distributed and wielded is a precondition for promoting just and efficient governance'. Dutfield and Suthersanen (2008: 22–23) repeat that these developments in intellectual property law began in Europe or North America and are spreading to the rest of the world through agreements such as the TRIPS or bilateral and regional free trade agreements. Consequently, national intellectual property, especially patent, regimes throughout the world are being increasingly held to standards of protection based on those of the most economically and politically influential countries. Based on these facts, Tansey (2002: 580) raises two issues: first, the way global rules are made reflects the clearly imbalanced nature of this process; and second, whether the content of these rules is appropriate (illustrating the 'one-size-fits-all' rule problem).

What was once plenty, available and free is now few, hardly accessible and costly. What was once 'commonly' held 'in trust' (mainly in situ in gene-rich developing countries) by 'Humanity' (Cooper, 1993) is now privately owned by monopolistic agrochemical companies (from developed countries) (Kloppenburg, 2004: 11). As Runge and Defrancesco (2006: 1722) have written 'there is no question that exclusive rights to private property, real or intellectual, confer tangible benefits to those who hold them'; but they continue on saying that this 'may require altering the balance in the bundle of these rights, and a reassertion of the private and public efficiencies gained from real and intellectual assets treated as common property'.

Revealing this tension and to force a rebalancing in the access to PGRFA, initiatives have bloomed worldwide within different stakeholder groups (farmers, academics, breeders, citizens, states and international research centres). These initiatives aim at an alternative path promoting a sustainable agriculture for the collective interest (Altieri and Nicholls, 2012), (i.e. to produce local, diverse, sustainable and healthy food)[6] *inter alia* by sharing, exchanging and conserving seeds 'freely', that is to say in an inclusive manner (which does not mean without any access and management rules). Examples of such initiatives are provided below.

La Via Campesina[7] is the most active and widespread farmers' organization worldwide (Claeys, 2014a, 2014b; Lambek et al., 2014). It was born in 1993 and defends small-scale sustainable agriculture as a way to promote social justice and dignity. It strongly opposes corporate driven agriculture and transnational companies that are 'destroying people and nature'. It comprises about 164 local and national organizations in 73 countries from Africa, Asia, Europe and the Americas. Altogether, it represents about 200 million farmers. Since its birth, one of its objectives is the fight against 'biopiracy'[8] of seeds and genetic information.[9]

Academics at Wisconsin-Madison University, along with breeders and farmers, have recently created the Open Source Seed Initiative (OSSI)[10] to 'keep the new seeds free for all people to grow, breed and share for perpetuity, with the goal of protecting the plants from patents and other restrictions down the line' (Miller, 2014). According to the OSSI website, OSSI is inspired

> by the free and open source software movement that has provided alternatives to proprietary software, OSSI was created to free the seed – to make sure that the genes in at least some seed can never be locked away from use by intellectual property rights. Through [their] Pledge, OSSI asks breeders and stewards of crop varieties to pledge to make their seeds available without restrictions on use, and to ask recipients of those seeds to make the same commitment. OSSI is working to create a pool of open source varieties, to connect farmers and gardeners to suppliers of open source seed, and to inform and educate citizens about seed issues.[11]
>
> (The Open Source Seed Initiative, https://osseeds.org/)

Citizens are also grouping themselves worldwide in associations to promote the free conservation, use and exchange pattern[12] for so-called 'traditional varieties'.[13] New movements on old seed varieties development and exchange seek to provide different, more diverse seeds to people to produce differently, more sustainably and to face climate change.[14] Association Kokopelli[15] created in France or the Garden Organic a UK-based association,[16] and its Heritage Seed Library are examples of such initiatives emanating from the ground. Networks of local and national associations of farmers, citizens, NGOs and other actors also group themselves to develop a collective action, *inter alia* in organic agriculture production and conservation with Réseau Semences Paysannes[17] for example (Demeulenaere, 2014, 2018).

Even states and international research centres collaborate to render more efficient the collective management of ex situ conservation and use of PGRFA. For instance, A European Genebank Integrated System (AEGIS) comprises 34 members and 58 associate members' agreements, manage a total of 25 291 accessions. AEGIS' objective (Maggioni et al., 2015: 14) is to conserve in a *collaborative way* and at agreed quality standards, the genetically unique and important accessions for Europe of all crops and *making them available for breeding and research* through [Standard Material Transfer Agreements] (SMTAs) (emphasis added).

These examples clearly show that at all levels (local, regional, national and international), and within different stakeholder groups (farmers, academics, breeders, citizens, states and international research centres), there is a need for 'dis-enclosing' and facilitating access to seeds, although it might not be for the same purposes overall. The consequences of enclosure resulting from the current property regime associated to PGRFA need to be overcome.

The tension between advancements in biotechnology led by mega-agri-businesses and small-scale farmers: raising an economic imbalance

A famous case law provides a good example of this tension: *Monsanto Canada Inc. v. Schmeiser.* Quoting Aoki (2010: 146), '*Monsanto Canada Inc. v. Schmeiser* established a troubling precedent, one that continues the North American legal trends commodifying seeds, plants and the genetic structures they contain, thereby favouring the interests of large agribusiness at farmers' expense'. Although Schmeiser cannot be associated with the representation one may have of a small-scale farmer in a developing country, this case law shows the imbalance of rights (occurring both in developed and developing countries) between multinational companies such as Monsanto (holding strongly recognized IPRs) and farmers (claiming the recognition and application of poorly recognized rights to save, sow and sell their own seeds), and the resulting growing economic divide between these different stakeholders. Besides the cost related to the law suit directly, Aoki denounces the economic pressure imposed on farmers and mentions that

> [f]armers may also face additional economic hardships such as the loss of the custom designed seed that they have cultivated over time, the loss of

> organic certifications (if contamination via genetic modification appears in the plant or seed), and replacement costs for purchasing new seed and new soil. Under this patent-maximalist view, farmers who do not want to use a patented genetically engineered agricultural technology system should either switch to another line of business or sell their farms to those who will use those systems.
>
> (Aoki, 2010: 159)

This section aims at explaining the economic disparity resulting from the imbalance in the recognition of these rights.

IPRs are closely related to advancements in biotechnology and have clearly contributed to the development of the current agricultural biotechnology business setting (Barton, 1999). While some authors argue that the progress made in biotechnology has been and remains the solution to hunger and poverty (Lenné and Wood, 2011) by producing more food (generally through mono-varietal large-scale cultivation), others remain sceptical as to the social, economic and environmental benefits of such technologies for the majority of the World's population (Tangley, 1987). Kloppenburg (2004: 11–15) argues that the privatization of seeds through the development of biotechnologies and the increase of IPRs led to the commodification of seeds, which in turn allowed for a massive industrialization of agriculture ruled by few monopolistic companies. He argues (ibid: 11) that initially, when the seed was 'unstable', 'the natural characteristic of the seed constitute[d] a biological barrier to its commodification'. He continues saying that the seed was 'rendered a commodity by legislative fiat as well as by biological manipulation'.

Farmers and breeders were once the same person. Today, breeding is a big business with high economic objectives. This massive industrialization changed the face of agriculture in developed and developing countries, propelling seeds to a purely economic value (Engels et al., 2011: 108; also see Charles, 2001a: 92–125). Thousands of family farms were turned into few immense monoculture farms (particularly in developed countries and specific countries in transition such as Brazil). The local hundreds of breeding businesses were swallowed by mega-agrochemical companies (Herdt, 1999: 8). Aoki (2010: 113) cites Janet Hope saying that 'the mergier-mania was driven primarily by the need to avoid high transaction costs associated with clearing multiple IPR, [. . .] and that most key enabling technologies are now in the hands of only a handful of firms'. According to Herdt,

> [t]he business plans of the mega-seed companies seem straightforward: control everything from genetic engineering of seeds to the selling of seeds to farmers, to marketing plant-grown drugs, modified foods, and industrial products. They aggressively employ patents to claim intellectual property and defend those claims equally aggressively.[18]
>
> (Herdt, 1999: 9)

This is confirmed by FAO data (FAO, 2001) showing that '270 patents related to genes of the soil bacterium Bacillus thuringiensis (Bt) granted from 1986 to

1997 in countries of the Organization for Economic Co-operation and Development (OECD), about 60% were owned by only six MNC [Multinational Companies]'. This trend takes away the autonomy of farmers, who become fully dependent upon these companies in every aspects of their work. An example is given with hybrid seeds, which farmers need to repurchase for every planting, as the lower quality of second-generation seeds diminish productivity and quality of the cropping.[19] With regards to the seed industry specifically, Anvar (2008: 433) argues further that it is the unicity of the offer, rather than the limited number of enterprises selling seeds, which constitutes the monopolistic scheme. She concludes that 'the sector proposes its products to customers, which technical and legal constraints have progressively rendered captive'.[20] Under these circumstances, the social and environmental values of seeds and of farmers' seed systems are overlooked,[21] while the social, economic and environmental impacts are high, in developing countries in particular.[22]

In developed countries, strong reactions (largely based on the fears about GMOs) occurred with the development of powerful civil society organizations (Mooney, 2011). Charles (2001a: 97; also see Charles, 2001b, 2001c) confirms that '[o]pposition to biotechnology became a way of opposing agribusiness and promoting an alternative vision to agriculture. The vision went by the name sustainable agriculture'. In his thesis, Brahy (2008: 134–136) clearly explains the public good problems related to the development chain in biotechnology (whether in the agricultural or pharmaceutical fields); i.e. that conservation and R&D activities create profits that are not totally appropriable by the 'conserver' or the developer. This generates externalities (part of the profits are captured by others), which diminish the incentives for conservation and R&D activities. Besides this public good dilemma, research and development activities based on genetic resources collected in developing countries during 'bioprospecting'[23] campaigns allowed big companies to generate huge profits, without compensating the country of origin (and in particular their local and indigenous communities) (Hamilton, 2006; Adair, 1997). Socio-environmental activists quickly made a link between IPRs as a tool for promoting innovation mainly in developed countries and the new concept of 'biopiracy'. Perhaps the most vocal criticisms of biopiracy have come from the Action Group on Erosion, Technology and Concentration (ETC Group) who define biopiracy (see their website) as: the appropriation of the knowledge and genetic resources of farming and indigenous communities by individuals or institutions who seek exclusive monopoly control (patents or intellectual property) over these resources and knowledge. Vandana Shiva (2001) states that biopiracy refers to the use of intellectual property systems to legitimize the exclusive ownership and control over biological resources and biological products that have been used over centuries in non-industrialized cultures (see also Hamilton, 2006: 160; Karjala, 1994; Rojahn, 2010: 29). Although these developments were not directly linked to the food and agriculture sector, the biopiracy claim spread by civil society organizations (McGown, 2006)[24] had an indirect impact on the international instruments developed to regulate these issues.[25]

In response, global public policies attempt to reconcile the tension between advancements in biotechnology by multinational companies and small-scale farmers. For example, the 1996 World Food Summit takes into account a more social dimension into global public food policies.[26] By using the benefits of biotechnology for the poor,[27] the CGIAR also attempts to mitigate this gap, in particular with the recent focus on the nutritional value of PGRFA[28] and the impact of climate change on local production (Bellon et al., 2011). In 1999 already, Serageldin was calling for 'a double shift in the agricultural research paradigm':

> The first involves integration of crop specific research into a broader vision that includes sound management of natural resources, as well as the productivity and profitability of smallholder farming; promoting synergies among livestock, agroforestry, food and cash crop, [. . .] and recognition of the socioeconomic realities of farmers. The second shift is to harness the genetic revolution. Cutting-edge work associated with genetic mapping, molecular markers, and biotechnology must be focused on benefiting poor people and the environment. It is vital to realize the promise of this revolution while avoiding the pitfalls.
>
> (Serageldin, 1999: 387–388)

This section aimed at describing the gap, which grew wider and wider, between a few mega agrochemical companies controlling the agricultural market and a vast number of small-scale farmers around the world (and mainly in developing countries), who encounter increasing difficulties in producing their food. This gap contributes to strongly politicize the sphere of agricultural negotiations.

The tension between 'informal' exchange networks and 'over-regulation' of access to seeds: raising a social sharing disruption

The tension between farmers' seed systems for the exchanges of PGRFA (Louwaars, 2008: 29–50) and national or international over-regulation on access to seeds (Serageldin, 1999) is at the heart of the current economic and political hurdle of seed management. Pautasso et al. (2013: 165) point that '[s]eed exchange is an important, yet poorly understood, factor shaping agrobiodiversity and helping its dynamic conservation'. They further (ibid: 152) recognize that 'the methods available to study the role of seed exchange networks in preserving crop biodiversity have only recently begun to be considered'. At all levels of use, accessing available PGRFA is a necessity, whether to grow one's food or to carry out breeding and research activities. However, the current over-regulation (resulting from the above-mentioned regime complex; see Chapter 2) disrupts the fluid exchange of seeds through various networks, at different levels. This section highlights the impact of such disruption in seed networks on social sharing.

For immemorial times, farmer and breeder communities have managed the exchange of PGRFA (Pistorius, 1997; Crucible II Group, 2000) and have accessed seeds through local markets or farmers' seed networks as a social practice (Visser and Smolders, 2003). As Pautasso et al. state (2013: 156), 'seed circulation is typically a social process: it is based on trust, may or may not be reciprocal, and is influenced by socio-cultural norms and practices'. Bertacchini (2008: 91) confirms that '[s]uch an exchange does not occur as a bilateral impersonal market transaction.[29] On the contrary, it is grounded on mechanism of reciprocity and cooperation. Indeed, traditional seed systems are mostly based on social and family relations, cast in the context of mutual interdependence and trust, often forming dynamic networks with a high degree of complexity'.[30] Until recently, it was common practice for breeders and researchers to similarly access seeds through farmers' seed networks and between collaborators from different laboratories, situated in different regions or continents. This common practice was made possible because there was no 'ownership' over PGRFA, controlling access under specific conditions.

However, the regulatory intensification of the past 30 years created a regime complex (see Chapter 2) where an 'over-regulation' of access to seeds has favoured only one type of situation: i.e. the dominant neo-liberal monopolistic seed market. Anvar (2008) confirms that the current regime hinders the mere existence of any other type of seed system, where flexibility and heterogeneity would allow for a plurality of situations to coexist, thereby unlocking the current monopolistic and enclosing system. Furthermore, the current system narrows the value of seeds and seed networks to merely an economic value, whereas it is recognized that this value is much wider and touches upon cultural heritage, social dimensions, ecosystems, etc (Cleveland et al., 1994; Ellen and Platten, 2011; Veteto and Skarbø, 2009; Sthapit, 2008). In his thesis, Bertacchini (2008: 90) argues that '[t]raditional farming systems – with practices of seed saving and exchange based more on reciprocity and cooperation – should be seen as a *form of social sharing* for germplasm production and distribution, which in turn enhances crop genetic diversity' (emphasis added). However, the current setting leads to a gridlock, a situation where 'formal' and 'informal' seed networks are opposed and competing. Whereas McGuire and Sperling (2016: 23) call for more integration of both formal and informal seed networks in order to 'deliver the types of products needed to catalyse smallholder advances: to encourage increased production; nutritional gains; and to foster farming system resilience'. They conclude *inter alia* that 'the seed sector strategy has to become more smallholder-focused'. Formal seed networks are locked by economic pressures (i.e. monopolies; conventional agriculture and increasing use of hybrid varieties) and by complex legal regulations (CBD and its contractual approach to accessing genetic resources, IPRs, UPOV, constraining national seed legislations),[31] which both hinder the social sharing dimension of seed networks. The 'formal' system grows in time and space, reducing the vital space for 'informal' networks to coexist, and therefore reducing the viability of their cultural, social and ecological purposes. Since neither market approaches nor neo-regulatory solutions are supposed to fit traditional farmers' interests within such a system, Bertacchini (2008: 90) explores the options and economic implications to support the traditional farmers' organization of germplasm production and distribution and refocus the debate on the social sharing value that characterize traditional agricultural systems. However, few studies exist on the role(s) of seed networks in preserving and sustainably using crop biodiversity.

As Pautasso et al. (2013: 157) argue, further (interdisciplinary) studies are needed to understand better the role of seed exchange networks in biodiversity conservation and use in a more holistic manner. They state indeed that 'with seed exchange itself, it is difficult to separate purely biological from social factors [. . .]; rather, these factors interact to a considerable degree, both in cause and effect'. This need for further studying seed networks is even more important in that 'local seed exchange networks are essential to agrobiodiversity conservation, because they permit access to seed and the maintenance of landraces in agroecosystems throughout the world, despite the trend towards more uniform seed material flowing through formal, commercial seed systems' (ibid).[32] Pautasso et al. suggest an exhaustive list of research methods that include: ethnographic fieldwork, participatory approaches, public good experiments, biogeography and landscape genetics, simulation models, scenarios, statistical analysis, indicators, life cycle assessments and impact evaluations, meta-analyses and finally network analyses. To complement this last research method proposed, nodal governance could be added as one method particularly useful in analysing the social sharing aspect of seed exchange networks. According to Burris et al (2005: 33) '[n]odal governance is an elaboration of contemporary network theory that explains how a variety of actors operating within social systems interact along networks to govern the systems they inhabit' (also see Drahos, 2004; Holley and Shearing, 2017). This approach could contribute to enhancing the continuum between formal and informal seed networks (favouring systems' heterogeneity, flexibility and pluralism) and between traditional and improved varieties rather than supplanting one system by the other.[33]

The North/South divide: a political stake

The purpose of this section is to highlight that the North-South divide renders reaching food security and sustainable agriculture objectives necessarily more complex to attain. The differences in political discourse and objectives between countries of the North and countries of the South[34] complicate the negotiation and implementation of international environmental treaties (Gonzalez, 2015: 408). The aim here is not to review the literature on North-South international relations (see *inter alia* Ravenhill, 1990; Gonzalez, 2015), nor to examine its origin in colonialism and post-colonial policies (ibid: 411–420). The analysis will be limited to what is directly relevant to the seed issues dealt within this thesis.

Regarding the international negotiations on biodiversity (in general) and on access and benefit-sharing in particular, De Jonge and Louwaars (2009: 28–40) recognize that there is a South-North imbalance in resource allocation and exploitation (see also Esquinas-Alcázar, 2005: xxv). Studying the perceptions of different stakeholders, they have identified six principles underlying ABS. The first three are 1) the South-North imbalance in resource allocation and exploitation, 2) biopiracy and the imbalance in IPRs and 3) an imbalance between IP protection and the public interest. These principles are driven by the perception of imbalance and a motivation to increase equity. The last three principles are 4) the need to conserve biodiversity, 5) a shared interest in food security and 6) the protection of the cultural identity of traditional communities. This second set of principles concentrates on other aims such as nature conservation, food security and the preservation of traditional cultures.

The above identified first three tensions confirm this analysis in that the mentioned tensions (between 'public seeds' versus IPRs; advancements in biotechnology versus small-scale farmers; and informal exchange networks versus over-regulation of access to seeds) are crystallized in the more general divide between the gene-poor but economically and technologically rich Northern countries and the gene-rich but economically and technologically poor Southern countries. This picture will be further analysed in this final Section.

Over the last century, parallel developments took place within three related fields (the economic, environmental and agricultural fields), which all led to a conviction of dispossession by and of the 'poor' and the 'small'. In the environmental field, bioprospecting activities – initially aimed at contributing to the conservation of genetic resources – were rapidly associated to biopiracy campaigns, which sole objective was seen as procuring more benefits to big companies from the North. People from developing countries voiced that these companies were stealing their resources and their associated traditional knowledge without any authorization or compensation.

In the agricultural field, the rise of GMOs and other modern biotechnologies, coupled with the expansion of patents over plants, has increased farmers' concern that they have been dispossessed of the seeds they had conserved, developed and improved over millennia, without any recognition for their role or compensation. The shift of the control over seeds and their access, from the collectively managed access by farmers in their fields, to individual and excluding rights of mega-companies in their laboratories has crystalized the North-South divide. Stoll writes that

> [t]he recognition of plant breeders' rights and the proprietary character of breeding lines were of comfort to the breeding industry – which in those days was mainly situated in the North. The so-called "farmers' rights" and the concept of a sovereign right on GRs can be roughly considered a counterclaim of the South.
>
> (Stoll, 2009: 7)

In the economic field and from a global perspective, recent studies published by Oxfam (2014; 2016; also see Hardoon et al., 2016) emphasizes the growing economic divide between the majority of the world's (poor) population and a very small percentage of it benefiting from most of the global wealth.[35] As regards to food, the lack of resources to grow or purchase sufficient food to satisfy their dietary needs plunges 805 million people in chronic undernourishment (FAO, 2014b: 11). There is today a wide recognition that current social and economic inequities, across and within regions and states, are a significant barrier to achieving development goals (IAASTD, 2009: 17–24). Gonzalez (2015: 433) strongly promotes a 'fundamental restructuring of international economic law' in order to live in 'a just and sustainable planet'. Throughout her analysis, she observes that '[i]nternational economic law systematically accelerates environmental degradation, subordinates the global South, and consigns environmental issues to the peripheries of legal discourse and policy-making' (ibid). Trade liberalization, which opened developing country markets to international competition too quickly or too extensively, further undermined the rural sector and rural livelihoods (IAASTD, n.d.). Indeed, the structural changes in the agricultural market (in production and consumption) took away autonomy from farmers, who became fully dependent on few monopolistic companies in every aspects of their work. Aoki goes further by adding that

> [t]he contemporary global agrifood system is capital intensive, centralized and consolidated, sourcing labor and natural resources where they are cheapest and selling where they bring the highest prices. In markets for other goods, if people can't generate a demand, they go without the particular good.
> (Aoki, 2011: 478)

These developments have strengthened the above mentioned ascertainment of inequity and widened the South/North divide in international negotiation and failure in the implementation of international environmental treaties. Gonzalez concludes her analysis with these words:

> [a] systematic examination of international environmental law from a North-South perspective can expose the historic and contemporary inequities that have compromised the effectiveness of international environmental law and hindered our ability to address the pressing environmental problems confronting the global community. This article has provided an overview of the origins of the North-South divide in colonial and post-colonial economic law and policy and the failure of sustainable development to remedy its social, economic and environmental consequences. The objective is to provoke further discussion and analysis about new approaches to international environmental law that will promote environmental justice in an era of growing economic inequality and looming ecological collapse.
> (Gonzalez, 2015: 434)

Conclusion

With Chapters 2 and 3, I drew a complete picture of the international seed regulatory system that developed during the twentieth century and of the tensions that arose from it. Understanding the past system and its weaknesses is a preliminary necessary step in order to assess the present regulatory setting (in Chapters 4 and 5) and suggest ways forward to develop a future more equitable and effective scheme (in Chapter 6). The hypothesis framed was that the historical evolution of PGRFA management has transformed what was previously considered as public goods available to all into overly privatized goods, accessible to few following strict (legal, economic and technical) access conditions. This evolution has crystallized an imbalance of rights pertaining to seeds and contributed to further limit access to and exchanges of seeds between all stakeholders. In turn, the limits in seed exchanges have weakened seed conservation and sustainable use objectives and rendered urgent the negotiation of a new international legally binding instrument to protect and promote seed conservation, sustainable use and fair and equitable exchange.

Notes

1 A note is made, however, regarding the physical possibility of 'privatizing' an open-pollinated seed (protecting it with an IPR), before specific innovation occurred in agricultural development allowing for the control in the use of the seed.
2 Depending on their background and disciplines, authors have associated various terminologies and concepts to the inclusive and openness characteristics of seeds such as 'public domain', 'global commons', 'common heritage of mankind', 'global public goods', 'pure global commons', etc. What is important to bear in mind at this stage is the 'public' dimension of seeds, as opposed to privately owned rights over seeds.
3 See Figure 5.1 for the quadrant of goods showing the non-rivalry and non-excludability characteristics of a public good, in Chapter 5.
4 Batur (2014: part II) explains exhaustively the different innovation contexts according to three different types of actors, and the impact of the enclosure of agrobiodiversity on the activities of these stakeholders.
5 See also Dutfield and Suthersanen (2008: 22–23) stating that the reason why commercial importance of intellectual property rights has grown considerably since the nineteenth century, and has accelerated since the 1970s, is because of the incessant and increasing pressure on businesses and national economies to be competitive.
6 Examples of such organisations are Arche Noah, Kokopelli, Pro Specie Rara, Red de Semillas, Réseau Semences Paysannes, Rete Semi Rurali, etc.
7 La Via Campesina claims to be an autonomous, pluralist and multicultural movement, independent from any political, economic or other type of affiliation. The main goal of the movement is to realize food sovereignty and stop the destructive neo-liberal process. It is based on the conviction that small farmers, including peasant fisher-folk, pastoralists and indigenous people, who make up almost half the world's people, are capable of producing food for their communities and feeding the world in a sustainable and healthy way. See http://viacampesina.org/fr/
8 For a recent press release on the matter see 'The Seed Treaty Undermined by the Gangrene of Biopiracy', posted on 15 October 2015, available at http://viacampesina.org/en/index.php/main-issues-mainmenu-27/biodiversity-and-genetic-resources-mainmenu-37/1886-the-seed-treaty-undermined-by-the-gangrene-of-biopiracy
9 The issue of digital sequenced information was a burning one during the eventh meeting of the Governing Body of the ITPGRFA in October 2017, Kigali, Rwanda. It will attract the attention of most of the Treaty community for the coming years. The use and appropriation of genetic information will be dealt with in chapter 5 below.
10 This initiative is led by Jack Kloppenburg, professor at the Department of Community and Environmental Sociology, Irwin Goldman centre, chair of the Department of Horticulture.

A similar open source initiative exists in India: the 'Centre for Sustainable Agriculture' (CSA), see CSA, 'Open Source Seed Systems', online document available at http://csa-india.org/wp-content/uploads/2014/11/Open_Source_Seed_Systems_1.0.pdf.

11 Open Source Seed Initiative website, available at https://osseeds.org/ (accessed April 2015).

12 An example in France: Réseau Semences Paysannes functions as a network of local and national associations of farmers, citizens, NGOs and other actors involved in organic agriculture production and conservation (see www.semencespaysannes.org/).

13 I call 'traditional varieties' seeds that are not registered in official plant variety catalogues, thereby seeds that do not fulfil one or several of the criteria for certification of seed i.e. Distinctness; Uniformity; Stability; and Value for cultivation and use – for agricultural crops. This notion covers 'non-conventional seeds', 'old / ancient / forgotten varieties', etc. as opposed to 'biotechnological seeds' (SEED, 2017).

14 E.g., varieties that do not need irrigation to grow. It should be noted that the objectives are the same as transgenic seeds resisting to drought, but the means to reach that objective are different as well as the related societal objective.

15 Association Kokopelli website, available at https://kokopelli-semences.fr/fr/ (accessed April 2015).

16 Garden Organic aims at conserving and making available to its members, through an annual catalogue, vegetable varieties, mainly of European varieties, that are not widely available (see www.gardenorganic.org.uk/).

17 Réseau Semences Paysannes website, available at www.semencespaysannes.org/ (accessed April 2015).

18 See also (Aoki, 2010: 151) the above mentioned case law *Monsanto Canada Inc. v. Schmeiser,* where 'Monsanto's utility-patent rights and licensing agreements put those who are accidental users due to natural forces in the position of being vulnerable to lawsuits for patent infringement'.

19 Another illustration of this dependence imposed on farmers is provided by the Genetic Use Restriction Technology (GURT). Also called 'terminator seeds', this method regulates gene expression and restricts the use of genetically modified plants by causing second-generation seeds to be sterile. Eventually however, GURT seeds were never commercialized. At the CBD-SBSTTA 4 meeting, which took place in Montreal, 21–25 June 1999, Recommendation IV/5 recommended a moratorium on field testing and commercial use of such technology 'until appropriate, authorized and strictly controlled scientific assessments with regard to, *inter alia*, their ecological and socio-economic impacts and any adverse effects for biological diversity, food security and human health have been carried out in a transparent manner and the conditions for their safe and beneficial use validated'. Available at www.cbd.int/recommendations/sbstta/?m=sbstta-04 At the Eighth Meeting of the Conference of the Parties to the CBD, which took place in Curitiba, Brazil in 2006, peasant and indigenous rights activists strongly opposed the technology and demonstrated outside the doors of the meeting venue in support of a complete ban on the sale and use of Terminator seeds.

20 Personal translation of '[l]a filière propose ses produits à une clientèle que les contraintes techniques et juridiques ont progressivement rendue captives'.

21 Kloppenburg (2004: 184–189; also see Coomes et al., 2015) points to the fact that it is generally 'asserted that only the application of scientists' labour [in breeding activities] adds value to the natural gift of germplasm' thereby ignoring any other sort of value the seed might have, i.e. cultural, social, or ecosystemic values.

22 Although Brush (2001: 162) argues that developing more GMO agricultural research may have a positive social impact in developing countries.

23 Initially, bioprospecting was seen as a way to access, collect and exploit plant and animal resources. The *Compact Oxford English Dictionary* defines bioprospecting as 'the search for plant and animal species from which medicinal drugs and other commercially valuable compounds can be obtained'. Bioprospecting activities were also justified by conservation

purposes, through the claim of collecting genetic resources to conserve them in ex situ gene banks.

24 However, there are also opponents to the biopiracy claim; see for example Chen (2006).

25 Brush (2004: 232) argues that the 'charge of biopiracy [. . .] epitomized the demise of the open collection of plants and the impoverishment of discourse by sloganeering. Few people bothered to ponder the nature of the ex ante system of common heritage. Rather, discourse on biological resources shifted rapidly to control and ownership by different actors: Nation-States, indigenous people, seed companies, international organizations, or research institutions'.

26 The Rome Declaration on World Food Security was adopted at the World Food Summit, 13–17 November 1996, Rome, Italy. The Summit was called by FAO and aimed at reiterating global commitments to fight hunger and react against widespread under-nutrition and growing concern about the capacity of agriculture to meet future food needs. The Rome Declaration calls for the members of the United Nations to work to halve the number of chronically undernourished people on the Earth by the year 2015. The conference produced a second key document: the World Food Summit Plan of Action. The Plan of Action sets a number of targets for government and non-governmental organizations for achieving food security, at the individual, household, national, regional and global levels. Full text available at www.fao.org/docrep/003/w3613e/w3613e00.htm

27 The Green Revolution is a controversial example of such a policy. Lappé et al. (1979) condemn the social and economic consequences of the Green Revolution (because the boost of food production in some developing countries replaced valuable traditional varieties with high-yielding new varieties of rice and wheat); Conway and Toenniessen (1999: 55–56) argue that there is a need for a second Green Revolution, which would avoid the pitfalls of the first one, and bring more benefits to the poor. They 'point to the need for a second Green Revolution, yet one that does not simply reflect the successes, and mistakes, of the first. In effect, we require a "Doubly Green Revolution", an agricultural revolution that is both more productive and more "green" in terms of conserving natural resources and the environment than the first. We believe that this can be achieved by a combination of: ecological approaches to sustainable agriculture; greater participation by farmers in agricultural analysis, design and research; and the application of modern biotechnology directed towards the needs of the poor in developing countries'.

28 Frison et al (2011: 246–247) believe that '[t]here is a new recognition of the profound challenges faced in increasing production to meet the needs of a growing population under changing climates and the need to do so in a sustainable manner. [. . .] While the temptation will always be to look for quick fixes, these are unlikely to be sustainable or to meet current concerns for an environmentally acceptable agriculture that responds to the needs of small-scale farmers throughout the world. Almost all of the approaches used to date in agricultural intensification strategies, for example the substitution and supplementation of ecosystem function by human labor and petrochemical products, contain the seeds of their own destruction in the form of increased release of greenhouse gases, water supplies depleted by mining, and degraded soils. We need to build production systems that deliver intensification without simplification'.

29 A wide-ranging study by McGuire and Sperling (2016: 13), based on a uniquely comprehensive data set of 9660 observations across six countries and covering 40 crops, list eleven different means of accessing seeds, that is to say through: the farmer's own stock, exchange, gift, bought, vouchers, direct seed distribution, seed loan, food aid, money credit, casual labour, and 'other'.

30 McGuire and Sperling (ibid: 20) demonstrate that 'farmers access 90.2% of their seed from informal systems with 50.9% of that deriving from local markets. In contrast, formal sector sources were modest, even though several decades of investment have largely focused on either the formal public or formal private sector. To review select findings tied to current models: a) the channels routinely supported supply an insignificant proportion

of seed sown by smallholder farmers; b) new varieties are not being accessed sustainably through supported channels; and c) the array of crops needed for production, nutrition and resilience goals will not likely be promoted via a commercialized formal sector approach alone. [. . .] At a minimum, our results suggest a need to address the imbalance in seed channel focus so as to give attention to the main seed systems smallholders use, including several informal channels'. Up to now, no other study has analysed and provided such a wide range of data and statistics on the topic. It should be noted however, that the analysis is mainly focused on African countries, and therefore the results cannot be generalized worldwide, calling for further studies

31 FAO accompanies African countries in adopting national seed legislations. 'With the support of FAO, the Southern African Development Community (SADC), the Economic and Monetary Union of West Africa (UEMOA) and the Economic Community of West African States (ECOWAS) have undertaken the harmonization of national seed regulatory frameworks in their respective Member States. Through a participatory process involving the key stakeholders in the countries, a legal framework for the harmonization of seed legislation is developed and subsequently adopted by these regional bodies'. See www.fao.org/agriculture/crops/core-themes/theme/seeds-pgr/seed_sys/rules/en/. See also the FAO (2015) Voluntary Guide for National Seed Policy Formulation available at www.fao.org/3/a-i4916e.pdf. However, one may question the result and impact of such capacity building project when reading the recently adopted African Regional Intellectual Property Organization (ARIPO) Arusha Protocol for the Protection of New Varieties of Plants. The Protocol was adopted by the four least developed countries of the 18 Member States present at the Diplomatic Conference held in Arusha, the United Republic of Tanzania on 6 July 2015. The Protocol remains open for signature by Member States of the Organization and other States, members of the African Union until 31 December 2015. Available on ARIPO website at: www.aripo.org/news-events-publications/news/item/81-the-united-republic-of-tanzania-signs-the-arusha-protocol#sthash.COyIQcmq.dpuf. The Arusha Protocol has been strongly criticized *inter alia* because the protocol compromises the implementation of the CBD and the Plant Treaty. A major point of contention regards farmers' right to save, use, exchange and sell farm-saved seed and other propagating material. Indeed, the Protocol is said to hinder member countries to adopt a *sui generis* plant variety protection system, thereby suppressing the possibility to establish more flexible farmers' rights, not to mention the fact that according to farmers' organizations, the Protocol is inappropriate for the African region, it undermines national sovereignty, and diverges from positions articulated by African nations at the regional and international levels. A recent German governmental study confirms what civil society organizations have recommended, i.e. that 'developing countries that have not yet joined UPOV should consider opting for alternative *sui generis* systems of PVP that allow for more flexibility in meeting the obligations of different treaties, for balancing the interests of diverse actors, and for protecting and promoting Farmers' Rights, compared with the UPOV system' (Christinck and Walloe Tvedt, 2015: 6; also see la Via Campesina and GRAIN, 2015; Shashikant, 2015).

32 An interdisciplinary approach to seed networks will also allow for the study of the multifunctionality of seed networks. According to the International Assessment of Agricultural Knowledge, Science and Technology for Development (IAASTD, 2009: 18) '[t]he concept of multifunctionality recognizes agriculture as a multi-output activity producing not only commodities (food, feed, fibres, agrofuels, medicinal products and ornamentals), but also non-commodity outputs such as environmental services, landscape amenities and cultural heritages'.

33 However, due to the scope limits of this work, this approach will not be searched further.

34 Rothstein (1984: 166) examines what is covered by the concept of 'North' and 'South' in the debate around the New International Economic Order negotiations in the 1990s. He discusses the utility of maintaining such dichotomy although one can clearly not easily

categorize such a diversity of countries in two, or even more groups. He states that 'issues such as the *global commons*, [. . .] are sensible candidates for North-South negotiations' (emphasis added).

35 Oxfam reports that in 2014, the richest 1 per cent of people in the world owned 48 per cent of global wealth, leaving just 52 per cent to be shared between the other 99 per cent of adults on the planet. The richest 20 per cent of the world's population own approximately 95 per cent of the plant's wealth, leaving just 5.5 percent for the remaining 80 per cent of people in the world.

4 The International Treaty on Plant Genetic Resources for Food and Agriculture

A legal analysis

The negotiations of the International Treaty on Plant Genetic Resources for Food and Agriculture were not alien to, but strongly influenced by the historical and geopolitical context in which they were developed (see Chapters 2 and 3 this book).[1] Safrin summarizes some of the major issues at stake regarding genetic resources management at the time of the negotiation of the Treaty:

> Some resources benefit from being shared [. . .]. The more the resources are shared, the more they are preserved. Genetic resources are this type of good. In contrast to engendering a tragedy of the commons, where a common resource is used to depletion, the sharing of genetic material under an open system increases the global genetic pool, as it ensures the maintenance of genetic material in multiple locations. The open system that predated the expansion of IP rights and sovereign rights over genetic material accounts for the widespread distribution and preservation of crops and crop varieties away from their places of origin. The maintenance of genetic material in multiple countries and locations has benefited all.[2]
>
> (Safrin, 2004: 670)

In the 1970s and 1980s, when an utopian socialism was still believed to be possible, the almost romantic concept of plant genetic resources, seen as 'heritage of [hu]mankind' to be made 'available without restriction', was defended with passion by most developing countries and some developed countries. This idealistic vision was reflected in the 1983 International Undertaking on Plant Genetic Resources (IU). After the fall of the Berlin wall and the start of an era of the so-called 'real politics', neo-liberal economic theories prevailed. Consequently to the increasing privatization of genetic resources, the concept of seeds as 'heritage of [hu]mankind' to be made 'available without restriction' was replaced by those of 'global concern', 'state's sovereignty' and 'facilitated access', as reflected in the Convention on Biological Diversity (CBD) and later in the Plant Treaty. The adoption of the CBD in 1992, and two years later of the TRIPS Agreement in the context of the WTO Uruguay Round, as binding international agreements, was a wake-up call for the agricultural sector. With compliance being voluntary, the IU lacked sufficient legal and political weight to defend the specificities and interests of agriculture. Increasing pressure from the commercial and environmental sectors made possible what seemed unimaginable at the beginning of the 'Seed Wars' in the 1980s (Aoki, 2008). Developing and developed countries, the seed industries and non-governmental organizations joined together with one common political objective: to transform the IU into a legally binding agreement that would (i) allow cooperation between trade and environmental sectors on an equal footing, and (ii) guarantee conservation, sustainable use and access to agriculturally important plant genetic resources for research and plant breeding through a fair system for access and benefit-sharing. Consequently, the new phase of the negotiations – specifically aiming at the development of the Treaty, as requested by Resolution 7/93 of the FAO Conference – commenced in a highly constructive atmosphere (Frison et al., 2011a: 9; also see § 108 of Resolution 7/93 adopted at the Twenty-seventh Session of the FAO Conference C 1993).

These formal negotiations took place between 1994 and 2001. The FAO Commission on PGRFA met in three regular sessions and six extraordinary sessions. In order to speed up negotiations by reducing the number of active negotiators, the Commission appointed a regionally balanced contact group composed of 47 countries. Between 1999 and 2001, the contact group held six meetings to discuss controversial issues and to pave the road for the Commission negotiations. The sixth extraordinary session of the Commission intended to conclude the negotiations, but delegates could not reach agreement on several points. These pending issues were resolved during the 121st session of the FAO Council in October 2001. In a euphoric atmosphere, the negotiations were completed during the thirty-first conference of FAO, on 3 November 2001, with the adoption of the Plant Treaty by consensus with only two abstentions: Japan and the USA (Thirty-First Session of the Conference of FAO, C 2001/PV: 73; also see Charles, 2001c). With an expression of disbelief and exultation after the vote, the Director-General of FAO, Dr Jacques Diouf, qualified the Treaty as a milestone on the North-South relationship (Frison et al., 2011a: 10).

The Treaty[3] adopted in 2001 entered into force on 29 June 2004, after it was ratified by more than 40 countries (Article 28.1). However, the following analysis will show that the Treaty has only really started to be operational around the years 2010–2011, once its main tools and mechanisms had been developed and adopted by the governing body (*inter alia* the SMTA, the Benefit-sharing Fund, the Compliance Committee, etc.). To date, 145 states are contracting parties.[4] Seven countries are signatory but not yet party to the agreement.[5] There remain 44 non-contracting parties including *inter alia* Bolivia, Botswana, China, Gambia, Israel, Mexico, Mozambique, New Zealand, the Russian Federation and South Africa.

By creating the Treaty and its MLS, contracting parties sought to strike an equitable balance between public and private interests in access to seeds but the Treaty tools and mechanisms do not necessarily fulfil this objective, and countries have difficulties in implementing the Treaty. As a result, I posit that a de facto imbalance of rights pertaining to seeds exists, which needs to be re-balanced to implement adequately the MLS and allow all stakeholders to reach the Treaty's objectives. To verify this hypothesis, I asked myself how do the Treaty and its mechanisms function, and are there data and evidence to assess the efficiency of the Treaty implementation. To answer these questions, a legal interpretation of the Treaty clauses is carried out, inspired from the classical method of public international law for Treaty interpretation (Corten and Klein, 2011).[6] This interpretative exercise is cross-checked with data and statistics on the implementation of the Treaty by its contracting parties (data come partly from the Treaty Secretariat website). The aim is to draw a precise picture of the limits in the Treaty text and in its implementation process by highlighting eight main topics, which are crucial for understanding and implementing the Treaty. The results of this in-depth legal analysis are to be published in extenso in 2020.[7] This chapter summarizes these results without entering into too many details of the technical legal interpretation and research methods. It presents the essence of the legal research under eight thematic sections: sustainable agriculture and food security; scope of the Treaty and its MLS; Farmers' Rights; access to seeds; benefit-sharing obligations and the Benefit-sharing Fund; the Third Party Beneficiary and other related legal procedures (monitoring, sanctions, conflict resolution); information and knowledge; and finally governance issues related to stakeholders' participation.

Sustainable agriculture and food security as Treaty overall goals

Sustainable agriculture and food security are the two overall goals of the Treaty (Article 1.1). These goals underpin and justify the Treaty's direct objectives of conservation, sustainable use and access and benefit-sharing. They also match well some important aspects of the theory of the commons, notably the central role of sustainability in common-pool resource management systems. Both objectives are intrinsically related to each other: there may be no sustainable agriculture without assuring food security, and food security will not be reached without achieving a sustainable agriculture (IPES-Food, 2016). These overall objectives represent the final reasons, the 'all-embracing' motives for contracting parties to act collectively in facing the global challenges of agricultural biodiversity erosion and repetitive food crises, which no country can face alone. These underlying objectives are critical in guiding contracting parties in their implementation of the Treaty. They remind us that conservation, sustainable use, or ABS are not meant for other purposes than promoting a sustainable agriculture and reaching food security. It is argued that sustainable agriculture and food security should be direct objectives rather than overall goals, matched with technical tools and concrete policy guidance to reach them.

Sustainable agriculture

Sustainable agriculture can be found mainly in Treaty Articles 1 (Objectives), 5 (Conservation), 6 (Sustainable Use) and 11 (Coverage of the Multilateral System), as well as in several paragraphs of the preamble. Their analysis shows that the overall goal of sustainable agriculture derives from specific characteristics of PGRFA and that it entails a number of specific obligations in the Treaty. Six specific characteristics serve as guiding compass when interpreting and implementing the Treaty provisions as they may enlighten on the intentions of the parties when they negotiated the text. These six elements explain why it is compulsory for contracting parties to address PGRFA management in a sustainable manner, through sustainable use practices to promote the development of a sustainable agriculture.

Six characteristics pertaining to PGRFA

The first characteristic is the risk of agrobiodiversity erosion related to the under user of PGRFA diversity. Contrary to other natural resources like water or wood, PGRFA do not suffer from overuse/exploitation, but rather from their underuse.[8] This reflects the opposite situation of the 'tragedy of the commons' (Aoki, 1998; Buchanan and Yoon, 2000: 2), that is to say a situation of 'anticommons' (Heller, 1998; Heller and Eisenberg, 1998). Articles 5 and 6 of the Treaty clearly reflect that there may be no (sufficient) conservation if there is no constant use of PGRFA diversity.

The second characteristic recognizes the special role of farmers in the conservation and sustainable use process, which implicitly strengthens the recognition of farmers' rights (Resolution 4/2015). Contracting parties' willingness to embed the conservation and sustainable use obligations in a broader social reality (Resolution 7/2013: §§ 1 and 3) implies a central role for farmers within a global interdependent network for a food secure world (Maxted et al., 2011).

The third characteristic relates to the special nature of seeds. The Treaty preamble § 1 recognizes the 'special nature of PGRFA, their distinctive features and problems needing distinctive solutions' (as recognized by Resolution 3 of the Nairobi Final Act;[9] also see the Conference of the Parties of the CBD in 1995 Decision II/15; UNEP/CBD/COP/2/19: 26; Hardon et al., 1994; Visser et al., 2003). The uniqueness of PGRFA is based on several criteria: they are crucial to satisfying basic human needs; they are man-made biological diversity being developed since the origins of agriculture; PGRFA are distributed throughout the world with greatest concentration in various 'centres of origin and diversity' of cultivated plants; countries' interdependence regarding PGRFA, i.e. the fact that 'all countries depend very largely on [PGRFA] that originate elsewhere' (Article 11.1), is much greater than for any other kind of biodiversity; finally, the target for conservation and use are not the species as such, but the genetic diversity within each species (Frison et al., 2011a: 9). This special nature of seeds justifies the mere existence of a new Treaty specifically dedicated to the question of PGRFA (FAO Conference C 1999 § 60).

The fourth characteristic links conservation to crop genetic improvement. Indeed, PGRFA are the 'raw material indispensable for crop genetic improvement' (Preamble, § 6), essential to adapt to unpredictable environmental and future human needs, whether by farmers' selection, standard plant breeding or modern biotechnology means. This justifies the broad, international efforts by contracting parties to conserve seeds. Breeding activities are positioned as a central means to realize food security and poverty alleviation.

The fifth characteristic highlights the intergenerational responsibility (Preamble §§ 4 and 13) of the present generation to past and future generations to conserve the world's diversity of PGRFA. This shows the temporal interdependence of generations for future sustainable livelihoods.

The last characteristic points to the necessity of a multi-sectoral approach to PGRFA management, involving agriculture, the environment and trade (Preamble § 12), i.e. a sustainable development law issue, requiring synergy among these sectors. As a reaction to the 'exclusion' of agriculture ministries by the environmental sector during the CBD negotiations, there is a growing recognition of the need for collaboration between the relevant international institutions (e.g. with the CBD representing the environmental sector, or the WTO representing the trade sector;[10] see Preamble § 9). An example of such a multi-sector approach is the Memorandum of Cooperation concluded between the Secretariats of the Plant Treaty and the CBD in October 2010 on a Joint Capacity Building Programme for the harmonious implementation of the Treaty and the CBD as well as the Nagoya Protocol (IT/GB-5/13/14: Appendix 1).

From overall objectives to concrete Treaty obligations for a sustainable agriculture

The above six characteristics of PGRFA justify the promotion of a sustainable agriculture policy, which is translated into specific obligations: Article 1 on the Objectives of the Treaty and Articles 5 and 6 specifying Contracting Parties' obligations relating to Conservation and Sustainable Use of PGRFA.

Article 1 of the Treaty is an essential provision as it sets the overall goals of the Treaty, i.e. sustainable agriculture and food security. These overall goals represent the ultimate reasons, the 'all-embracing' motives for contracting parties to act collectively in facing the global challenges of agricultural biodiversity erosion and repetitive food crises, which no country can face alone. These overall goals are critical in guiding contracting parties in their implementation of the Treaty. They imply that the objectives of conservation, sustainable use, and ABS are not meant for other purposes than promoting a sustainable agriculture (and reaching food security).

Articles 5 and 6 on conservation and sustainable use of PGRFA were non-contentious during the negotiations of the Treaty and offer a modern framework for states to act for the benefit of conservation and sustainable use of PGRFA

(Cooper, 2002). This framework is based on earlier provisions of the IU (Articles 3 and 4) combined with text and suggestions derived from the CBD (Decision III/11) and from the Global Plan of Action for Plant Genetic Resources for Food and Agriculture (GPA – a voluntary non-legally binding instrument).[11] Article 5 deals with the 'Conservation, Exploration, Collection, Characterization, Evaluation and Documentation of PGRFA'. Article 6 is entitled 'Sustainable Use of Plant Genetic Resources'. They both support an integrated approach to conservation and sustainable use, suggesting actions the joint implementation of which increase their mutual effectiveness. The details of the technical, scientific and political actions promoted, such as: promoting collection of PGRFA, promoting on-farm conservation and management by farmers and local communities, strengthening research for the benefit of farmers, broadening the genetic base of crops and increasing the range of genetic diversity available to farmers, etc., will not be dealt with here. What is especially interesting to note from a legal perspective is the reference to the principle of national sovereignty in Article 5 when stating that each contracting party shall, 'subject to national legislation' promote the above mentioned integrated approach. This reference marks the overall influence of the CBD in the 'privatization' of biodiversity management through the (state) control over genetic resources.

Both articles are quite specific in their content (as opposed to similar articles in the CBD), which might facilitate a benchmarking when monitoring the effective implementation of these articles. Articles 5.2 and 6 jointly promote rather an environmentally friendly and diverse agricultural policy, which mitigates the threats to PGRFA identified through the actions proposed under Article 5.1. The lists of actions to be taken under both Articles are not exhaustive (as the terms 'in particular', and 'such measures as' suggest). These actions are driven from the GPA (Treaty Article 14). What is worth noting is that Articles 5 and 6 are interrelated and cover concrete actions to promote the overall goal of 'sustainable agriculture'.

Evaluation of the implementation of the conservation and sustainable use provisions

Once the special characteristic of PGRFA have been translated into concrete obligations for contracting parties, it is worth looking at the implementation of the sustainable agriculture overall goal in the Treaty. Four elements are analysed highlighting the discrepancy between the importance of the sustainable use of PGRFA item on the agenda of every Governing Body meeting and the reality of its implementation at the national levels.

First, it should be noted that sustainable use of PGRFA systematically constitutes a core agenda item of every Governing Body meeting (IT/GB-6/15/12: § 1) and several Resolutions have been adopted by the Governing Body (Resolution 7/2011; 7/2013; 4/2015; 6/2017). This is not the case for Article 5 on Conservation because ex situ conservation has been dealt with by states for several decades already through other instruments at the national (national policies and genebanks for PGRFA conservation)[12] and international levels (such as the establishment of the CGIAR and their genebanks, the GPA,[13] the State of the World's PGRFA,[14] and more recently with the Global Crop Diversity Trust[15] (GCDT)). A considerable amount of money is dedicated to the implementation of national and international ex situ conservation activities,[16] contrary to in-situ and on-farm conservation, which have received little attention and funding up to now (Frison and Hodgkin, 2016). Moreover, conservation activities are implicitly and intrinsically included within Article 6, as sustainable use of PGRFA is implemented through in-situ and on-farm conservation. Indeed, the direct complementarity between in-situ, on-farm and ex situ approaches to conservation of PGRFA and the resulting PGRFA information has been recognized by the Ad Hoc Technical Committee on Sustainable Use of Plant Genetic Resources for Food and Agriculture (ACSU, established by Resolution 7/2011; also see IT/GB-6/15/12: § 8).

Second, notwithstanding the importance of this topic in the GB agenda, national implementation lags behind (IT/GB-3/09/Inf.5: § 44). The Fourth Session of the GB, recognized that most countries do implement explicit policy and legal measures for different aspects of the concept of sustainable use through a variety of policies, but that an integrated and coordinated approach towards the promotion of sustainable use of PGRFA is lacking. For this reason, contracting parties established the Ad Hoc Technical Committee on Sustainable Use[17] to assist them in this task (by developing a toolbox, convening a stakeholders' consultation,[18] elaborating a Work Programme on Sustainable Use (WP-SU), etc.). At GB5, Resolution 7/2013 requests contracting parties: (1) to promote access of all farmers to PGRFA in the Multilateral System; (2) to implement measures and activities of the WP-SU and (3) to report on the implementation activities of the WP-SU to the Governing Body using the indicators of the second GPA. Besides, this resolution also requests the Treaty Secretariat to take various actions to promote the implementation of Articles 5 and 6.[19]

Third, it is hard to assess the degree of implementation of Articles 5 and 6 as few national reports[20] and submissions by other stakeholders[21] (i.e. civil society organizations and international research centres) have been submitted to the Treaty Secretariat. However, it is clear that having established the Ad Hoc Technical Committee on Sustainable Use and having systematically set on the Governing Body agenda the issue of the implementation of Article 6 shows that there is an urgent need to promote the implementation of these obligations. Besides, the implementation of Articles 5 and 6 remains critical for the implementation of all other Treaty obligations. In particular, it is very closely related to the implementation of Farmers' Rights at the national level (Resolution 7/2013: Goal 1 and 4). The work undertaken by the Ad Hoc Technical Committee on Sustainable Use further strengthens this very close link between Articles 5 and 6 and Article 9 of the Treaty.

Fourth, a final comment relates to the relation between conservation and sustainable use of PGRFA and national seed legislations. When coupled to IP legislation, many variety registration and seed certification legislations that only recognize commercial varieties create obstacles to farmers' rights in the exchange and selling of farm-saved seed, and in the marketing of landraces and of many farmers' varieties (Winge, 2015; Wattnem, 2016: 850–867). This constitutes a serious hurdle to on-farm conservation and sustainable use of crop genetic diversity. Andersen (2015: 5) notes that '[i]t is a paradox that rules originally intended to protect plant health in fact contribute to removing the basis for ensuring plant health in [the] future, namely the diversity of genetic resources'. Such seed laws together with strict plant breeders' rights restrict traditional farming practices and traditional seed conservation, thereby contributing significantly to the further erosion of the primary material for all plant breeding and farming activities. Andersen (ibid) questions the possibilities that exist to make such laws more compatible with these customary rights of farmers – which are so crucial to the maintenance of agrobiodiversity for food security, today and in the future (ibid; also see Winge, 2015; Prip and Fauchald, 2016; Girard and Frison, 2018b).

Food security

Evolution of the concept

Although in 1996 the Rome Declaration on World Food Security emphasized that 'the primary responsibility rests with individual governments', food security, touching upon so many different aspects of nations' government (agriculture policy, social and economic policies, war, climate change, etc.), cannot be dealt with by individual states but necessarily calls for global action (Sen, 1981). The concept of food security (Hadiprayitno, 2014: Cadieux and Blumberg, 2014) has been on the political agenda of FAO for more than 50 years (Shaw, 2007; McDonald and Matthew, 2009: 5–7). However, the concept has significantly evolved, from a simplistic view of producing more food in order to supply more foodstuffs to people, to the more complex approach encompassing the nutritional value of the food supplied, as well as the political, social and economic aspects of locally producing foodstuff in accordance with cultural preferences and traditions (Sen, 2003; Shaw, 2007: 383–386; IPES-Food, 2016). Food security can be envisaged through different lenses as a technical concept (food security), a legal right (right to food) or a political objective (food sovereignty) (Windfuhr and Jonsen, 2005). The FAO World Food Summit Plan of Action (point 1) provides a simplified definition stating that:

> Food security exists when all people, at all times, have physical, social and economic access to sufficient, safe and nutritious food to meet their dietary needs and food preferences for an active and healthy life.
>
> (FAO World Food Summit Plan, 1996)

This rather technical approach to food security has pushed civil society movements and farmers' communities to design a wider political concept encompassing *inter alia* the defence of farming practices or cultural aspects: food sovereignty (Edelman, 1999). Reaching international spheres through proactive civil society movements (Claeys, 2014b, 2015), food sovereignty was asserted as a right at the NGO Forum to the World Food Summit held in Rome in 1996 (Claeys and Lambek, 2014: 34–35). To further fight against the 'neo-liberal agriculture policies [which] have led to the destruction of our family farm economies and to profound crisis in our society' (ibid: 35), 500 delegates from more than 80 countries adopted the 'Declaration of Nyéléni' in 2007 providing and extensive definition of food sovereignty.

Following this trend, the International Assessment of Agricultural Science and Technology for Development (IAASTD), an intergovernmental process under the sponsorship of the United Nations and the World Bank, defined food sovereignty as being 'the right of peoples and sovereign states to democratically determine their own agricultural and food policies' (IAASTD, 2008: 15). Although quite vague, this definition recognizes at the global level a much wider concept based on the importance of local control and self-sufficiency (Hadiprayitno, 2014).

Besides the evolution of the concept of food security to food sovereignty, it has also gained recognition as a Human Right, through the development of the Right to Food (Universal Declaration of Human Rights: Article 25; International Covenant on Economic, Social and Cultural Rights: Article 11; International Covenant on Civil and Political Rights: Article 6; Convention on the Rights of the Child: Articles 24 and 27), in particular in Article 11 of the International Covenant on Economic, Social and Cultural Rights (Morten Haugen, 2014). Notwithstanding the development of the right to food within the UN fora (Human Rights Council Resolution 13/4, 24 March 2010) and the recognition of its role in reaching food security (De Schutter, 2010a), its implementation and realization has been slow (Claeys, 2015). To progress on this question, in 2012 civil society pushed for the negotiation of a UN Declaration on the Rights of peasants and other people working in rural areas, which is still currently being discussed (Human Rights Council Resolutions 21/19, 26/26, and 30/13).

Many other commitments to reduce hunger and poverty have been taken by states, such as the Declaration on World Food Security in 1996, the Millennium Development Goal (MDG) in 2000 (Leiserowitz et al., 2006), the 2008 Cordoba Declaration on the Right to Food and the Governance of the Global Food and Agriculture System or the 2015 Sustainable Development Goals (SDGs) and their second goal 'Zero Hunger'. These instruments highlight the growing interdependence between nations and regions for food security, and the need for international cooperation and solidarity between areas experiencing different levels of development to achieving food security for all. International aid in times of crisis is put forward as an emergency solution to critical food insecurity, but the past and current crises show that this response is not sufficient and that it does not solve the problem. Despite the fact that Food Security seems to be high on the international agenda, voices argue that access to affordable and sufficient food is not recognized as a universally guaranteed human right, at the same level as access to water, health and education (Vivero Pol and Schuftan, 2016), thereby rendering it difficult to reach the goal of food security set in the SDGs.

Literature has stressed the importance of taking into account food security in the design of scientific research but also of national, regional and international policies (De Schutter, 2009b), thereby recognizing the tight link between PGRFA and the criteria of interdependence and food security (Thrupp, 2000). Fowler and Hodgkin (2004: 144) consider for example that 'genetic resources underpin plant breeding and agricultural production and are thus essential to food security, livelihoods and the development aspirations of every country on Earth'. This is reflected during the negotiations of the Treaty, where negotiators listed food security as one of the objectives to be attained by facilitating access to PGRFA through the creation of a multilateral agreement (CGRFA-Ex3/96/Rep: 50). They considered that access to and the sharing under fair and most favourable terms of both genetic resources and technologies are essential for meeting world food security and the needs of the growing world population, and must be facilitated. By adopting this language, stakeholders have voluntarily linked food security to the idea of a multilateral mechanism facilitating access to – and sharing benefits of – PGRFA. Indeed, in the preamble and in Article 1, food security is matched with sustainable agriculture, as being the two overall goals of the Treaty. In Article 11.1, contracting parties use food security as one of the two criteria, with interdependence, for determining what PGRFA should be part of the Treaty Annex I list of crops and forages.

Implementing the concept of food security in the Treaty: interdependence as a key

Interdependency of states and food security are the two intertwined criteria (Article 11.1) for determining what PGRFA should be covered by the MLS as part of Annex I to the Treaty.

PGRFA interdependence between countries results from long run human collaboration in the exchange of food and feed plants across the world. The degree of dependence of a country's main food crops on genetic diversity in areas of origin and primary diversity located elsewhere has been established (FAO, 1996a; Palacios, 1997) and later on confirmed (FAO, 2010) by several worldwide studies (Khoury et al., 2015). They show that all regions in the world are highly dependent upon resources originating for another region, North America being the highest dependent region, and Asia and the Pacific region being the least dependent region. As Palacios says (1997: 3), '[m]odern agriculture is heavily dependent on plant genetic resources from practically all countries. [. . .] There is therefore an ongoing need to exchange plant genetic resources' to allow countries to produce their food and reach food security. This argument explains the rationale for establishing the multilateral system, an innovative governance scheme, where rights and obligations for seed management are collectivized.

Up to recently, the way the MLS was implemented (see below same chapter) did not seem to contribute significantly to reaching food security. As I argue in Chapter 6, new ways of envisaging and implementing the Treaty's overall goals are necessary to enable their realization. It is argued that sustainable agriculture and food security should be direct objectives rather than overall goals to the Treaty, matched with technical tools and concrete policy guidance to reach them. Gladly, a shift seems to have occurred at GB7. With the adoption of Resolution 1/2017 (being the first agenda item shows the importance contracting parties devoted to the matter), the Governing Body has decided to materialize much the Food Security objective by linking it with the achievement of the goals of the 2030 Agenda for Sustainable Development. This mind (and language) shift might be the start for a much more concrete and delineated policy guidance to reach global food security. Language in Treaty documents and decisions shifts more and more to food security 'and nutrition' and links food security (and sustainable agriculture) to climate change, thereby integrating the two Treaty overall goals into the wider political objectives and challenges that the international community is faced with. Resolution 1/2017 emphasizes

> that the effective implementation of the International Treaty contributes to the achievement of the 2030 Agenda for Sustainable Development (2030 Agenda) and the Sustainable Development Goals (SDGs), in particular SDG Targets 2.5 and 15.6, relating to conservation, and access and benefit-sharing of genetic resources, while also contributing indirectly to SDGs 1, 12, 13 and 17.

It also highlights 'contributions made by the Multilateral System to the achievement of SDG Target 2.5 by facilitating access to over 4.2 million samples to breeders, farmers and stakeholders all over the world', thereby relating the Treaty policy and objectives to other international instruments. It further decides

> to take stock, at its Ninth Session in 2021, of progress made in the achievement of SDGs 2 and 15, in particular targets related to PGRFA' (ibid: point 12), thereby setting, for the first time, some sort of evaluation of the reaching of the food security goal of the Treaty. Finally, it requests 'that the new Funding Strategy and its result framework emphasize the contribution of the International Treaty to the 2030 Agenda.
>
> (ibid: point 16)

This move allows the dedication of the necessary funds to more concretely set objectives encompassed in wider international instruments and political commitments (Agenda 2030 and SDGs), for which progress made in the implementation by the contracting parties will be evaluated for the first time. This might contribute to rendering the overall food security goal more tangible.

To conclude on the Treaty's overall goals of food security and sustainable agriculture, it can be said that they justify the mere existence of the MLS (and even perhaps its enlargement to all PGRFA), and that their relevance is likely to increase in the future due to environmental, economic and other hazards. The fact that food crises continue to be chronical, that hunger and malnutrition are far from being eradicated (SOFI, 2017), shows that food security remains a key objective to be attained through the promotion of sustainable agriculture(s). It also requires taking into account a wider approach to food policy, where economic power and geopolitical interests are acknowledged. The very recent shift in policy towards food security of the GB might show that new ways of envisaging and implementing these overall goals are necessary to enable their realization. It is argued that sustainable agriculture and food security should be direct objectives rather than overall goals to the Treaty, matched with technical tools and concrete policy guidance to reach them (see Chapter 6).

Scope of the Treaty

The scope of the Treaty refers to two different aspects: the resources the Treaty relates to and the boundaries of the Treaty application. A description of specific rules and procedures relating to the scope of the MLS will be provided, before a short description of where contracting parties stand regarding the implementation of these provisions.

Defining what material is covered by the Treaty and its MLS

Defining what should be understood by PGRFA is crucial, as it determines to what resources the Treaty (and its objectives of conservation, sustainable use and access and benefit-sharing) should be applied. Generally speaking, PGRFA are crops and forages used as nutrients for humans and animals. The notion of 'PGRFA' encompasses both the physical material (i.e. the plant, the seed, the tuber, cuttings, bulbs, etc.), and the related information and data accompanying the material. Indeed, PGRFA would be useless to farmers, scientific researchers and breeders if they had no information about the characteristics of a specific variety compared to another, or about the environment (soil, weather, ecosystem conditions, etc.) where this variety comes from. PGRFA should therefore be approached as being both a tangible and an intangible good, as these two sides of the same coin are intrinsically linked (Brahy, 2008). This dual character complicates the management of seeds significantly, especially as far as the intellectual protection of the intangible aspect of the PGRFA is concerned (Correa, 1999).

The definition of 'PGRFA' and 'genetic material' are determinant in understanding the scope of application of the Treaty and of the MLS and consequently in delimitating its implementation. Article 3 specifies that the 'Treaty relates to PGRFA', that is to say to 'any genetic material of plant origin of actual or potential value for food and agriculture' (Article 2 § 4). The Treaty further defines genetic material as 'any material of plant origin, including reproductive and vegetative propagating material, containing functional units of heredity'. Requiring that the PGRFA must be of actual or potential value leaves the door wide open to include as many crops and forages as possible, as one cannot predict the potential use of genetic material for future needs. However, this wide scope is narrowed by the consideration that the said value should be related to food and agriculture purposes. This is consistent with the lex specialis characteristic of the Treaty as opposed to the CBD.

Regarding the genetic material, its definition mentions 'functional units of heredity'. There has been a debate during the negotiation of the Treaty as to specifying in Article 2 whether the functional units of heredity defined as 'genetic parts or components' (understood as being individual genes or gene sequences) would be considered as PGRFA by themselves (Moore and Tymowski, 2005: 35). As a compromise, contracting parties left the definition ambiguous, although it might mean leaving the Governing Body to interpret the matter at a later stage. However, if the term 'functional units of heredity' is in the end not specified, according to Moore and Tymowski (ibid), it 'would include at least all genetic elements containing DNA, i.e. genes'. The lack of precision on contentious terms was voluntary to allow moving on with the negotiation. However, it has not erased the contentious views but only postponed the agreement on difficult questions to the implementation phase, making it more complex.

A good example arose during GB7 with the touchy question of digital/genetic sequence information (DSI)[22] (Resolution 13: 2017; also see 9/2017: § 12; 10/2017: § 14). As a phenomenon of dematerialization, DSI covers 'the information and knowledge content of genetic material [that could be] extracted, processed and exchanged in its own right, detached from the physical exchange of the plant genetic material' (FAO, IT/GB-5/13/4). The topic was recently introduced on the agenda of the CBD (CBD decision XIII/16; Nagoya Protocol decision 2/14) and the sixteenth Regular Session of the CGRFA in February 2017 established a new work stream on digital sequence information on GRFA. Noting this ongoing work, many Treaty contracting parties (from developing countries) wanted to include the topic on the agenda of GB7 and of the forthcoming biennium, as DSI constituted the flesh of most discussions during GB7 (ENB, 2017: 3; TWN, 2017; GB7 Report). Indeed, defining precisely whether the information contained in the 'functional units of heredity' is part of the 'genetic material' or not has many important implications. In Resolution 2/2017, the co-chairs of the contact group on the SMTA proposed that the 'consolidated text for the revised SMTA made provision for the concept of digital sequence information (DSI) to be reflected in the revised SMTA through a new definition of "genetic parts and components"' (Resolution 2/2017: Annex 2, § E.14). However, there was no consensus on this point, and contracting parties finally decided, *inter alia*, to

> consider at its Eighth Session the potential implications of the use of "digital sequence information" on genetic resources for the objectives of the International Treaty, and to consider it for inclusion in the [Multi-Year Programme of Work 2018–2025] at that meeting.
>
> (Resolution 13/2017: point 2)

Furthermore, contracting parties and other relevant stakeholders are requested to provide information on the

> relevance of "digital sequence information" on PGRFA for food security and nutrition, in order to facilitate consideration by the Governing Body, at its Eighth Session, of the potential implications of the use of "digital sequence information" on PGRFA for the objectives of the International Treaty, including exchange, access and the fair and equitable sharing of the benefits arising from their use.
>
> (Resolution 13/2017: point 4)

The Treaty Secretariat shall compile this information for the next GB.

In an area where DSI gain more and more value in plant breeding biotech, defining whether DSI is considered part of the 'genetic material' or not will have huge implications in terms of benefit-sharing and IPRs. It has been argued that, as coded information, DSI were not genetic material as covered by the Treaty (EU, Japan and Seed Industry declarations made during GB7) even though the genetic sequence originates from the coding of physical PGRFA originating from the MLS. As argued by Halewood et al. in a very recent article (2018: 2), laws regulating the access and use of PGRFA 'require users to share monetary benefits in exchange for access to, and use of *material* genetic resources in the development of new commercial products; they do not extend in scope to genetic sequence information' (emphasis added). This emphasis on the 'materiality' of PGRFA is questionable, because it is written nowhere in the Treaty that PGRFA refers to the 'material/physical' shell of the genetic resources. According to their interpretation, DSI falls outside the scope of the MLS obligations, in particular the benefit-sharing obligation, but also (and for now this is not highlighted in the debate) the IPRs obligations (Article 12.3d). According to these stakeholders, DSI should be freely available for further use without falling under the scope of the MLS.

I do not share this view and I strongly question this interpretation of the Treaty clauses. Article 12.3.c) of the Treaty states that '[a]ll available passport data and, subject to applicable law, any other associated available non-confidential descriptive information, shall be made available with the plant genetic resources for food and agriculture provided'. Halewood et al. in the same article explain that DSI technology enables

> the extensive characterization and traceability of the stocks that are currently held by gene banks. [. . .] Moreover, extensive genotyping, linked to measured traits, allows repositories to be searched for strains containing desired genetic elements and/or trait characteristics.
>
> (Halewood et al., 2018: 3)

Therefore, Halewood et al. confirm that the purpose of DSI technology is to characterize PGRFA in a more efficient manner. According to me, this characterization process is precisely what Article 12.3.c) refers to, i.e. the passport data and descriptive information of the PGRFA. And this is no more than logic. Contracting parties know very well that accessing the physical part of the PGRFA on its own is of no use. This is precisely why, since the beginning of the design of the Treaty, they included the obligation to provide the related information with the PGRFA. This is also the reason why the definition of the 'parts and components' terms was so contentious during the negotiations. It is hypocritical to state that the genetic information extracted from a variety (i.e. the DSI) is not defined as its 'parts and components' simply because one can now use the coded information to synthesize a gene and insert it in a variety without needing to access to the original physical material (whereas this was not so obvious at the moment the Treaty was negotiated and signed). Validating such interpretation would render the whole MLS void. The fact that we can now provide more precise genetic information related to the PGRFA using genomics does not change the purpose of the exchange or the (material and informational) object accessed. The coded information derives from the physical material and is used to design a new variety. Doing so, the DSI will regain physicality in an improved variety to be commercialized and has no value outside of its physical shell. Not including the DSI in the definition of the genetic material is simply a means to avoid being obliged by the MLS benefit-sharing provisions as well as the IPRs clauses. This is clearly not fair, nor equitable. This interpretation does not respect the spirit or the objectives of the Treaty and it does not implement properly the interpretation principles of the Vienna Convention on the law of Treaties.

Coverage of the Multilateral System

The scope of the Treaty applying to all PGRFA (Article 3)[23] encompasses a more limited coverage for the MLS (Article 11), similar to a Russian doll (Frison, 2016: 123–148). Indeed, within the Treaty's general scope of application, the specific clauses on Access and Benefit-sharing obligations (the MLS) only relate to the Annex I list of 64 crops and forages. It is estimated that these crops, combined, provide about 80 per cent of our food from plants. However, crops important to food security like soybean, groundnut or tomato remain excluded from the system.

Some authors also argue that the scope of the Treaty is too limited because the facilitated access only applies to PGRFA used for breeding, research and training purposes (Lawson, 2009: 252). However, the scope of the MLS – i.e. material listed in Annex I for the uses prescribed by the Treaty – is a reflection of the political climate during which the Treaty was negotiated (Frison et al., 2011a: chapter 2), and which, in many ways, prevails until the present day. It is clear from the history of the negotiations, and from the way in which the list of materials included in the MLS fluctuated during the negotiations (Lim and Halewood, 2008: 249), that the Treaty might never have been finalized if some delegations would have insisted on the MLS covering *all* PGRFA for *all* purposes. Thirteen years after its entry into force, a proposal to enlarge the scope of the MLS to all PGRFA, and therefore to suppress the nested scopes in the Treaty and MLS, was put on the table by Switzerland at GB7. While this proposal has not been adopted by the GB, it shows that mentalities have evolved on this matter, and perhaps in a near future such harmonization of scope will occur.

Coupled to this issue is the question as to whom are the Treaty obligations applicable? All providers and recipients of PGRFA in all contracting parties' territories? Do they extend to recipients in non-contracting parties, as the SMTA contains a viral clause of application imposing any further transfer of the material to occur using the SMTA? The geographical scope of the Treaty is an important question to tackle. It raises the difficult problem of porous boundaries of the whole system. As long as countries with major genebanks, such as China are outside the Treaty, it is relatively easy to bypass the MLS system by accessing identical PGRFA accessions outside the MLS (i.e. the economic concept of free-riding). Expanding the scope of the Treaty to all PGRFA and to all countries would be a way to circumvent the bypassing problem.

Article 11 enumerates five types of PGRFA which form the MLS basket of material: 1) PGRFA listed in Annex I, which are under the management and control of contracting parties and in the public domain (Article 11.2); 2) contributions from all other holders of PGRFA listed in Annex I, upon invitation by contracting parties (Article 11.2); 3) PGRFA included voluntarily by natural and legal persons within the jurisdiction of the contracting parties, who hold material listed in Annex I (Article 11.3); 4) PGRFA listed in Annex I and held in ex situ collections of the CGIAR (Articles 11.4 and 15.1a); and 5) PGRFA listed in Annex I and held in ex situ collections of other international institutions (Articles 11.4 and 15.5). Moreover, there is only one case in which the obligations under Article 11 can be limited, that is to say when the PGRFA is 'under development' (including material being developed by farmers). In this case, the material 'shall be at the discretion of its developer, during the period of its developments' (Article 12.3(e)). Yet, in practice, it is difficult to identify which material is effectively included in the MLS, and the boundaries of the MLS basket of accessions seem to be rather blurred. The question of the coverage of the MLS has been one of the most contentious issues during the Treaty negotiations: what material is to be included in the MLS by contracting parties? And indeed, it is an important question as the access will be facilitated only to those specific PGRFA, and as the effectiveness of the benefit-sharing mechanism will partly depend on the material included in the system. The recent discussion about DSI and their inclusion or not in the MLS exemplifies this important question.

Rules and procedures related to the coverage of the MLS

This section reviews for what type of use the materials included in the MLS can be accessed. It also details the specific rules and procedures applicable for products under development.

Article 12.3(a) states that 'access shall be provided solely for the purpose of utilization and conservation for research, breeding and training for food and agriculture, provided that such purpose does not include chemical, pharmaceutical and/or other non food/feed industrial uses'. This means that PGRFA that are used for another purpose, such as the production of biofuels, cosmetics or pharmaceuticals are not considered as PGRFA under the MLS (Treaty Secretariat, 2012a, 2012b: §§ 19–21). This distinction is important because the exchange mechanism and applicable law will differ when the subject matter is used for food/feed or for other purposes. Indeed, for the latter, the CBD and its access and benefit-sharing obligations under the Nagoya Protocol are applicable (Garforth and Frison, 2007: 4; Treaty Secretariat, 2012b: §§ 27–31). In fact, the same species or even the same variety can be both within and outside of the MLS, depending on the intended use. For example, this is the case for some yam (Dioscorea sp.) species which are in the MLS when intended to be used for food but are not included if they are intended for a pharmaceutical purpose, in making the contraceptive pill for example. The same applies for a maize variety that can be used for animal feed or for biofuel production.

Furthermore, the Treaty creates a special regime for PGRFA under development where the material under development is part of the MLS but is subject to certain discretionary powers of the developer as long as it is under development. This special regime will be applied for a limited period of time and could be assimilated to a sort of trade secret protection mechanism.[24] Indeed, the developer's discretionary powers include the authority not to transfer any information to the MLS Third Party Beneficiary, to restrict access to the material being developed (not to the original material) and to set special conditions (including financial) on the transfer of that material during the period of development. Upon commercialization of the final product on the open market, the normal MLS regime applies again, and the material sold shall be subject to Article 13.2(d)(ii) regarding benefit-sharing obligations.

Implementing the provisions on the scope of the Treaty

Implementation of the provisions relating to the scope and boundaries of the Treaty cross-cut with the implementation of the provisions on access to PGRFA under the MLS, addressed below in this chapter. Therefore, only brief information will be provided here.

Agreeing on a common understanding of what PGRFA are covered by the Treaty and by the MLS is important to implement Treaty obligation in a coherent and harmonized manner. In order to operationalize the ABS provisions of the MLS it has been necessary to identify and designate the accessions that are meant to be part of the MLS. During the implementation phase, states have had difficulties in doing so. Indeed, few information on the collections and accessions to be part of the MLS have been communicated to the Treaty Secretariat.[25]

Today, following the proposal made by Switzerland at GB7, contracting parties are envisaging reopening the debate on the coverage of the MLS. However, the G-77 and the African groups are still reluctant to discuss this option before significant progress is made on other issues, such as the implementation of the ABS mechanism (and in particular the benefit-sharing aspects), or the promotion of

FRs. However, some stakeholders hope that enlarging the coverage of the MLS might contribute to increase the monetary returns to the Benefit-Sharing Fund. Furthermore, it is argued that it would simplify significantly the administrative burden related to the access procedures, as it would allow the use of one unique material transfer agreement.

Up till recently, the majority of holdings brought into the Multilateral System are those of the CGIAR Centres or other international organizations (e.g. the Tropical Agricultural Research and Higher Education Centre or the International Coconut Genebank for Africa and the Indian Ocean). The experience of the CGIAR Centres with the implementation of the Treaty seem to be positive (IT/GB-3/09/Inf. 15; IT/GB-4/11/Inf. 4, point 14; and IT/GB-4/11/Inf. 5), as the Treaty considerably simplifies the task of the Centres in making PGRFA available and notably reduces the administrative costs involved. Even more so as the Governing Body, at its second meeting, recognized that the Centres should use the same SMTA for both Annex I and non-Annex I material.[26] However, some difficulties have arisen regarding the tracking obligation when transferring PUDs. This question is being addressed by the Governing Body (Resolution 1/2015: §§ 22–25) and will hopefully be resolved once the review process of the MLS and SMTA is over. Using the SMTA for non-Annex I materials (Article 15; GB-2/07/Report, §§ 66–68) is a means that de facto enlarges the scope of the MLS (Frison et al., 2010: 5–6). The 11 Centres holding such collections in trust signed agreements with the GB in 2006, placing the collections under the Treaty. The Second Session of the Governing Body in 2007 recognized that the CGIAR Centres should use the SMTA when distributing non-Annex I materials. All recipients of those materials from the Centres are receiving them on legal terms and conditions identical to those applying to materials in the multilateral system, and they are contractually bound, when passing them on (or new PGRFA incorporating those materials), to use the SMTA. This may represent a significant de facto expansion of the Treaty's MLS. Some countries have also decided to use the SMTA when distributing non-Annex I PGRFA. Likewise, countries participating in the European Genebank Integrated System (AEGIS) have agreed to use the SMTA for distributing important PGRFA designated as European Accessions, whether they are listed in Annex I or not (AEGIS, 2009).

De facto enlargement of the scope of the MLS also occurs by transfers to recipients in non-contracting parties. On this matter, the Treaty is silent, but it appears there is nothing to prevent a provider in a Contracting Party to send materials to a recipient in a non-Contracting Party using the SMTA. Since the recipient in a non-Contracting Party would then be bound by the terms and conditions of the SMTA through the viral clause of the SMTA, he should use the SMTA for subsequent transfers of the same material and new PGRFA incorporating the material received. This viral clause is similar to a

> *copyleft* feature of open-access schemes, [which] endeavors to attach the sharing norm to the intellectual resource itself-not only to the contractual

> parties. The copyleft feature can lead to a particular provision or regulation applying to any user of the work or invention-almost equating the contract to a right against the world.[27]
>
> (emphasis in original, Dusollier, 2007: 1395; also see Van Overwalle, 2018: 95)

By doing so, the 'reach' of the MLS can de facto expand beyond the territories of contracting parties (Hope, 2008: 79), although this can serve as a disincentive for states to ratify the Treaty as they can access the material without having to contribute to the MLS with their own genetic resources (Halewood et al. 2013: 13).

In sum, today some people still consider that the list is too extensive (Lim, 2011), while others believe the MLS should apply to all PGRFA (van den Hurk, 2011). Nonetheless, more and more Treaty stakeholders believe that modifying and/or expanding Annex I will be necessary in the medium term, especially because very important crops, such as tomatoes, soybeans or peanuts are not included (Coradin and Sampaio, 2011;[28] Cubero, 2011; Fernandez Diaz-Silveira, 2011; Fraleigh and Harvey, 2011; Mozafari, 2011; Visser and Borring, 2011) and because external factors, such as climate changes, might impact on what crop is or has become crucial for food security. Echoing the views of developing countries, some authors do not reject the idea of a modification of the list, potentially to all PGRFA, but do not support such a development before it is clear that the MLS functions efficiently, in particular with respect to its benefit-sharing provisions (Fernandez Diaz-Silveira, 2011; Egziabher et al., 2011; Lim, 2011). This analysis is even stronger since the passionate debate on DSI that took place during GB7.

Farmers' rights

The concept of farmers' rights is an important topic in light of the theory of the commons because it is based on several important principles embedded in the theory: the notion of (farmers') community, the notion of self-organization, the idea of access and use of resources rights, etc. What does this concept cover under the Treaty and what impact does its implementation have?

Defining farmers' rights

Farmers' rights have been one of the most contentious Treaty provision to be negotiated. The increasing 'hyper-ownership' (Safrin, 2004) over seeds during the end of the twentieth century has fed a rising debate among FAO Member states about the 'asymmetric benefits accruing to farmers and breeders. Farmers, whose efforts over centuries in improving and selecting farmers' varieties have made an immense contribution to modern agriculture, and the producers of commercial varieties that take these farmers' varieties as a starting point and reap the benefits from what were characterized as relatively small improvements' (Moore and Tymowski, 2005: 67). Farmers' rights (FR) were initially conceived to be a formal recognition of farmer communities' conservation and selection efforts over millennia and a formal means allowing them to participate in the benefits derived from the use of improved seeds. Has the Treaty provision, as drafted in its Article 9, succeeded in this purpose?

During the negotiation of the Treaty, agreeing on a definition of FRs was extremely difficult (Moore and Tymowski, 2005: 67–78). Stakeholders had different understandings of the concept (Brush, 2005: 90; Oguamanam, 2007; De Jonge and Louwaars, 2009; Aoki, 2010), which made it extremely difficult for contracting parties to agree on a coherent bundle of rights to define the legal concept of FRs in the Treaty. At the end of the day, Article 9 resulted in vague obligations at the discretion of national governments. First, paragraph one recognizes the immense contribution of farmers to the conservation and development of PGRFA. This formal recognition is crucial to justify the establishment of the international benefit-sharing mechanism. Second, in paragraph two, the Treaty provision does not provide an exhaustive list of rights under the concept of FRs. Rather, it mentions three important issues: the protection of traditional knowledge related to seeds (CBD Article 8J; Brush, 2005: 108; Chiarolla, 2011: 128; Srinivas, 2008); the multilateral benefit-sharing concept (De Jonge and Louwaars, 2009; De Jonge and Khortals, 2006); and the right to participate in national decisions on PGRFA. However, Article 9.2 leaves considerable discretion to states to specify these rights in further details within their national legislations. This was exactly the purpose for some countries like North American countries to include the following mentions that the realization of FRs '*rests with national governments. In accordance with their needs and priorities, each Contracting Party should, as appropriate, and subject to its national legislation*' take suggested measures (emphasis added). In this way, the potential of Article 9 to act as a 'farmers' privilege-type' clause (i.e. space for a recognized farmers' exemption whether for PVP or patents) has been counteracted. Finally, with paragraph three, the concept is further downgraded in terms of the 'universal recognition' of FRs and the central role of the international community in their realization, from the initial intention in the IU text (FAO Conference Resolution 9/83, CPGR, 1989 report: 4) to the current Treaty text (Brush, 2005: 87). The fact that the right to save, use, exchange and sell farm-saved seeds is not formally recognized at the international level maintains FRs at a de facto 'lower level' than breeders' rights, which are expressed in the UPOV international agreement. Indeed, recognizing farmers' past role in the use and conservation of PGRFA does not equate to recognize their continuous (current and future) role in innovation and breeding – and therefore the rights attached to that role too. The idea that IPRs is an effective manner to promote research and innovation in the breeding field can sound strange if one looks back at history. IPRs are very recent in human history. Farmers and breeders did not wait for IPRs to innovate by developing and creating new varieties adapted to local conditions, culinary and cultural preferences or religious beliefs. Farmers have always been innovators and breeders. The current text does not effectively recognize this and puts farmers in a passive position of beneficiary (through the BSF).

This downgrading has created a legal imbalance between strong internationally recognized proprietary rights over improved seeds and internationally weak – not to say non-existent – FRs over the vast majority of the World's PGRFA. In that sense, I disagree with Moore and Tymoswki who state that the Treaty text is 'neutral' in this regard (Moore and Tymowski, 2005: 75). By explicitly refusing to recognize FRs at the international level, in a similar way to breeders' rights, the Treaty favours the position of the breeding sector and does not take advantage of its position to promote seed exchanges between farmers, which is a critical practice for food security and agrobiodiversity conservation (Coomes et al., 2015; Pautasso et al., 2013). Indeed, the implementation of FRs is 'subject to national legislation' 'in accordance with contracting parties' needs and priorities' and takes place 'as appropriate'. On the contrary, IPRs are explicitly recognized in Article 12.3(f) and SMTA Article 5(d) and particularly Article 6.10 and contracting parties are requested to respect and enforce these rights. Furthermore, by creating a benefit-sharing mechanism mostly focused on the collection of financial outcomes originating from the enforcement of IPRs in the MLS, it necessarily reinforces the international recognition of IPRs to the detriment of FRs. This point raises the question of the 'mutual supportiveness' of the MLS and FRs. A careful reading of other Treaty provisions mentioning farmers allows to point to clear obligations, which indirectly promote FRs (e.g. Articles 13.3, 18.5). These other very specific (and more easily implemented) obligations may function as a vehicle for supporting farmers and promoting FRs (Andersen, 2008: 111).

However, there is a significant nuance in the weight recognized for both these rights. The nuance resides in the fact that IPRs provide a property right for a certain period of time while FRs recognize a stewardship right with no time limit. The current international regime complex strongly recognizes individual property rights for the benefit of one owner versus collective use practices for the benefit of humanity. Why, in our present system of value, does an individual property right have more weight than collective stewardship? The justification for this imbalance in rights' recognition can be questioned (Morten Haugen, 2014). Both farmers and breeders need seeds to work. Both stakeholders select varieties, improve and conserve them. They both need access to a diversity of genetic material (although not necessarily the same type of material). However, farmer communities are the ones which, through the ancestral practice to save, use and exchange PGRFA, developed all existing PGRFA which breeders use daily (with the exception of crop wild relatives). Why should this not be formally recognized and enforced at the international level (i.e. in the MLS)? And above all, why should these traditional practices, of which breeders benefit every day, be rejected at the benefit of biotechnology and dematerialization of breeding technics? I believe it is contrary to interests of humanity, including breeders' long-term interests. I believe a diversity of systems and rights could coexist.

Implementing farmers rights

Although attention is growing on the need to help countries in designing FRs policies and legislations (Resolution 7/2017), FRs are still poorly implemented in national legislations (Andersen and Winge, 2009). According to the 'Farmers' Rights Legislation & Policy Database', only ten countries in the World have enacted a legislation dealing with the matter (out of which eight are Asian countries).[29] The purpose here is not to examine the few existing legislations in detail (Ramanna, 2003; Singh et al., 2011) but rather to show that leaving the recognition of FRs subject to national laws has not permitted to enforce these rights (and has further increased the imbalance with IPRs). Indeed, added to the difficulty of agreeing on the substance of these rights, their enforceability within national legislations is hypothetical. Even in the case where farmers' rights are recognized as constituting human rights, which should be enforceable, it is another challenge to make sure that there is an effective judicial protection accompanying these rights and that people can access this judicial protection (Alston and Weiler, 1998: 668).

Against this background and to mitigate this weakness, contracting parties have systematically included farmers' rights on the agenda of the Governing Body, since its Second Session, and six Resolutions have been adopted (2/2007; 6/2009; 6/2011; 8/2013; 5/2015; 7/2017). This systematic inclusion of the matter in the Governing Body agenda and in resolutions clearly shows the desire of a majority of contracting parties to the Treaty to render FRs effective and up-grade it to a wider recognition at the international level. But most of the resolutions had little substance. Only the recently adopted Resolution 7/2017 represents a big step forward through the establishment of an Ad Hoc Technical Expert Group on Farmers' Rights whose role is to 'produce an inventory of national measures that may be adopted, best practices and lessons learned from the realization of Farmers' Rights, as set out in Article 9 of the International Treaty'; and 'based on the inventory, develop options for encouraging, guiding and promoting the realization of Farmers' Rights as set out in Article 9 of the International Treaty'.

With the same objective to advance the realization of FRs and following recommendations made in several Treaty resolutions, three important workshops were held: the Informal International Consultation on Farmers' Rights held in Lusaka, Zambia, in 2007; the Global Consultation on Farmers' Rights that took place in Addis Ababa, Ethiopia, in 2010; and the Global Consultation on Farmers' Rights held in Bali, Indonesia, in 2016. These consultations encourage contracting parties in developing national action plans for the implementation of Article 9, and to share information regarding progress with the development and implementation of such action plans. Following the call to share information (Resolution 8/2013: point 5), in-depth reports have been sent to the Treaty Secretariat by several stakeholders at GB 6 (Greiber et al. (IUCN), 2015; Shashikant and Meienberg, 2015; La Via Campesina and GRAIN, 2015; ESA, 2015). These activities contribute to raise awareness and highlight the importance of capacity building activities (Article 13.2(c)), but due to lack of funds these activities remain limited (IT/GB-3/09/Report Appendix A: 41; Semences Paysannes, 2015), even when taking into account the BSF projects (the Treaty Secretariat launched four calls for projects, in 2011, 2013, 2015 and 2017).

Besides the clear craze for the Benefit-sharing Fund round calls[30] by which the Treaty Secretariat aim at funding benefit-sharing projects in developing countries, stakeholders also manifest their interest in the question of FRs implementation by other means. A Joint Capacity Building Programme was set up between the Plant Treaty, the Global Forum on Agricultural Research (GFAR) and other organizations including Bioversity International to respond to the limited capacity of governments to implement FRs, with the hope (depending on funding) to lead to a stable and systematic capacity building programme similar to the one set up by the UNEP-GEF for the implementation of the Biosafety Protocol in more than 100 developing countries (UNEP-GEF, n.d.; Frison and Joie, 2006a, 2006b).

Implementation of FRs could also significantly benefit from an explicit recognition of farmers' direct access to MLS seeds. The MLS facilitates seed exchanges solely for the purpose of research, training and breeding (Treaty Article 12.3 (a)) with no mention of direct use by farmers, while the SMTA has a clear focus on commercialization of improved seeds, and defines what a 'product' or a PGRFA 'under development' is and what is understood under the terms 'sales' and 'commercialization' (SMTA Article 2). These terms are far away from what constitute the daily lives of millions of farmers feeding a majority of the population around the world. As mentioned earlier, the fact that Article 9 is not integrated within the MLS (keeping farmers aside from the innovation line as passive users) has created an imbalance of rights, which may be partly mitigated by reinforcing the mutual supportiveness of the MLS and Article 9 through their common implementation. Point 7 of Resolution 8/2013 recognizes explicitly for the first time that farmers, local and indigenous communities should have direct access to MLS seeds. Recognizing the right for farmers to directly access MLS seeds (even in 'soft' terms such as 'invites contracting parties to promote access [. . .] by farmers') constitutes another further step, which consolidates the link between Article 9 and the MLS.

A related aspect deals with the viability of farmers' seed networks. Studies have shown the importance – including for the formal seed sector – of the informal seed networks in the conservation, sustainable use of and access to PGRFA (Louwaars and De Boef, 2012; Coomes et al., 2015; McGuire and Sperling, 2016). Much flexibility will be needed in maintaining the diversity of existing seed systems in order to create a mutually supportive framework, rather than the current international framework largely favouring breeders' rights and strict, formal variety release and seed certification legislations.

A final comment is made regarding the interrelationship between FRs and WIPO/UPOV. As the analysis in Chapter 3 has demonstrated, international and national formal instruments promoting breeders' rights and dealing with seed release and commercialization (negatively) impact informal seed networks and the maintenance of the common practice of farmers to save, exchange and use seeds (Anvar, 2008; Louwaars et al., 2005, 2006; Louwaars, 2008; Andersen, 2015). The Governing Body has acknowledged the need to further understand the relationship between the Treaty MLS and UPOV/WIPO international instruments

(Resolution 10/2015). A timid collaboration was set up (Resolution 7/2017: §§ 11–12), *inter alia* through the organization of a 'Symposium on possible interrelations between the International Treaty on Plant Genetic Resources for Food and Agriculture and the International Convention for the Protection of New Varieties of Plants' by UPOV. However, these two institutions have not yet meaningfully taken up the subject into their agendas, although Treaty stakeholders[31] have clearly highlighted the difficult cohabitation of both FRs and breeders' rights systems,[32] and call for further collaborative work to promote a truly mutually supportive global system. This is clearly not an easy task. It will need to overcome the almost 30 years of competition on the primacy of either concepts of breeder's rights or FRs (Halewood, 2014: 300). However, such a mutually supportive system is a desirable and feasible option (Bertacchini, 2008; Bocci et al., 2011; Chiarolla and Jungcurt, 2011) and sui generis options for plant variety protection in developing countries have already been put on the table (Correa, 2015). What is needed now is for states to effectively embrace the matter, which is mainly a political decision.

To conclude on the analysis of farmers' rights, it is clear that the lack of formal recognition of these rights at the international level – in contrast with the strong recognition of IPRs – create an imbalance of rights which prevents the Treaty from functioning effectively. By denying a formal recognition of rights protecting the present role of farmers (in breeding innovation and in producing our food) at the same level of IPRs, the effectiveness of the Treaty is imperilled. However, throughout every Governing Body Resolution on FRs, the subject has been dug further and further down to concrete matters in order to render the abstract and vague obligations under Article 9 more tangible. Every step taken by contracting parties individually, or collectively through the Governing Body, and by other institutions and stakeholders upon the Governing Body's request, contributes to clarifying states' conduct in implementing FRs and the outcome to be reached. De facto, FRs become more substantial, more specific, and therefore more easily implementable, even though there is no agreement on one precise definition as to what is covered under the concept of FRs. The outcome of the forthcoming 'UN Declaration on the rights of peasants and other people working in rural areas' might also bring some flesh to the recognition and implementation of FRs. Promoting a stronger collaboration with existing international instruments such as the Nagoya Protocol, which recognizes right to communities and aim at protecting their innovation, may also contribute to reaching this objective.

Facilitated access to PGRFA

The fourth theme addressed in the Treaty analysis is the facilitated access to PGRFA, set up by the Multilateral System of access and benefit-sharing (MLS). In Section 1, it was shown that using seeds sustainably was a crucial element for their conservation and for reaching food security. The compulsory prior step to this process is accessing the necessary genetic material. Access to Annex I list of crops and forages is facilitated for a range of specific uses. Several Treaty provisions deal with the facilitated access concept: Articles 1, 10, 12, 13.1 and 15 and the obligations deriving from the Standard Material Transfer Agreement (SMTA). In this section, a presentation of the facilitated access concept will be followed by an assessment of its implementation.

Defining facilitated access under the MLS

The MLS functions as a virtual common-pool of PGRFA for which access is facilitated and which triggers benefit-sharing obligations upon their use (Manzella, 2013: 150–163; Stannard, 2013: 243–263). The Treaty (Articles 10–13), complemented by the provisions of the SMTA, establish the rules regarding the material included in the MLS, the terms and conditions of its access and use, and how (monetary and non-monetary) benefits shall be shared. Managing access and benefit-sharing in a multilateral way and providing for a facilitated access to many important crops is considered as a major benefit in itself (Article 13.1). Under the MLS, access is free (or under limited processing and shipping fees) and facilitated through the use of the SMTA. The fact that no contract needs to be negotiated between providers and recipients puts every party on an equal footing and constitutes a major achievement in the exchange of genetic resources. For now, the CGIAR has been the most important provider of material on the basis of the SMTA, mainly to developing countries.

The prevalence of State's sovereign rights going against the intrinsic logic of PGRFA

Following the adoption of the CBD (CBD Article 3; Rio Declaration on Environment and Development, Principle 2; Stockholm Declaration on the Human Environment, Principle 21), access to genetic resources has been hooked on to states' sovereign rights to control the use of 'their' genetic resources (Treaty Article 10). FAO Conference Resolution 8/83; FAO Conference Report 1985 § 294; FAO Conference Report 1989 § 105; (Halewood, 2014: 301–307) This 'principle means that a state has the power and jurisdiction to establish the manner in which the resources will be shared and used as well as whether they are the object of property rights (private or public) and the conditions under which this may occur' (Correa, 2013: 182).

To match the primacy of the sovereign rights principle (Aalberts, 2012; Nagan and Hammer, 2004: 141–187) with the idea of creating a common-pool of seeds (to be accessed without any systematic prior informed consent by contracting parties), negotiators proposed to 'twist' the principle (ibid). By stating in Article 10.2 that the MLS is established through the exercise of states' sovereign rights, contracting parties turned around the necessity to obtain prior informed consent for each request to access genetic resources on a state's territory. By doing so, unrestricted availability of germplasm was ambitioned without renouncing to the primacy of national sovereignty (Halewood, 2014: 301–307).

However creative this solution might have been, it has only built around the difficulties that the sovereign rights principle entails on access to genetic resources, i.e. enclosure due to appropriation. Furthermore, it has not explicitly acknowledged the necessity of creating a mechanism where common management adapted to the resource would prevail. Although clearly reflecting a majority of the contracting parties' intention, this strong formal recognition of sovereign rights as a founding principle of the MLS is, in essence, contradictory with the intrinsic logic of PGRFA conservation and sustainable use and with the founding principle of unrestricted availability of seeds under the IU. I believe that this constitutes a fundamental mistake, which might explain why the Treaty does not function well and why the contracting parties have difficulties in implementing the MLS obligations.

The facilitated access to Annex I PGRFA in the MLS

The Governing Body of the Treaty has adopted a specific contract, i.e. the Standard Material Transfer Agreement, in order to facilitate the exchange of PGRFA material through its MLS. The SMTA is a contract between two parties (a provider and a recipient) defining the terms and conditions for the transfer to take place. As a contract, the SMTA operates at the level of private contract law, rather than public international law (even if one of the parties, or both, are public entities). On this basis, the SMTA imposes rights and obligations only to its parties. However, there is a major difference between the SMTA and a 'normal' material transfer agreement: the origin of the material transferred is the provider and the destination of the benefits is the MLS (through the BSF) rather than the provider of the material. There is therefore a tripartite relation between the usual provider, the recipient and the MLS as a third virtual 'entity'. In order to recognize the latter, the SMTA has created the 'Third Party Beneficiary' (3PB, see below in this chapter), as formal representative of the MLS.

This SMTA is used by all providers and recipients willing to exchange material under the scope of the MLS (for a detailed explanation of its functioning see Manzella, 2013). This scheme is applicable to PGRFA covered by the MLS, that is to say material listed in the Annex I of the Treaty and all PGRFA held in trust by the CGIAR, but also in the case where the contracting parties (i.e. providers and recipients) agree to use the SMTA for non-Annex I PGRFA, or if a Contracting Party (like Germany and The Netherlands) decides to impose at the national level the use of the SMTA for all PGRFA under its control and management. The purpose of the access must remain for research, breeding, or training for food and agriculture (Article 6.1 SMTA), otherwise the general rules of the CBD (Article 15) and its Nagoya Protocol (Article 6) should be applied. When PGRFA are accessed under the MLS (including after being developed further), they are to be made available by the recipients for further accesses under the terms of the SMTA (SMTA Article 6.4). When a product that incorporates material accessed through the MLS is commercialized, the recipient must pay an equitable share of the commercial benefits to the Benefit-sharing Fund (Treaty Article 13.2d(ii)) if further access to the material commercialized is restricted by the recipient, for instance through IPRs (Chiarolla and Jungcurt, 2011). Payment is otherwise voluntary.

Finally, material can be transferred between contracting parties and non-contracting parties with the SMTA when the material originates from the MLS (Article 6.4 SMTA). However, a provider from a non-member country may provide access to non-MLS PGRFA to a recipient in a Treaty Contracting Party under whatever contract he chooses (in accordance with its national legislation), even if the material is a duplicate from material covered by the MLS. In this manner, a recipient (i.e. the seed industry) may avoid using MLS material in order to escape the benefit-sharing related obligations.

Intellectual property rights and the Multilateral System

Article 12.3(d) – one of the most sensitive provision of the MLS – requires that recipients shall not claim IPRs that would limit facilitated access to the PGRFA or 'their genetic parts and components, in the form received from the Multilateral System'. Besides, Article 12.3(f) recognizes existing IPRs. These provisions can be interpreted in very different manners. Some parties fear that recipients will patent their products and keep them outside the MLS (Helfer, 2005). In order for IPRs to support the Treaty's objectives, IPRs should not limit the facilitated access to PGRFA in the MLS and IPRs should be a means to capture value from the development and commercialization resulting from the facilitated access to crops in the MLS. The Treaty specifies that Annex I materials that are 'under the management and control of the contracting parties and in the public domain' are included in the MLS. If, as a result of IPRs on PGRFA, the latter is not in the public domain, then one of the conditions for being automatically included in the MLS is not satisfied. That is not to say that IPRs owners could not elect to place their materials in the MLS; they could. Further, if a public research organization chose to seek patent protection for a new PGRFA, it would not be automatically included in the Multilateral System. In this way, IPRs can function to limit what is included in the MLS.

It is clear that a recipient cannot take IPRs that prevent others from obtaining, from the MLS, a PGRFA in the same form in which it was originally sent to the first recipient, e.g. a seed or a cutting (Article 6.2 SMTA). It is still not clear, however, if a recipient can seek IPRs over isolated parts and components of those seeds or cuttings from materials within the MLS, such as genes. In any case and importantly, independently of the outcome of this debate, such property rights must not prevent future recipients from obtaining the same seeds or cuttings (Halewood and Nnadozie, 2008: 129). In the future, a Governing Body decision or a dispute settlement decision might provide a common interpretation on the terms 'parts and components, in the form received'. For now, it is considered that IPRs do not hinder too much the facilitated exchange of PGRFA, although voices heard from developing countries in particular put forth that IPRs limit the efficient implementation of the MLS (Frison et al., 2011b: general conclusion; Mwila, 2013).

The Treaty does not prevent recipients from seeking IPRs over improved products that incorporates materials received from the MLS. If the material is restricted or protected in a way that does not allow further recipients to use it for further breeding and research, the recipient has the obligation to share a percentage of the commercial benefits back to the MLS (see below). The restriction may be technological or legal (SMTA Article 2), such as for example, with a patent. The IPR provision of the SMTA was designed for the MLS to receive funds in order to be able to implement its plans and programmes under the Treaty (Article 18.4(a)).

Patents may limit what enters the MLS but not alter the material already included in the system as long as it is in the system. However, material may be taken out of the MLS in two different ways. First, anyone can simply decide to stop conserving an MLS accession; if there is no duplicate, the material is lost. Second, a Member State may step out of the Treaty (Article 32). In this case, MLS material from that country, which has not been transferred to another country, would leave the MLS and be available for patenting.

Implementing the facilitated access to the Multilateral System

Since June 2004, contracting parties have been busy creating all the necessary tools and instruments to operationalize the Treaty and implementing all self-standing obligations from the Treaty. Implementation of the MLS has certainly concentrated most of contracting parties' attention since the entry into force of the Treaty (together with the Funding Strategy). This constitutes Phase One of the implementation covering the period 2004–2013. By the end of Phase One, contracting parties came to the observation that the MLS was not fulfilling their expectations in terms of facilitated access and in terms of financial outcomes for benefit-sharing activities (Galluzzi et al., 2016: 5). As a reaction, contracting parties initiated a review process in order to 'increase user-based payments and contributions to the [BSF]', and 'enhance the functioning of the [MLS]' (Resolution 1/2015: § 2 and point 4). This review process launched the implementation Phase Two, from 2013 to nowadays, which will attract more attention below.

Implementation Phase One 2004–2013

During the first period, contracting parties devoted their efforts to building an efficient MLS. Through numerous Governing Body Resolutions, they adopted the necessary tools and instruments, such as the SMTA (Resolution 2/2006), the Third Party Beneficiary entity (Resolutions 5/2009; 5/2011; 9/2013), the mechanism to promote compliance and address issues of non-compliance (Resolutions 3/2006; 1/2007; 2/2009; 2/2011; 9/2013), as well as the Funding Strategy (Resolutions 1/2006; 3/2009; 3/2011; 2/2013) and the Benefit-sharing Fund, etc. to fulfil their obligations. This busy process was facilitated by the work of the Secretariat of the Treaty and its Secretary, who organized six Governing Body meetings and numerous inter-sessional meetings of various working groups/contact groups/ad hoc working groups, expert groups, etc. In the following sub-section, various aspects of this implementation process will be analysed. While the efficient institutional functioning of the Treaty will not be addressed (for details see Frison, 2016: 178–187), I will focus on the actual (much more difficult) implementation of the MLS obligations by contracting parties. An evaluation of the collections and accessions in the MLS, as well as data on the use of SMTAs and flows of PGRFA will be addressed.

One should not underestimate the theoretical breadth of what is included in Annex I. As a matter of fact, all PGRFA, from the moment they are part of the Annex I list and under the management and control of the contracting parties and in the public domain, are 'automatically' included in the MLS without any declaration or notification. It is clear that the Treaty provisions impose no specific procedure to include material in the MLS. This is what happens in theory. In reality however, 'actual use of material depends on information being made public about what materials are available and where they may be accessed, along with related non-confidential information' (IT/AC-SMTA-MLS 1/10/Report: 11; IT/AC-SMTA-2 MLS 2/10/Report: point 12; IT/AC-SMTA-MLS 3/12/Report: points 12 and 15; IT/AC-SMTA-MLS 4/12/Report: point 9). Although no specific procedure is explicitly defined in the Treaty (except from the fact it should be as simple as possible with not tracking obligation; see Treaty Article 12.3(b) and SMTA 5(a)), the concrete need to know what accession is part of the MLS constitutes a logical prerequisite for users to be able to identify potential material to be accessed. To this end, the Treaty Secretariat has made available specific tools to help contracting parties in this process of identification. First, a 'Letter of Notification of Inclusion of Material in the Multilateral System' can be downloaded on the Treaty website. 'Notification to the Secretary, or an equivalent public statement, creates a legitimate expectation on the part of potential recipients that the materials in question will be made available under an SMTA, on request' (IT/AC-SMTA-MLS 2/10/Report: 5). The Treaty Secretariat also established a website to publish the notifications of material included in the MLS. In some cases, these notifications contain a link to web portals where the samples are documented or where the material can be ordered online. Following a recommendation of the Ad Hoc Advisory Committee on the SMTA and the MLS (IT/AC-SMTA-MLS 2/10/Report: 9 and in particular point 65), a 'Handbook to the Implementation of the Multilateral System of the International Treaty on Plant Genetic Resources for Food and Agriculture' (IT/AC-SMTA-MLS 3/12/Inf.22/Inf.2) was drafted to help countries implement their obligations, including that of including PGRFA in the MLS. Following this guidance, the MLS has been furnished with material designated by contracting parties (Article 11.2), by the CGIAR Centres and other international organizations (Articles 11.5 and 15), by other natural and legal persons (Article 11.4), as well as the material resulting from BSF projects and included in the MLS.

As of March 2018, 43 out of 145 contracting parties notified one or several lists of accessions to the Treaty Secretariat. One can objectively say that this participation could significantly be improved. This clearly shows that the 'automatic' inclusion of material in the MLS is not such an easy and straightforward process. Indeed, it requires that national repositories maintain updated and complete list of the material with all the necessary information (passport data, etc.). This is clearly not the case for many countries (IT/AC-SMTA-36 MLS 2/10/Report: 36). Furthermore, the fact that the negotiation of the Annex I list of PGRFA impose on contracting parties to distinguish what seeds are covered and what seeds are not covered by the MLS creates a significant administrative burden on Member states, in particular for developing countries. Indeed, many contracting parties need more support to operate this technical identification of the material covered by Annex I, in the form of capacity building. There is no doubt that the system would be much easier if all PGRFA were covered by the MLS and contracting parties would only have to provide access to their national databases of collections, without the need to investigate, almost accession by accession, whether the content of their collections enters the MLS or not. Some countries have adopted this philosophy by developing a wide national policy providing access to all the PGRFA under their control and management using the SMTA. The Netherlands and Germany decided to do so and already effectively use the SMTA for all the crops and forages under their management and control, thereby de facto widening the MLS.

In accordance with Treaty Articles 11.5 and 15, the CGIAR centres have included their PGRFA in the MLS, through a letter of agreement signed on 16 October 2006 between the eleven IARCS of the CGIAR and FAO, as legal entity representing the Governing Body. This collaboration is the logical continuation of the previously International Network of Ex Situ Collections under the Auspices of FAO held in trust for the international community since 1994. Other international research centres have also signed such agreements with the Treaty. In furtherance of the CGIAR policy of widest diffusion of PGRFA, contracting parties agreed that CGIAR centres commit to using the SMTA for the transfer of all material held by CGIAR collections, not only Annex I material (IT/GB-2/07/Report: §§ 66–68). This aimed at simplifying the exchange procedures by avoiding to use different MTAs and hence reduce costs. Transfers are made with the same SMTA, complemented by an interpretative footnote which clarifies that the SMTA provisions should not be interpreted as precluding the use of the SMTA for transfers of non-Annex I material, collected before the entry into force of the Treaty (IT/GB/2/07/13 rev.1: § 9). Following this new enlargement of the material covered under the MLS, material originating from the CGIAR constitutes an important percentage of the accessions made available under the MLS,[33] with more than 665,684 accessions notified to the Treaty Secretariat on a total of 1, 561,638 materials notified by the end of 2017 (IT/GB-7/17/Inf.4: 4; for a detailed overview of the material per country see 'Appendix: Plant Genetic Resources Available in the Multilateral System' of the same document).

Article 11.3 enables natural and legal persons within the jurisdiction of contracting parties and who hold PGRFA listed in Annex I to include such PGRFA in the MLS. Similarly to the notifications by contracting parties, very few natural and legal persons have done so (only six between March 2009 and 2016).

Finally, one last option contributes to furnish the MLS with accessions: incorporating the material resulting from the activities under the calls of the Benefit-sharing Fund (IT/AC-SMTA-MLS 4/12/Report: point 9). While there is no such obligation stated in the Treaty or in its operationalizing instruments documents, including material resulting from BSF funded projects follows the intentions of contracting parties to have as wide an MLS as possible. It is reasonable to interpret this fact as an implicit strong encouragement to voluntarily contribute material to the MLS (Treaty Articles 11.3, 11.5, 15; SMTA Article 6.9).

At every Governing Body Session and in inter-sessional meetings, contracting parties have tackled the issue of the inclusion of PGRFA in the MLS and have integrated it in Resolutions related to the implementation of the MLS (Resolution 4/2009: points 7–11; Resolution 4/2011: points 3–5; Resolution 1/2013: points 14–17; Resolution 1/2015: point 11). By doing so, states recalled the importance of the 'availability of plant genetic resources for facilitated access [as] the foundation of the Multilateral System' (IT/AC-SMTA-36 MLS 2/10/Report: 36), thereby insisting upon inclusion being a primary precondition for the system to be effective.

The adoption of the text of the SMTA at the first meeting of the Governing Body enabled contracting parties to start implementing the facilitated access obligation as from January 1st, 2007. After an enthusiastic beginning when figures (from the CGIAR) rapidly showed that SMTAs were used to exchange PGRFA all around the world, the enthusiasm lowered significantly after a few years. Indeed, SMTAs are signed by recipients and providers to exchange seeds under the Treaty, but these users seem to be those very same users accessing PGRFA through the previously existing CGIAR collections. A report (IT/GB-4/11/Inf. 5) showed that from 1 January 2007 to 31 December 2009, the Centres distributed a total of 1.15 million samples of PGRFA. Approximately 84 per cent of the samples were sent to developing countries or countries with economies in transition, 9.5 per cent to developed countries and 6.5 per cent to CGIAR Centres. 18 per cent were sent by the Centres' genebanks, and 82 per cent from the breeding programmes'. Few other data were available before 2013,[34] i.e. on the number of SMTAs signed (other than those with CGIAR centres) or on the type of providers and recipients, or on the type of material exchanged. A study stated that by the end of June 2013, there were 261 users[35] registered in the Treaty Data Store.[36]

These figures show that important companies from the Seed Industry (except from the European Seed Industry) tend to avoid accessing seeds using the SMTA. Besides the cost related to potential benefit-sharing obligations, the administrative burden related to the SMTA might explain the reluctance in its use to access PGRFA, as well as the fact that they own important private genebanks. Although the initial aim was to render access as simple as possible, the fact that a contract (even a standard one) has to be filled in with specific information (notably on the pedigree of the material), signed and sent, and that relevant information needs to be transmitted to the Third Party Beneficiary, is quite a burden for the providers and recipients (IT/AC-SMTA-MLS 2/10/Report: §§ 21–29 and Appendixes 4 and 5; IT/AC-SMTA-MLS 2/10/3). Issues about traceability and control of transfers are also regularly mentioned by Treaty stakeholders. Even for the CGIAR Centres' using the SMTA for all their material exchanges (including non-Annex I material) it is a challenge. SMTA Article 6.5(b) requests the provider to identify the material received from the MLS and specify that the transferred PUD derives from that material. This obligation poses a problem to the CGIAR, because all improved material are transferred with an SMTA, thereby creating enormous tracking obligations with most of the material exchanged (Resolution 1/2015: §§ 22–25).

To enhance the use of the SMTA, the Treaty Secretariat created the 'Easy-SMTA' online tool, which aims at facilitating the signature and handling of SMTAs. However, this does not seem to have dramatically increased its use and contracting parties came to the conclusion that the SMTA had to be reviewed in order to 'increas[e] user-based payments and contributions to the [BSF]', and 'enhance[e] the functioning of the [MLS]' (Resolution 1/2015: § 2 and Point 4).

Implementation Phase Two 2013 – nowadays (end 2017)

The realization that it is difficult for contracting parties to implement their obligations and that this does not allow all users of PGRFA to effectively benefit from the MLS has functioned as a wake-up call at GB5. This was translated into an official review process through the establishment of the Open-ended Working Group on the MLS and the Advisory Committee on the Funding Strategy. In this section, the purpose is not to analyse paragraph by paragraph the work conducted by the above mentioned working group and advisory committee;[37] the review is not complete yet and options evolve from one GB to the other. Rather, a brief explanation of the main solutions sought for the enhancement of the MLS by contracting parties at the last GB (GB7, held in Kigali, Rwanda, 30 October – 3 November 2017) will be provided and commented upon. Funding issues are definitely crucial to the functioning of the system. However they are not legal instruments per se, and therefore they fall out of the scope of this analysis.

The work carried out by the various working groups, advisory committee, contact groups and co-chairs has evolved from the initial mandate to 'increase user-based payments and contributions to the BSF in a sustainable and predictable long-term manner, and enhance the functioning of the MLS by additional measures' (Resolution 2/2013: point 23), to the lately renewed mandate to 'develop a proposal for a Growth Plan to attain the enhanced Multilateral System, taking into account *Annex 1* to this Resolution, as appropriate;

revise the Standard Material Transfer Agreement [. . .]; and elaborate criteria and options for possible adaptation of the coverage of the Multilateral System'. (Resolution 2/2017)

To help contracting parties progress in the negotiation, various study-topics were requested by the Treaty Secretariat, for which background studies (Moeller and Stannard, 2013), synoptic studies (IT/OWG-EFMLS-2/14/3; IT/OWG-EFMLS-2/14/4; IT/OWG-EFMLS-2/14/5; IT/OWG-EFMLS-2/14/6) and research studies were outsourced. These documents allowed the negotiators to progress on the matter, but not sufficiently to adopt firm proposals.

Contracting parties came to the conclusion that one option to enhance the functioning of the MLS was to revise the financial benefit-sharing mechanism included in the SMTA (mainly Articles 6.7 and 6.11) to render the SMTA more attractive. One of the solutions explored is to create a 'subscription payment' mechanism (IT/OWG-EFMLS-4/15/4; IT/GB-6/15/06 Rev 2; IT/OWG-EFMLS-3/15/Inf.5), which would become a real alternative to the Articles 6.7/6.8 standard benefit-sharing mechanism. The rationale behind this proposal is the same as the one behind SMTA Article 6.11 but enhanced with the recognition that both payment options are closely interrelated and that both options should be equivalent in terms of costs and benefits for the users, for them to be able to actually make a choice between using one or the other (and therefore be encouraged to use the SMTA rather than acquiring material by other means). To widen up the range of situations where payments would be due and to follow on the example given by Norway, the Subscription System proposed at GB7 would entail payments at different (not yet negotiated) rates, i.e. when a recipient commercializes a product that would remain available without restriction as well as when commercializing a product that is not available without restriction to others for further research and breeding (Resolution 2/2017: Annex 2, point 1 and 2; Annex 3, Article 3).

Article 6.11 functions as an upfront payment scheme. This option is called the 'crop-based alternative payment scheme' because benefit-sharing is triggered by the sales of products belonging to one specifically identified crop or related products (other products that are PGRFA belonging to the same crop). This alternative payment scheme is rarely chosen by SMTA users (Correa, 2011: 249–256). As mentioned above, the payments are to be made whether the product is under restricted access or not. The rate is of 0.5 per cent of the sales of the product or related products during a ten-year period of time. The seed industry has repeatedly highlighted that this rate was too high (Guiramand et al., 2012). Furthermore, it is not clear what happens once the ten-year period is over: Do parties fall back on the 6.7 regime? And if so, would this mean that tracking obligation would need to be complied with since the signing of the SMTA?

In the last proposal, the revised SMTA would provide

> the option to subscribe to a Subscription System. Companies and other users interested in becoming subscribers of the Subscription System would

> be given a period of one year after a date to be set by the Governing Body to declare their willingness to sign on to the Subscription System. The new and revised SMTA containing the Subscription System as an access mechanism would automatically become effective, once companies that represent a to be agreed amount of money corresponding to approximately [30%] of global seed sales of crops contained in the Multilateral System (i.e. listed in Annex I) sign on to the Subscription System.
>
> (Resolution 2/2017: Annex 1 Growth Plan)

The proposal further suggests that 'at the same time that the Governing Body adopts the new and revised SMTA, it should decide on amending Annex I of the Treaty, including a new list of crops. The amended Annex I would (1) delegate the power of future expansion of the Multilateral System to the Governing Body, and (2) include a list with a first set of specific crops, thereby once again linking coverage of the MLS with the conditions of access. Furthermore, to facilitate the notification of accessions by contracting parties, the Treaty Secretariat encourages to use the new tool they developed for that purpose, i.e. the Digital Object Identifiers (DOIs) of the Global Information System (see below), as one of the methods for the identification of material available in the MLS (Resolution 4/2017: point 4).

While the issue of adapting the coverage of the MLS remained 'taboo' until the 2011–2013 biennium, contracting parties are now much more open to discussing this option (Resolution 2/2017). This is definitely a good thing in my view. Indeed, in an ideal world, if the MLS covered all PGRFA everywhere, for all food and agriculture purposes, and if the payment scheme was of an upfront payment type, users would have no other choice than acquiring material through the system. The system would be unique, uniform and simple. There would be no tracking obligation and potentially no need for any contractual (formal) agreement. Of course we are far from this utopia, however, there are voices expressing their willingness to move in that direction (Stannard, 2013).

While it is undeniable that a review is needed, the way in which negotiators have tackled this issue sounds partly unproductive to me. The highly technical studies, projections and scenarios proposed at GB6 in the numerous studies on which negotiators have based their discussions have added confusion rather than simplified the proposed mechanism. The proposal for a Subscription System at GB7 seems simpler and more realistic, but the tsunami caused by the vivid discussion on the DSI as 'a cross-cutting matter affecting the entire Treaty agenda' (ENB, 2017: 14), does not predict a rapid decision on these matters. It remains to be seen what will come out of the review process at GB8 in 2019.

In this section, the purpose was to understand what covers the facilitated access to Annex I PGRFA obligation and to evaluate its implementation. The analysis above shows that contracting parties have difficulties in providing access to their resources; that accessing seed by all PGRFA beneficiaries is not straightforward; and that working on reviewing the existing system will necessitate to leave some space to all stakeholders to participate in the review process as well as in the implementation of the forthcoming system. I support a simplification of the facilitated access procedure not for the sake of homogenization, which I believe is detrimental to the diversity of resources and their management mechanism, but to render the system implementable. A subscription system might be one way of rendering the process easier to implement and more attractive to some stakeholders. However, I question the utility of remaining in the same appropriation paradigm regarding seed management. I argue that a shift needs to operate from considering the physical resources as exclusively appropriable, the dematerialization of resources openly accessible under no rules, and the facilitated access rationale based on IPRs, to a holistic approach to seed management, where interdependence between seeds and their custodians is put at the core of the system, with an inclusive approach to the right to use those resources for the benefit of humanity, present and future (see below final chapter).

Benefit-sharing, the Benefit-sharing Fund and the touchy issue of money

The negotiations on the benefit-sharing provisions (Article 13) of the MLS were closely related to those on facilitated access (Article 12) and on the Treaty's Funding Strategy (Article 18). This section addresses the touchy issue of money, indispensable to fund the benefit-sharing activities and allow contracting parties to fulfil their Treaty obligations. Notwithstanding the importance of this issue (and the enormous amount of Treaty documents on the subject), the facts can be summarized very simply: there is no (sufficient) money. For this reason and because of the limits of the scope of the present work, financial resources for the Treaty will be addressed very concisely, and only when it directly relates to the concept of benefit-sharing. On the contrary, more details are provided to explain what benefit-sharing obligations cover and how they are implemented.

Defining benefit-sharing

Benefits should be fairly and equitably shared by way of the exchange of information (Article 13.2(a)), access to and transfer of technology (13.2(b)), capacity building (13.2(c)), and the sharing of monetary and other benefits of commercialization (13.2(d)). According to Article 13.1, the primary benefit of the MLS is the facilitated access to the material in itself. To uncover major issues related to benefit-sharing, three points will be dealt with: the Funding Strategy and financial resources of the Treaty; monetary and non-monetary benefit-sharing obligations; and the Benefit-sharing Fund.

Financial resources of the Treaty under the Funding Strategy

Funding is needed for two kinds of expenditures: money to allow for the functioning of the Treaty's Governing Body, Secretariat and other bodies (i.e. the Core Administrative Budget), and funds to implement the Treaty obligations (i.e. the Special Fund for Agreed Purposes, the Benefit-sharing Fund and the Fund to support the Participation of Developing Countries to Treaty meetings). The Funding Strategy (Article 18) is an agreed upon strategy mobilizing funding from multilateral, bilateral and voluntary sources.[38] It is not a fund or a financial mechanism per se (contrary to the Global Environmental Facility, which was assigned as the permanent financial mechanism of the CBD; CBD Articles 20 and 21. CBD-COP Decision III/8). Rather, it constitutes 'an agreed strategy for mobilizing funds primarily from existing channels, though it will also cover financial resources provided for in the Treaty itself, such as the mandatory and voluntary payments to be made under Article 13.2(d)(ii)' (Moore and Tymowski, 2005: 139). We are far from the initial 'International Fund' imagined under the International Undertaking in the 1980s.

Article 18 provides the 'skeleton' of the funding strategy, which includes 'the decision for its establishment, a general objective, a list of possible funding sources and some basic priorities' (Treaty Secretariat, 2013: 9). Article 18.5 insists that priority will be given to 'funding activities that target farmers who conserve and sustainably use PGRFA, especially in developing countries'. The Funding Strategy is therefore a tool to finance Treaty activities and obligations, including those related to benefit-sharing, whether monetary or non-monetary. Indeed, implementing non-monetary benefit-sharing provisions also requires financial resources.

Monetary and non-monetary benefits

While Article 18 provides the general rules guiding the establishment and functioning of the Funding Strategy, Treaty Article 13 and the related provisions of the SMTA define the procedures to collect and share the monetary and non-monetary benefits.

Monetary provisions address how to obtain money from the use of PGRFA in order to implement the fair and equitable sharing obligations of the MLS. Monetary benefit-sharing are referred to in several Treaty Articles: Article 13.2(d); Article 15.1(b)(iii) in relation to the CGIAR; and Article 18.4(e) on Financial Resources. Financial benefits are directed to the Treaty Trust Account (Article 19.3(f)): i.e. the Benefit-sharing Fund. According to Treaty Article 13.2(d), a recipient of Annex I who commercializes a PGRFA product that incorporates material accessed from the MLS must pay to the BSF an equitable share of the benefits arising from such commercialization. This obligation is mandatory only if further access to the material or resultant product is restricted by the recipient, for instance through IPRs; otherwise payment is voluntary. The terms and conditions of the MLS, including details on the type and level of payments to be made, are set up in the SMTA, which is currently under review. Notwithstanding the review process, a brief explanation of the existing system is made for both the SMTA Article 6.7 payment scheme and the Article 6.11 alternative scheme.

Under Article 6.7 of the SMTA, the recipient is the only SMTA contracting party who bears monetary benefit-sharing obligations. Whenever the recipient commercializes a product derived from material originating from the MLS, which availability is restricted to others for further research and breeding, he/she has to pay 1.1 per cent of the gross sales of the product, minus 30 per cent (i.e. 0,77 per cent of the gross sales) for the period equivalent to the duration of such restriction (mainly 20 years for IPR-based restrictions). The 30 per cent is a lump sum that includes the administrative costs, taxes, shipping costs, etc. that would probably be subtracted from the gross income anyway (i.e. the difference between gross sales and net sales).

The alternative payment scheme was proposed by the African Group (Frison et al., 2011a: 53) in the hope to guarantee a faster monetary flow back to the BSF. According to SMTA Article 6.11 and its Annex 4, the recipient may pay 0.5 per cent on the sales of all PGRFA products of the same crop accessed (regardless of whether the products include material obtained from the MLS or not). This payment is made 'regardless of the restrictions for further research and breeding on the product, and for a period of ten years, which is renewable' (Manzella, 2013: 156). Furthermore the recipient may access more material of the same crop using other SMTAs, but will not have to pay the share twice.

Non-monetary benefit-sharing are of two types. This first one is the facilitated access to PGRFA itself covered by Article 13.1. The second covers the exchange of information and results of technical, scientific and socio-economic research on PGRFA (Article 13.2 (a)); the access to and transfer of technology (Article 13.2 (b)); and capacity building (Article 13.2 (c)). The second type of benefits is not possible without the realization of the first, as transfer of technology related to a specific improved material, for example, is useless if the recipient has no access to the improved material first. However, this raises the question about the direct access by farmers of Annex I material. The Treaty does not explicitly provide a direct access to Annex I material to farmers, but farmers are positioned as the first benefit-sharing beneficiaries, and facilitated access to seeds is viewed as the first benefit. . . . Something is going wrong in this loop.

The Benefit-sharing Fund

The Benefit-sharing Fund is the main Treaty instruments implementing ABS obligations. The BSF (i.e. the Trust Account) was established following Resolution 1/2006, when the Funding Strategy of the Treaty was adopted (Treaty Secretariat, 2013: 42–45), to administer the financial resources over which the Governing Body has direct control. The BSF has three priorities: information exchange, technology transfer and capacity building (i.e. non-monetary benefits); on-farm (in situ) management and conservation of PGRFA; and sustainable use of PGRFA. The BSF is therefore the most important tool through which contracting parties implement their benefit-sharing obligations. The BSF is the recipient of the financial benefit-sharing originating from contracting parties, or of voluntary payment made by stakeholders. It is the main administrative instrument distributing the benefits for the implementation of plans and programmes under this Treaty. Priority beneficiaries are 'farmers in developing countries, especially in least developed countries, and in countries with economies in transition, who conserve and sustainably utilize PGRFA' (Article 18.5). The BSF fund activities which fulfil the three priority areas, when assessing what project is eligible for the call for proposals to its projects portfolio. Up to date, three calls have taken place, and the fourth call is ongoing (Treaty Secretariat, 2013). The BSF may be (and up to now has been mainly) financed by voluntary payments from contracting parties or international organizations.[39] These voluntary contributions are the result of innovative approaches to provide money to the Funding Strategy. Implementing benefit-sharing obligations in the MLS with an empty purse.

In 2005 already, Brush had anticipated that

> [t]he weakness of that [T]reaty, however, is that it does not give proper emphasis to the obligations of industrial countries and developing countries alike to support conservation of crop resources *beyond funds raised in connection to commercializing improved crop varieties.*
>
> (emphasis added; Brush, 2005: 108–109)

Indeed, the SMTA mechanism was created around the hope that financial returns would flow back to the system upon access, use and commercialization of improved Annex I material. Following this scenario though, benefits would only arise 5 to 20 years (depending on the technology used) after initial facilitated accesses were granted, due to the long research and development period prior to the commercialization of a new product (Drucker and Caracciolo, 2012: 57). Shortly after the adoption of the SMTA, countries, especially developing contracting parties, started to claim that the breeding development cycle was too long for the Benefit-sharing Fund to receive enough income within a reasonable period of time. Additionally, contracting parties might have not sufficiently anticipated the slow pace of implementation of the MLS (i.e. slow inclusion of material in the MLS, and the resulting limited SMTA exchange, which itself implies limited benefit-sharing returns), and the consecutive serious financial problems the Treaty has been faced with (Moeller and Stannard, 2013; Stannard et al., 2012; Srinivasan, 2012; Guiramand et al., 2012). In the following subsection, the implementation processes of the funding strategy, of non-monetary benefit-sharing obligations, and of the Benefit-sharing Fund will be detailed.

Implementation of the financial resources provisions

Contracting parties have designed the Funding Strategy mainly between 2006 and 2009. With Resolution 1/2006, a notable aspect of the strategy was the recognition of the Global Crop Diversity Trust (GCDT)[40] as an essential element for ex situ conservation and availability (IT/GB-1/06/Report: §§ 35–40). The GCDT is an endowment fund, which provides funds in perpetuity to support long-term conservation of PGRFA and ensure conservation and availability of PGRFA which are most relevant for food security and sustainable agriculture.

Although the GCDT is an independent international organization, it operates within the framework of the Treaty as it receives policy guidance from the Governing Body. Its focus is limited to ex situ conservation activities. Therefore, contracting parties and the Treaty Secretariat should continue to enhance direct efforts for the implementation of the benefit-sharing and other conservation and use provisions, in particular on farm and in situ activities.

Unfortunately, Brush's prediction materialized into lack of funding. As a consequence, contracting parties explored other sources of income within the Funding Strategy (e.g. additional voluntary contributions[41] not derived from SMTA-related monetary benefit-sharing), to implement their benefit-sharing obligation with 'farmers in all countries, especially in developing countries, and countries with economies in transition, who conserve and sustainably utilize [PGRFA]' (Treaty Article 13.3). Contracting Parties committed to raise US$ 116 million between July 2009 and December 2014 through the 'Strategic Plan for the Implementation of the [BSF] of the Funding Strategy' (Resolution 3/2009). The plan was to secure cumulative targets over a five-year period to launch the calls for project cycles under the BSF. By the end of 2016, the target was far from been reached (Resolutions 2/2013; 2/2015).[42] This shortage has slowed down the pace of the calls for project cycles under the BSF (see below). Following this failure, the GB decided to change strategy and to adopt a dynamic and synergistic programmatic approach, i.e. a 'new vision' as follow: 'The Funding Strategy enables the Governing Body, contracting parties, funding agencies, farmers and other relevant actors to secure funding and other resources for the programmatic implementation of the International Treaty in a long-term, coordinated, synergistic and effective manner' (Resolution 3/2017: point 4). It remains to be seen whether this new vision, which embeds the Funding Strategy in a wider map by connecting with other existing funding sources and partners (i.e. the GEF, the 2030 Agenda target, etc.), will succeed in securing long-term sufficient funding.

Implementation of non-monetary benefit-sharing obligations

Besides Article 13, non-monetary benefit obligations are explicitly expressed in several Articles throughout the Treaty: Articles 7 and 8 on international cooperation and technical assistance; Article 9.2(b) on FRs; Article 14 on the Global Plan of Action; Article 17 on the Global Information System on PGRFA; Article 18 on Financial Resources through the design of the Funding Strategy. They were not at the core of contracting parties' attention during the early implementation of the Treaty. However, their rapid implementation constitutes, in my view, an essential element to facilitate the smooth and effective application of most Treaty's obligations. Indeed, some parties, especially from developing countries, do not consider access to PGRFA as a major benefit of the MLS because they have limited financial and technological capacity to utilize PGRFA. They find more important to ensure that benefits derived from the use of genetic resources reach those who need them most and that capacity building and transfer of technology and information is effectively implemented. They consider the exchange of information and results of technical, scientific and socio-economic research on PGRFA as more important benefits to be shared.

After 2013, collaborations with other international organizations were set up to implement specific capacity building activities as part of the non-monetary benefit-sharing obligations (Resolution 2/2013: point 15; Resolution 10/2015). Few examples of technology transfer and capacity building projects outside the BSF activities exist (IT/ACSU-2/15 /Inf.2; Resolution 7/2013).[43] Notwithstanding these initiatives, non-monetary benefit-sharing under the Treaty remains limited and many contracting parties voice their discontentment (Frison et al., 2011a: chapter 4; Mwila, 2013), resulting in numerous Governing Body Resolutions on the subject (Resolution 4/2009, Resolution 4/2011, and Resolution 1/2013 for aspects related to the implementation of the benefit-sharing obligations in the MLS; Resolution 1/2006, Resolution 3/2009, Resolution 3/2011 and Resolution 2/2013 for aspects related to the financial issues in the Funding Strategy). All of these illustrate and confirm that by the end of the biennium 2011–2012 a number of important Treaty obligations were still not complied with and that the MLS did not produce the excepted outcomes (see the four reports of the Ad Hoc Advisory Technical Committee on the Standard Material Transfer Agreement and the Multilateral System of the Treaty). This led to the decision taken at GB5 to review the MLS and the Funding Strategy (Resolution 1/2013; 2/2013).

Implementation of the Benefit-sharing Fund

The BSF became operative in 2008 with voluntary contributions from a few contracting parties. The Strategic Plan for the implementation of the BSF aimed at improving the funding set a funding target of US$ 116 million to be raised by 2014 (Resolution 3/2009; IT/GB-3/09/Report: §§ 26–30). Moreover, Norway pledged to implement a national policy where an amount corresponding to 0.1 per cent of all seed sales are to be transferred by the Norwegian government to the fund (Treaty Secretariat, 2013: 17). Notwithstanding this positive initiative and the four calls for BSF projects that have already been launched, the total amount of money remains far below the US$ 116 million target set in 2009. Since 2013, regular decisions are taken to speed up the review process and accelerate the resource mobilization (Resolution 2/2013; IT/GB-5/13/Report: §§ 22–26 and Appendix A.2; Resolution 1/2015: point 2; Resolution 2/2015: point 3; IT/GB-6/15/Report, Appendix A.2). Although contracting parties have welcomed the financial contributions made by Indonesia, Italy, Austria, Norway and Sweden during the 2014–2015 biennium in support of the fourth round of the BSF project cycle as well as the contribution made by the European Seed Association at Governing Body 6 as the first collective contribution from actors of the European seed sector (Resolution 2/2015), the financial crisis in which the Treaty is stuck remains critical. By GB5, only UD$ 20 million had funded projects involving 340,000 farmers in 55 countries. Major efforts will have to be made in order to comply with Treaty Article 18.2 on the importance of the availability, transparency, efficiency and effectiveness of the provision of financial resources. Transparency does not seem to be a major obstacle, but compliance with the obligations of efficiency, effectiveness and above all availability of funds is clearly not met.

Information and knowledge related to PGRFA

The supporting components (Part V of the Treaty) play an important role in facilitating the implementation of the Treaty by its contracting parties. The four interdependent components are: the Global Plan of Action (Article 14), the ex situ Collections of the IARCs and other International Institutions (Article 15), the International Plant Genetic Resources Networks (Article 16), and the Global Information System (GLIS) on PGRFA (Article 17). Deep technical analysis, spreading beyond the boundaries of the Treaty, would be required in order to provide a comprehensive assessment of the implementation of these four elements. However, taking into account the limits (in scope, time and length) of the present book, Article 17 on the Global Information System will be the only subject analysed. Indeed, information and knowledge are crucial elements inseparable from the PGRFA physical material and enabling its use. Moreover, the GLIS has received increased attention over the last few years within the Treaty forum and various tools have been created to allow for its implementation. This highlights the strategic importance of designing and implementing a PGRFA information and knowledge policy. In the present section, an explanation of the concept is first provided, followed by an assessment of the implementation process of Article 17. After the analysis of the GLIS, a short explanation on traditional knowledge related to PGRFA will be provided. I recognize that this very important topic would require a whole book on its own, and that the description made here is very limited. However, it falls outside the scope of the present study and it is left for other researchers specialized in topics related to local and indigenous communities and traditional knowledge to tackle the important questions at stake.

A note is made regarding the tsunami that occurred at GB7 on DSI. Not much more can be said besides what has been written above in this chapter. Indeed, the topic has only recently been recognized as requiring attention and discussion by contracting parties (Resolution 13/ 2017; 2/2017), and we must wait the next GB before a deeper analysis can be carried out.

Defining the Global Information System on PGRFA

Article 17 states that contracting parties shall cooperate to develop and strengthen a GLIS to facilitate the exchange of information, based on existing information systems, on scientific, technical and environmental matters related to PGRFA. In this section, the GLIS will be briefly outlined and the diversity of existing information systems will be highlighted. A final note is made regarding traditional knowledge.

The GLIS is aimed at enhancing the documentation on PGRFA, including crop wild relatives, on-farm and ex situ material, as well as promoting its distribution in particular among developing countries. Article 17.1 highlights the necessary cooperation with the up and running CBD Clearing House Mechanism.[44] Article 17.2 stresses the role of the information system in early warning about hazards threatening PGRFA. Article 17.3 focuses on the role of the GLIS in the periodic reassessment of the state of the World's PGRFA and in updating the rolling GPA (IT/GB-2/07/Inf.7: §§ 22–26; IT/GB-3/09/Inf. 7: §§ 15–20).

Several types of information systems are of relevance for the development of the GLIS. Information can relate to *inter alia* 1) the accession itself (passport and characterization data); 2) the existence, location and access conditions of material in national, regional or international collections or networks; 3) related technology, tools or research results; 4) publicly available genomic data; etc. The galaxy of existing information systems makes it difficult to have a clear vision of the current situation. The diversity among these systems makes it difficult to coordinate them together (with the view *inter alia* to avoiding too much duplication).

Examples of this variety of information systems and related initiatives are: 1) The World Information and Early Warning System on Plant Genetic Resources for Food and Agriculture (WIEWS) in FAO is the monitoring approach of the GPA adopted since 2004, which establishes National Information Sharing Mechanisms (NISM). 2) The System-wide Information Network for Genetic Resources (SINGER) of the CGIAR. 3) The American Germplasm Resources Information Network (GRIN) supports the national PGRFA collections, collectively called the National Genetic Resources Programme (NGRP). 4) The European Plant Genetic Resources Search Catalogue (EURISCO). 5) Genesys is a plant genetic resources portal that gives breeders and researchers a single access point to information of about a third of the world's genebank accessions. 6) The Global Biodiversity Information Facility (GBIF) is a multilateral initiative aiming at making the world's biodiversity data freely and universally available via the Internet. 7) The DivSeek initiative aims at characterizing crop diversity and developing a unified, coordinated and cohesive information management platform to provide easy access to genotypic and phenotypic data associated with genebank germplasm. 8) The Global Open Genome Sequence Data Framework proposes to establish a 'public license for genomic information on crop germplasm' as the first mechanism to ensure that 'such data will be systematically treated as a public good for the benefits of mankind'. 9) The Multi-Crop Passport Descriptors (MCPD V.2.1) is the globally adopted international standard for passport data of ex situ genebank accessions. It is compatible with WIEWS and Genesys. 10) Work on Digital Object Identifiers (DOIs) is undertaken to identify genetic material with permanent unique identifiers, also indicating whether the material is included in the MLS or not.

> Once a DOI is assigned to the material, an accession can be referenced easily and unambiguously forever, even across organizations, with the advantage that users and modern information systems will be able to discover and

access the information associated with the material and also to add value to it to through automatized web links.[45]

11) The Capfitogen Programme and the Platform for the Co-development and Transfer of Technologies are initiatives to support national and regional programmes in the development and transfer of information technologies for, and data analysis of, PGRFA (IT/ACSU-2/15 /Inf.2). Many countries also have national information systems on PGRFA.

Implementing the Global Information System

Notwithstanding this diversity of systems, the Second Report of the State of the World on Plant Genetic Resources for Food and Agriculture highlighted a significant imbalance among regions and countries within regions in maintaining an integrated national information system on germplasm holdings (FAO, 2010: chapter 4). Therefore, contracting parties requested the Secretariat to develop a Vision paper which takes stock of existing information systems, outlines a process for the development of the GLIS (Resolution 4/2011: point 13), and identifies the gaps and needs for information of providers and users (IT/GB-4/11/19: §§ 46–47; and IT/GB-4/11/Inf. 4: §§ 20–22). Following the development (IT/GB-5/13/17; Resolution 10/2013; IT/GB-5/13/Report: point 41) and adoption of the Vision paper and the Programme of Work on the GLIS (Resolution 3/2015; IT/GB-6/15/Report: §§ 30–32 and Appendix A.3; T/GB-6/15/7), the first global data library of genetic data of food crop seeds was set up.[46]

Resolution 3/2013 states that GLIS should integrate and augment existing systems to create the global entry point to information and knowledge for strengthening the capacity for PGRFA conservation, management and utilization. The idea is to promote inter-connectivity among systems and provide overarching mechanisms to ensure ready access to the information and services provided (IT/GB-6/15/Report: Appendix A). This translates into the following objectives for the period 2016–2022: 1) to create a web-based platform with use-oriented entry points to PGRFA information; 2) to provide a comprehensive overview and facilitate access to sources of PGRFA and associated information; 3) to promote and facilitate interoperability among existing systems by providing clear principles, technical standards and appropriate tools to support their operations in accordance with the principles and rules of the Treaty; 4) to promote transparency on the rights and obligations of users for accessing, sharing and using PGRFA associated information and to establish ways to exercise those rights and obligations within the Global Information System; 5) to create and enhance opportunities for communication and international and multidisciplinary collaboration to increase knowledge about and add value to PGRFA; 6) to provide capacity development and technology transfer opportunities for the conservation, management and use of PGRFA and associated information and knowledge paying special attention to the needs of developing countries;

and 7) to create a mechanism to assess progress and monitor effectiveness of the Global Information System.

Digital object identifiers project

To progress on this task the Treaty Secretariat launched the Digital Object Identifiers (DOI) project, following the advice of the Scientific Advisory Committee, to unambiguously and permanently identify PGRFA being exchanged across organizations. DOI allows to digitally provide a Permanent Unique Identifiers (PUID), i.e. a common standard, to facilitate the management and exchange of information between parties using all the above-mentioned tools.[47] The GB encourages contracting parties to use DOIs to accelerate implementation of GLIS (Resolution 5/2017). The objective is to significantly simplify and reduce the administrative burden of tracking PGRFA exchanges. It is also presented as a tool to facilitate the inclusion of accessions in the MLS (Resolution 4/2017: point 4).

In order to attract more users to the MLS, it was sought to increase the visibility and accessibility of the type of material in the MLS, including by improving the access to technical information on genetic resources (Resolution 1/2013, points 20–25). Focus is therefore set on the type of material included in the MLS, on the related (genetic) information available, and on the means of making it available. Indeed, having a clearer view on what is part of the MLS can definitely incite stakeholders to use it. Two specific recent initiatives are presented to contracting parties as ways to enhance the availability and use of genetic information associated with MLS material: the DivSeek Initiative and the Global Open Genome Sequence Data Framework.

The DivSeek initiative

The DivSeek initiative is a global 'big data' partnership to 'harness genetic resources for food security' (DivSeek website). To date (January 2018) there are 68 partnering institutions: genebanks, breeders, plant and crop scientists, universities, database and computational experts but also the Global Crop Diversity Trust, the CGIAR Consortium and the Global Plant Council. The Treaty Secretariat was initially involved in the partnership, but retrieved from a direct collaboration following the discontent of many contracting parties and other Treaty stakeholders (Third World Network, 2016). DivSeek functions as an umbrella organization in 'creating a well-coordinated, international effort, based on experience and knowledge of relevant stakeholders' (DivSeek, n.d.). The aim is to provide easy access to genotypic and phenotypic data associated with genebank germplasm (Unknown, 2015; Piotrowski, 2015) in order to enhance the use of genebank materials, promote effective utilization of genetic variation in plant improvement and better understand how components of genetic variation contribute to plant performance in diverse climatic environments. DivSeek wants to 'unlock the potential of crop diversity stored in genebanks around the globe and make it available to all so that it can be utilized to enhance the productivity, sustainability and resilience of crops and agricultural systems' (DivSeek, n.d.).

The Global Open Genome Sequence Data Framework

In the 2015 Global Sustainable Development Report,[48] researchers pledged for a 'universal access to genome information, needing nothing more than a web browser' to 'transform plant breeding [. . . and] spawn innovation around the world' (Warthmann and Chiarolla, 2015). Responding to the call for 'a New Data Revolution' for sustainable development expressed by the UN Secretary General's High Level Panel on post-2015 development goals (United Nations, 2015b: 23), the rationale behind this initiative is that 'using genetic data can improve the speed and efficiency of plant breeding compared with classic trial-and-error practices'. Do to so, Warthmann and Chiarolla (2015: 2) propose to establish a 'public license for genomic information on crop germplasm' as the first mechanism to ensure that 'such data will be systematically treated as a public good for the benefits of mankind'. This license would ensure that genome sequence information and related data remains free and is made available as a public good, as long as it originates from publicly funded repositories and other stakeholders. This means that anyone sequencing germplasm in the public domain should be able to render the resulting information publicly available as is, 'without prior curation, and without being exposed to the risk of infringing someone else's rights on the material and/or related information, including patents or copyright' (ibid: 3). Article five of the draft public license imposes a reach-through obligation that is to say that 'the genomic information of descendants of material subject to this license is again subject to the same terms and conditions' (ibid: 3–4).

Similarly to the DivSeek initiative, the Global Open Genome Sequence Data Framework is certainly to be welcomed by many stakeholders. Promoting an open access to PGRFA-related genetic information could work around the increasing enclosure of seeds and allow public scientists and breeders to develop the new varieties to face our current and future needs. However, because of the type of information that is currently covered (i.e. genomic data), one understands easily that it will be useful mainly for a specific type of users, i.e. professional plant breeders, and not for the end users, i.e. the smallholder farmers. This situation reveals at this level of data management the imbalance of rights and capacities in the Treaty between breeders and farmers.

Therefore, a caveat is placed on the above-mentioned initiatives which mainly focus on one specific type of PGRFA user, i.e. the (high-tech) breeder, leaving aside the majority of seed users, i.e. the (smallholder) farmer. It is understood that the end-beneficiary of this type of initiative is to be the farmer, but it does not recognize the fact that the majority of the seeds used by smallholder farmers does not originate from commercial breeders but from informal seed exchange systems between farmers (McGuire and Sperling, 2016), where traditional knowledge related to PGRFA is an important asset.

Traditional knowledge

While there is not yet an accepted definition of traditional knowledge (TK)[49] at the international level, within the WIPO fora, TK is presented as 'knowledge, know-how, skills and practices that are developed, sustained and passed on from generation to generation within a community, often forming part of its cultural or spiritual identity' (WIPO webpage on TK). For smallholder farmers, seeds (the material) and traditional knowledge (the information related to the material) are indissociably related. A seed without its associated traditional knowledge is of no value, of no use. Therefore, this type of information is crucial for the conservation and sustainable use of PGRFA. This is recognized in the Treaty, where Article 9.2 stipulates that 'each Contracting Party should [. . .] take measures to protect and promote Farmers' Rights, including: a) protection of traditional knowledge relevant to plant genetic resources for food and agriculture'. Farmers' traditional knowledge is undeniably considered as 'information' relating to PGRFA (Onwuekwe, 2004; Brush, 2005, 2007; Biber-Klemm and Cottier, 2006; Jarvis, 2011), which could be covered by the development of the GLIS. Addressing issues related to TK through the GLIS could be a way to limit misappropriation, which have enflamed polemics on cases of biopiracy (Mgbeoji, 2001; Ebhojie, 2013). The issue of documenting and digitizing TK and related intangible cultural knowledge is not an easy one (Oguamanam, 2007). Current negotiations on an international legal instrument to ensure the effective protection of TK, traditional cultural expressions and genetic resources are taking place within the WIPO Intergovernmental Committee on Intellectual Property and Genetic Resources, Traditional Knowledge and Folklore (WIPO-IGC). Besides, the development taking place within the ambit of the Nagoya Protocol (Morgera et al., 2014; Lakshmi Poorna et al., 2014) will also very likely influence the way TK is handled in the implementation of the Treaty (Girard and Frison, 2018b: 12; Chiarolla and Sand, 2013).

There are two ways of protecting TK through IPRs: defensive and positive protection. Defensive protection refers to

> a set of strategies to ensure that third parties do not gain illegitimate or unfounded IP rights over TK. [. . .] Some countries and communities are also developing TK databases that may be used as evidence of prior art to defeat a claim to a patent on such TK.[50]
>
> (Wendland, 2002: 101–138; see also Van Overwalle, 2002, 2005)

These databases could be integrated in the GLIS under the Treaty. Defensive protection include means such as maintaining secrecy, or documenting TK. Positive protection rather focuses on the protection of TK by IP rights through 'preventing unauthorized use, and active exploitation of TK by the originating community itself' (Wipo website). Positive protection are mechanisms such as patents, geographical indications, sui generis systems, ABS systems or contracts.

While (Article 9.2a) stipulates that Traditional Knowledge is important and should be protected, the fact that this responsibility is recognized only 'as appropriate, and subject to [. . .] national legislation[s]' has not promoted a general move towards TK protection at the international level. According to the WIPO database, 17 countries adopted sui generis legislation.[51] Nevertheless, after several years of negotiations at the IGC, countries still have not come to an agreement and the debate does not seem to be close to an end. Relationships between the Treaty Secretariat and the WIPO-IGC seem rather distant, and there is no concrete collaboration foreseen in the coming agendas of both bodies as to common work on the protection of information and knowledge related to PGRFA (Resolution 12/2017).

Legal rules and procedures supporting compliance with the Treaty

What rules and procedures support the functioning of the Treaty? What can be done when there is a situation of non-compliance? By whom? What happens in case of a dispute regarding the SMTA or a Treaty obligation? What happens when contracting parties want to change or adapt a Treaty provision? Various rules and procedures are established by the Governing Body to answer these questions and allow for an efficient implementation of the Treaty.

These rules and procedures intervene at two levels: 1) at the level of the MLS and its SMTA, where PGRFA users and the Third Party Beneficiary are involved; and 2) at the level of the Treaty, where contracting parties (i.e. states) are the major stakeholders at play. These procedures will each be described and assessed in four sub-sections: the Third Party Beneficiary; procedures and operational mechanisms to promote compliance; dispute settlement; and amendment to the Treaty.

The Third Party Beneficiary

The Third Party Beneficiary (3PB) was created to address some issues related to the multilateral character of the system and the enforcement of the SMTA (Moore 2013: 164–176). This innovative concept bridges the gap between the private contractual law relationship between parties to the SMTA and the public international law setting where contracting parties (i.e. states) to the Treaty and its MLS are anchored.

Defining the third party beneficiary

The 3PB is the virtual entity representing the MLS, designed to assert its rights and to allow for the enforcement of the SMTAs. As a matter of fact, the SMTA creates a triangular relationship between the provider, the recipient and the MLS. Unlike in a classical MTA, benefits flow back to the MLS (and in particular the

BSF) rather than to the provider of the material. Therefore, the provider has no incentive to ensure that the terms of the SMTA are respected by the recipient. As Moore explains

> the MLS is the real beneficiary of the benefit-sharing provisions of the SMTA [. . .]. The SMTA recognizes this fact by providing for the appointment of a third party beneficiary *to represent the rights of the MLS* and by giving this third party beneficiary the power *to initiate dispute settlement action*, including arbitration, in the event of a breach of the terms and conditions of the SMTA affecting those rights.
>
> (emphasis added, Moore, 2013: 164)

While the concept of 3PB will be found nowhere in the Treaty text, I agree with Moore when he says that it is implicitly provided for in the Treaty, in particular in its Article 12.4, which establishes the triangular relationship between the provider, the recipient and the MLS. Moore (ibid: 168) further analyses the concept of 3PB rights in national law and concludes that

> the possibility for a contract to provide for enforceable third party beneficiary rights is expressly and unambiguously recognized in the Principles of the International Commercial Contracts, which were adopted by the International Institute for the Unification of Private Law (UNIDROIT) in 2004.
>
> (ibid: 167)

For this reason, the SMTA refers explicitly to the UNIDROIT principles when dealing with the choice of applicable law (SMTA Article 7). In the SMTA, explicit mention of the 3PB is made in Articles 4.3, 4.4 and 8. Article 4.3 requests parties to the SMTA to recognize the 3PB as acting on behalf of the Governing Body and its MLS. Article 4.4 provides the 3PB with the right to request information as required by SMTA Articles 5e, 6.5c, 8.3 and annex 2 § 3, in particular where the provider or the recipient failed to provide this information to the Governing Body. Article 8 states that dispute settlement may be initiated by the provider, the recipient or the 3PB. Mechanisms for dispute settlement include amicable negotiation, mediation and arbitration (3PB Procedures Articles 5, 6 and 7).

Another difficulty was to be resolved with the concept of 3PB: the fact that, in representing the Governing Body and its MLS, it does not have the legal personality necessary to act under international or national laws. Therefore, the entity which would embody the 3PB would necessarily need to have 'its own legal personality and capacity to take legal action to protect [its] rights' (ibid: 170). In 2006, FAO has accepted to represent the 3PB, acting under the direction of the Governing Body (Circular Letter of 22 December 2006). This proposal for FAO to act as the 3PB was endorsed by the Governing Body (Resolution 5/2009: point 2).

Implementation of the Third Party Beneficiary

The 'Procedures for the Operation of the Third Party Beneficiary' were adopted by Resolution 5/2009 (Annex I), and further developed with the 'Rules for Mediation of a Dispute in Relation to a Standard Material Transfer Agreement' (Resolution 5/2011). A list of experts to serve as mediators and arbitrators can be found on the Treaty website.

At GB5 (IT/GB-5/13/Report: points 42–43; Resolution 11/2013), a potential case of non-compliance related to two agreements signed by two CGIAR centres (CYMMIT and ICARDA) with private sector entities was reported (IT/GB-5/13/19; IT/GB-5/13/19 Add.1). These agreements 'may have resulted in the transfer of barley germplasm without the required SMTA, and/or in violation of other Treaty provisions concerning the availability of germplasm' (IT/GB-5/13/19, point 6). The case was spotted by the Action Group on Erosion, Technology and Concentration (ETC Group, 2012: 6–7) in a Communiqué, then through a letter to the Treaty Secretary (IT/GB-5/13/19: point 5). For the first time, FAO acting as the 3PB initiated the alternative dispute resolution process (Procedure for the Operation of the Third Party Beneficiary, Article 5; SMTA Article 8) and requested clarifying information to both centres during 2012–2013. Upon subsequent informal consultations, a number of clarifying and corrective actions have been initiated to comply with the Treaty obligations, particularly Article 15, including that new SMTAs covering retroactively all the transfers to the recipients were to be signed (IT/GB-6/15/10: point 6). Resolution 1/2015 (points 14 to 21) dedicates significant space to the 3PB agenda item and closed the case. Since then, no new possible cases of non-compliance have arisen (IT/GB-7/17/10). The fact that contracting parties have focussed significant attention on this mechanism might sound as an implicit threat to non-complying parties, or at least as a recall of the fact that 'FAO [now has the right] to initiate legal action through dispute settlement procedures to protect the integrity of the [MLS]' (Moore, 2013: 172).

Compliance

Compliance procedures are distinct from dispute settlement (Article 22) and from the possibility to seek recourse under Article 12.5 regarding contractual disputes under the SMTA. They deal with general issues of compliance or non-compliance, are non-adversarial and can be raised by any contracting party and the Governing Body ('Procedures and operational mechanisms to promote compliance and address issues of non-compliance' Rule VI. 1).

Although Treaty Article 21 requires the Governing Body to approve compliance mechanisms at its first meeting, negotiations were long and difficult, and the GB only adopted the 'Procedures and Operational Mechanisms to Promote Compliance and Address Issues of Non-compliance' at GB4 (Resolution 2/2011). At the same meeting, the Compliance Committee was requested to develop further rules of procedure relevant to its work and submit them to the next GB Session and to develop a succinct standard reporting format to allow contracting parties assessing where they stand with the implementation of the Treaty. Both elements were approved at GB5 (Resolution 9/2013: Annex 1 and 2). Thereafter, contracting parties were requested to submit their standard reports in a timely manner, so that the Compliance Committee could consider them before the Seventh Session of the Governing Body. An Online Reporting System (ORS) to streamline the process through electronic means was set up to support contracting parties in fulfilling their reporting commitments under Article 5 of the Compliance Procedures.

The adopted procedures and operational mechanisms consist of ten sections covering respectively: objectives; principles; institutional mechanisms; committee functions; monitoring and reporting; procedures regarding submissions relating to issues of non-compliance; measures to promote compliance and address issues of non-compliance; information; other procedures regarding the promotion of compliance and review of the procedures and mechanisms. Compliance mechanisms may potentially play a key role in the Treaty implementation. However, countries have difficulties complying with all Treaty obligations and it is too early to provide an assessment of their implementation of the compliance procedures. The above procedures could significantly contribute to a better implementation of the Treaty but it remains to be seen if and how contracting parties will use this tool as a complement to other initiatives taken by the GB to enhance the Treaty's implementation.

Settlement of disputes

Article 33 of the UN Charter obliges states to settle disputes in a peaceful manner; it provides a list of potential procedures: i.e. negotiation, enquiry, mediation, conciliation, arbitration, judicial settlement, etc.

Article 22 of the Treaty spells out almost word by word the settlement of dispute Article 27 of the CBD. It provides for gradually more intrusive and formal procedures to settle disputes regarding the interpretation or application of the Treaty between two or more contracting parties. When a country becomes party to the Treaty, or any time thereafter, it may declare in writing to the depositary of the Treaty that a dispute which was not resolved under Article 22.1 (negotiation) or 22.2 (mediation) is to be compulsorily settled either by arbitration (Article 22.3(a)), or by submission to the International Court of Justice (Article 22.3(b)). Alternatively, conciliation (Article 22.4) is made available to contracting parties who have not accepted the 22.3 procedure. Arbitration and conciliation are further developed in Annex II to the Treaty. Regarding mediation, a specific procedure has been developed under the Third Party Beneficiary Procedures.

All EU countries as well as Myanmar have accepted the Article 22.3(a) (arbitration) procedure. None of the 145 contracting parties have accepted the ICJ as a means of compulsory dispute settlement under Article 22.3(b) (even countries that have accepted ICJ resolution under the CBD). The time-consuming and costly procedure probably explains this situation. Since the entry into force of the Treaty no dispute has arisen between contracting parties.

Amendments to the Treaty

Modifying Treaty provisions is sometimes necessary when contracting parties agree that the existing obligations no longer serve the purposes of the Treaty, or that they are not implementable. Amendment (Article 23) is the classic way of reviewing or adapting a Treaty. Other procedures, such as adding a Protocol to the Treaty, will also be addressed below.

Defining amendment procedures

Contracting parties mandated the Ad Hoc Open-ended Working Group on the Multilateral System (WG-MLS) to explore what legal procedures would allow a review of the MLS and Funding Strategy (Resolution 4/2011: Appendix, point 6). The FAO legal office and the WG-MLS identified three main options: 1) minimal changes to the SMTA; 2) the adoption of amendments to the Treaty; 3) the development of a possible protocol, supplementing and/or improving the existing legal framework of the Treaty (IT/OWG-EFMLS/14/Inf. 5; IT/OWG-EFMLS-3/15/Inf.4 Rev.1).

Amendments to the Treaty can be proposed by any Contracting Party (Article 23.1), and Switzerland has done so at GB7 by proposing to enlarge Annex 1 to all PGRFA (Resolution 2/2017: point 3). Amendments shall be adopted by consensus at a Session of the Governing Body, which equates to a right of veto for any Contracting Party. The potential right of veto makes it very difficult to amend part of the Treaty text, including a potential expansion of the Annex I list of crops and forages. As a comparison, within the CBD amendments may be adopted, as a last resort, upon a two-third majority (CBD Article 29(3)). According to Moore and Tymowski (2005: 161), this 'provision was viewed as essential by some countries during the negotiations as a way of ensuring that their essential interests would be taken into account in all aspects of the functioning of the Treaty, including its amendments'. Amendments shall come into force ninety days upon the deposit of the ratification/acceptance/approval instrument by a contracting party (Article 23.4). This means that they may enter into force at different dates for different states.

Contracting parties may decide to enhance the functioning of the MLS by adopting a Protocol to the Treaty, in accordance with Article XIV, paragraph 2(b), of the FAO Constitution. In international law, two types of Protocols are identified by the Working Group (IT/OWG-EFMLS-3/15/Inf.4 Rev.1): amending Protocols (which alter the wording of provisions of an existing Treaty) and supplementary Protocols (which complement an earlier convention with additional provisions and obligations). A supplementary Protocol is a stand-alone instrument, often designed to implement an earlier Convention and occasionally to broaden substantive provisions of the Treaty it supplements.

Implementing the amendment procedure

At its Sixth Session, the Governing Body requested the Working Group to 'elaborate a full draft revised SMTA focusing especially on the development of a Subscription System and aiming to avoid the necessity of any other legal instrument primarily through a revision of Article 6.11 of the SMTA' for the next GB Session (Resolution 1/2015: point 3). The Resolution adds that

> [i]f a legal instrument would nevertheless be deemed necessary to develop an effective subscription system, [the Working Group is requested] to elaborate a complete proposal for an appropriate legal instrument (including an amendment of or a Protocol to the Treaty).
>
> (ibid)

Contracting parties therefore seemed to prefer avoiding going through a formal amendment process, without shutting that door completely. Rightly so, as by GB7, the GB had received the Swiss proposal to amend the Treaty and decided to 'consider it in conjunction with the agenda item on *Enhancement of the Functioning of the Multilateral System of Access and Benefit-sharing*, for which it adopted Resolution 2/2017' (IT/GB-7/17/Report: § 27). Contracting parties therefore seem to openly decide to consider the MLS and Funding Strategy review process as an amendment to the Treaty. The same report states that the 'Governing Body will approve a new and revised SMTA, providing the option to subscribe to a Subscription System' (Resolution 2/2017: Annex 1, point 1), and that it should decide, at the same time on

> amending Annex I of the Treaty, including a new list of crops. The amended Annex I would (1) delegate the power of future expansion of the Multilateral System to the Governing Body, and (2) include a list with a first set of specific crops.
>
> (Resolution 2/2017: Annex 1, point 2)

The GB justifies this choice by arguing that the

> declaration of companies and other users of their willingness to sign on to the Subscription System, following the decision of the Governing Body to adopt a new and revised SMTA, would create confidence on the part of contracting parties that more user-based income will eventually accrue to the Benefit-Sharing Fund. The decision of the Governing Body to adopt a process for the amendment of Annex I would generate confidence on the part of prospective subscribers of the clear intention of contracting parties to increase the coverage of the Multilateral System.
>
> (Resolution 2/2017: Annex 1, point 3)

Finally, the GB limits the breadth of the amendment procedure by stating that 'all new provisions would be added in Annex I, whereas no other provisions of the Treaty would need to be revised, and the amendment would follow the procedures foreseen in Articles 23 and 24 of the Treaty' (Resolution 2/2017: Annex 1, point 6). It remains to be seen how this will develop by GB8, in 2019.

Treaty governance and stakeholders' participation

Like for any international convention, the contracting parties of the Plant Treaty have established its administrative bodies (Treaty Articles 19–20), which care for the administrative functioning of the instrument. Under this section, the functioning of these bodies will be explained. Then, an assessment of this governing scheme will be provided, with a particular focus on the (non-)role of other (non-state) Treaty stakeholders.

Administrative bodies governing the Treaty

Two administrative bodies are responsible for the institutional functioning of the Treaty: the Governing Body and the secretary. As these institutional provisions are common in international environmental treaties, comments will remain concise.

The Governing Body

Article 19 establishes the highest organ of the Treaty, the Governing Body, which is composed of representatives of all its contracting parties (Article 19.1). This Article defines the role, procedures, rules, etc., which contracting parties have to apply for the Treaty's main organ to be effective. The Governing Body's function is 'to promote full implementation of this Treaty, keeping in view its objectives' (Article 19.3). Sessions of the Governing Body must be regular, at least one meeting every two years (Article 19.9) and preferably held to back with CGRFA meetings. Special sessions may be held, if necessary and agreed upon by contracting parties (Article 19.10). Decisions are taken by consensus (unless decided otherwise) (Article 19.2). However, consensus is compulsory for the adoption of amendments to the Treaty (Article 23) or to its Annexes (Article 24). Consensus implies that every state may potentially exercise a veto over Governing Body decisions. So far, even if a contracting party does not agree with the totality of every decision content, a veto has never been exercised explicitly. Each Contracting Party has one vote, but its delegation may comprise several representatives: i.e. one (voting) delegate, eventually accompanied by alternates, experts and advisers (Article 19.4).[52] Furthermore, Article 19.8 specifies that there must be a quorum of members at every session for Governing Body decisions to be adopted. In case there is not a majority of contracting parties attending a session (i.e. 50 per cent of its members plus one), decisions cannot be adopted.

Observers

Observers may attend sessions of the Governing Body (Article 19.5) as long as they are 'qualified in fields relating to conservation and sustainable use of [PGRFA]'. Candidate entities must inform the secretary, which transmits the request to the Governing Body. The Governing Body may deny access to the observer if one third of the contracting parties object to its participation. This procedure seems fairly straightforward and difficulties for candidates to acquire the status of observer within the Governing Body forum have not been reported.[53] Usual observers are NGOs, including farmers' organizations, Universities[54] or other research institutions. CGIAR centres which have signed an agreement with the Governing Body are automatically accepted as observers without other formalities. Representatives of the United Nations and specialized agencies may also participate in Governing Body sessions as observers (Article 19.5 & Rules of Procedure, VII).[55] Finally, states that are not contracting parties (signatory members or non-parties) may also attend Governing Body sessions as observers (ibid).

The functions of the Governing Body

Article 19.3 specifies non-exhaustively the functions of the Governing Body, which relate to providing policy directions (19.3(a)), adopt plans and programmes (b), adopt a funding strategy to be periodically reviewed (c), adopt the Treaty budget (d), consider and eventually establish subsidiary bodies when necessary (e), establish a Trust Account (f), cooperate with other relevant international organizations and treaties (g), envisage amendments to the Treaty (h) and its annexes (i) if necessary, encourage voluntary contributions (j), perform any other functions necessary for the fulfilment of the Treaty's objectives (k), take note of CBD-COP and other relevant international organizations' decisions (l) and inform the CBD-COP and other relevant international organizations of relevant Treaty implementation issues (m), and finally approve the CGIAR centres' term of agreement relationship (n) referred in Treaty Article 15. All these functions have been exercised throughout every Governing Body sessions.

Rules of Procedures

Article 19.7 states that the Governing Body shall adopt and amend, as required, its own rules of procedures, which were adopted at GB1 (IT/GB-1/06/Report; § 9 and Appendix D). Rule 1 clarifies that the 'rules of procedure shall apply to all sessions of the Governing Body and the activities of its Secretary. They shall also apply, mutatis mutandis, to subsidiary bodies of the Governing Body unless the Governing Body should decide otherwise, in accordance with Rule 9.2'. Basically the Rules of Procedures builds on Article 19 provisions and details more specific functioning rules to allow for the Governing Body meetings to be effective. Such specific rules are for example how to introduce an item in the agenda of the following Governing Body, the role of the secretary and its secretariat in compiling and rendering available all necessary documents and information for the functioning of every Treaty meeting, etc.

The Bureau

The Bureau (Article 19.11) is the administrative entity without which Governing Body sessions would not function. The bureau is constituted by the chairperson of the Governing Body, six vice-chairpersons (one per FAO region, except from the region of the Chairperson) and a rapporteur. They are elected by the Governing Body at the beginning of every Governing Body meeting and they serve during the whole biennium. The rules of procedures of the Treaty detail the role of the bureau in Rule II. The bureau facilitates the operation of Governing Body sessions and all inter-sessional meetings. The bureau meets regularly in between sessions and during Governing Body meetings. As the chairperson of the bureau is also the person chairing Governing Body meetings, he/she functions as the link between the formal negotiation activities and the administrative endeavours necessary to support the operation of official meetings. The chairperson conducts the Governing Body meeting by going through each item to be discussed, following the agenda point by point; by giving the floor to people; by taking note of requests from the floor; etc. When conducting the meeting, the chairperson may consult the bureau to respond to requests from contracting parties, for example if there is a need for clarification on the procedure to follow. While it remains an administrative body, the bureau, in fact, may also have some sort of political power. This can take the form of facilitating the dialogue between representatives of FAO regions within a small and confidential setting; unlocking difficult situations when discussions are blocked in the Governing Body meeting; or providing guidance to the secretary during the inter-sessional period. Although this political role is referred to nowhere in official documents, it cannot be denied to such point that some contracting parties have already contested actions taken by the bureau, stating that it was overriding its role and rights.

The Secretary

The secretary (Article 20) of the Governing Body is appointed by the director-general of FAO with the approval of the Governing Body. He/she may be assisted by the required staff. Its role is to provide administrative support for Governing Body sessions, assist the Governing Body in carrying its functions, and report on its activities (Article 20.2). He/she has the duty to communicate to all contracting parties and to FAO director-general all decisions of the Governing Body and information received from contracting parties (Article 20.3). He/she shall do so in the six official languages of the UN (Article 20.4). Finally, he/she shall cooperate with other organizations and treaty bodies, in particular the Secretariat of the CBD (Article 20.5).

Following FAO Resolution 3/2001, the Secretariat of the CGRFA has acted, since November 2001, as the Secretariat of the Interim Committee for the Treaty, with the Secretary of the Commission (Josè (Pepe) Esquinas-Alcázar) acting as the secretary of the Interim Committee, until a new secretary would be appointed. At GB1, contracting parties decided to commence the process for the appointment of the secretary, and adopted the procedures for doing (IT/GB-1/06/Report: §§ 16–17; IT/GB-1/06/11: Appendix 1). Following the procedure under Appendix J.2 of the Report, the Governing Body agreed that '[e]xceptionally, on a specific mandate from the Governing Body, the chair of the Governing Body shall propose the candidate to the director-general of FAO, on the basis of the recommendation of the Screening Committee, *without the prior approval of the Governing Body*' (emphasis added). At GB2, Dr Shakeel Bhatti was hired by the director-general of FAO as the Treaty's secretary.[56] The duration of the term of office of the secretary is four years, renewable without limitation (IT/GB-1/06/REPORT, Appendix J.1). However, no rules were foreseen to establish a procedure for the possible renewal of the appointment of the secretary, which created quite some uncertainty at the end of Bhatti's second mandate (see below).

Implementing governance procedures

Treaty governance is not an easy item to assess, as the Treaty follows classical rules of international law regarding laws of treaties, but also many informal practices that can only be understood by participating to the process from the inside (and which have no legal value, but strong political impact). In the following sub-section, brief information is provided regarding the functioning of the Governing Body, the Secretary's mandate, and most of all on the attempts to include stakeholders in different ways during the review of the Treaty process.

The Governing Body meetings

Since the First Governing Body in 2006, contracting parties have been very active in crafting the legal and technical apparatus to apply Treaty obligations. This dynamism transpires from the many meetings that took place. The Governing Body meetings allow the 145 contracting parties to identify and discuss their needs and to negotiate and adopt the resulting instruments and mechanisms. The Treaty secretary and its secretariat play a crucial role in the efficiency of the administrative process. Governing Body meetings have systematically been held within the two-year schedule (Article 19.9) and numerous inter-sessional meetings have been organized (upon availability of funds) in order to provide negotiators with the necessary time and space to address the issues under discussion and be ready to negotiate and adopt measures during the Governing Body meetings. While negotiations have not necessarily always been easy, the negotiating 'mood' has most of the time remained positive and constructive.

Renewal of the Secretary's mandate

In 2010, the FAO secretariat followed an ad hoc procedure of consultation of the bureau of the Treaty, as a result of which the Director-General of FAO extended the appointment of the executive secretary for a term of four years. In 2014, as the second term of office of the secretary was drawing to a close, the FAO secretariat proposed an interim extension of the appointment of the secretary, pending review of the matter by the Governing Body and approval of a procedure to that effect. At the Sixth Session of the Governing Body, a document was presented to establish such a procedure (IT/GB-6/15/26.). With Resolution 12/2015, contracting parties decided to set the matter on the provisional agenda of the Seventh Session of the Governing Body and to extend the appointment of the current secretary until the designation of a new secretary, following approval by the Governing Body at its Seventh Session in 2017. At GB7, contracting parties attempted to remedy this procedural gap by considering the document 'Procedures for the Appointment and Renewal of the secretary of the Governing Body of the International Treaty' (36 IT/GB-7/17/30). However, the GB 'failed to reach a consensus on the proposed procedures due to concerns raised regarding the shared responsibility of the FAO Secretariat and the Governing Body on the Procedures for the Appointment and Renewal of the secretary of the Governing Body of the International Treaty. Therefore, it decided to revisit this issue at its Eighth Session and requested the FAO Secretariat to reconsider, in close consultation with the Bureau, the proposal for the selection and appointment of the secretary of the Governing Body of the International Treaty to more appropriately reflect the concerns raised at this Seventh Session' (IT/GB-7/17/Report: § 47). Notwithstanding this delay, the GB appointed Dr Kent Nnadozie, who has been exercising the function ad interim for more than a year already, as secretary of the International Treaty for a period of two years, consistent with the decision of the Council of FAO at its 155th Session (IT/GB-7/17/Report: § 45).

Consultation with stakeholders, in particular the seed industry, during the review process of the MLS and funding strategy between 2012 and 2015

It is the first time that an official sub-organ of the Treaty conducts an official consultation with a non-contracting parties stakeholders' group outside the GB, other than their usual participation as observers (or within country's delegation) to Treaty meetings. Negotiators realized that if they wanted the seed industry to use the SMTA, they would need a good understanding of its expectations. It should therefore be applauded that such consultation of members from the seed industry took place in 2012–2013. The review (Moeller, 2014) seems to have been well-received by the private sector, whose interviewees have frequently expressed support to the Treaty. The results of this review are available on the Treaty website and have been integrated in the working documents of the WG-MLS.

The myth of the financial solution by the seed industry

To this positive note, some remarks will be added. First, this consultation is of primary importance, as one of the issues that render the Treaty so difficult to implement is the fact that stakeholders other than contracting parties are not sufficiently integrated in the Treaty institutional functioning (as part of the governance scheme). However, I believe it is a myth to imagine that obtaining the participation of the seed industry in the MLS will solve major financial issues in the Treaty. It is unlikely that the private sector will voluntarily pay for the overall cost of conservation and sustainable use activities worldwide. This is not to say that they should not contribute. On the contrary, they should pay their fair share. However, unless governments manage to impose the system to the 'Big Five' who have been most reluctant to access material from the MLS,[57] it will not work. Indeed, while these companies have the financial means to pay a significant, fair share, they also can perfectly live without accessing material from the MLS for the near future, relying on their significant private collections. The sole participation of the small-to-medium-size private sector industry would not be sufficient to create a stable, long-term secured and financially sustainable funding source for the Treaty's Funding Strategy. Efforts in convincing the Big Five to participate in the system should be enhanced. Dialogue with the CEOs could be an opportunity to better share views and identify common ground for participation. However, up to now, such dialogue has not been possible. A solution might lay in informal consultation with these CEOs. The Informal Multi-stakeholder Dialogue initiative (see below)[58] that took place in 2014 could have provided for such open and frank debate and could have fertilized the soil for finding an innovative agreement (out of the box). Unfortunately, the initiative has not been pursued. It is therefore crucial that efforts towards finding other means of obtaining financial support are maintained and come to a positive outcome.

The absence of other stakeholder groups' consultations

A second remark is made regarding the fact that other stakeholder groups outside contracting parties have not been consulted. While it is easy to understand that there are no specific stakeholder groups, other than the seed industry, whose participation in the MLS would potentially bring significant financial benefits back to the Treaty, it is nonetheless a pity that the voice of these groups were not heard on the same basis. Indeed, other PGRFA users might have interesting opinions and concrete proposals to make in ways of enhancing the functioning of the MLS, regarding both monetary and non-monetary aspects. Farmers' Consultations have taken place (see above same chapter, Farmers Rights' section), but not at the same time, and not with the same purpose.

'Silent observers' at the WG-MLS

Third, an interesting note is made regarding the composition of the WG-MLS. Besides the usual experts and negotiators representing the FAO Regions in the Treaty Governing Body,[59] the secretariat staff and two official observers representing main Treaty stakeholder groups (civil society organizations, the seed industry, farmers' organizations, and the CGIAR centres; Resolution 2/2013: point 23), additional 'silent observers' were admitted to the meetings (IT/OWG-EFMLS-1/14/Report). Their participation could be considered as an indirect additional way of consulting stakeholder groups, along the line of the WG-MLS' mandate. These silent observers included people from NGOs (such as the South Centre, the Third World Network or the Centre for Research, Information, Action in Africa), the Seed Industry (Syngenta International AG or the European Seed Association), as well as representatives from FAO and other international bodies (CBD, CGRFA) (IT/OWG-EFMLS-1/14/Report; IT/OWG-EFMLS-2/14/Report; IT/OWG-EFMLS-3/15/Report; IT/OWG-EFMLS-4/15/Report). This shows the clear need and will of contracting parties to collaborate with the direct users/stakeholders of the MLS in finding ways to enhance its functioning. This sounds logical, as these are the real people who actually use the SMTA and MLS material. However, from an international law point of view, such an initiative is not the usual practice, as official representatives of governments are the only negotiators with recognized power of decision, engaging the state they represent. The fact that these technical meetings have been opened to more stakeholders confirms the realization that decisions need to take into account and include the voice of the stakeholders on which the decisions will apply. This note might sound simplistic and quite logical, but from the viewpoint of negotiation practices within the Treaty forum, it may be considered as a 'little revolution'. This is probably why negotiators, very cautiously, specified in the first meeting report that 'this decision would set no precedents for other inter-sessional bodies of the Governing Body' (IT/OWG-EFMLS-1/14/Report). This practice should be encouraged and widened. Yet, it should not be interpreted in an overly optimistic manner. 'Normal' observers may only talk when they are invited to do so (contrary to contracting parties who may request the floor at any time), and as their title says, the fact that these additional observers are 'silent observers' certainly limits their real involvement in the (formal) discussions.

The multi-stakeholder dialogue initiative: a missed opportunity

Finally, a last comment is made regarding the 'Informal Multi-stakeholder Dialogue' initiative that took place in 2013–2014. At the second High-level Roundtable on the International Treaty on Plant Genetic Resources for Food and Agriculture (United Nations Conference on Sustainable Development Rio de Janeiro, Brazil, 21 June 2012), a 'Rio Six-Point Action Plan for the International

Treaty on Plant Genetic Resources for Food and Agriculture' (the 'Rio Six-Point Action Plan') was adopted by consensus. The third point of the Rio Six-Point Action Plan had the objective to 'facilitate a new Keystone-type dialogue, to complete the governance of all plant genetic resources for food and agriculture under the Treaty'. This was to take the form of an informal multi-stakeholder dialogue (the Informal Dialogue), as adopted by Resolution 2/2013, where contracting parties welcomed 'the organization of an informal multi-stakeholder dialogue to enhance the functioning of the Multilateral System and increase contributions to the Benefit-sharing Fund, which may provide input to the Ad Hoc Working Group' (Resolution 2/2013, point 7). Similar informal consultation had taken place between 1994 and 2001 under the Keystone Dialogue sessions, and between 1999 and 2002 with the follow-up Crucible Group, with the same facilitators Michael Lesnig and Timothy Mealey (see below Chapter 5).

Bioversity International and the Meridian Institute jointly convened the Informal Dialogue which took place in September 2014 in the form of a two-day workshop (IT/OWG-EFMLS-2/14/Inf.4.1).[60] The rationale for organizing such dialogue was mainly to allow all stakeholders to talk to each other and express their views and perspectives more freely, applying Chatham House rules, in parallel to the formal negotiating process. Indeed, the formal process does not necessarily provide space for investigating options and solutions outside the limited scopes of official positions. Allowing representatives of a variety of stakeholder groups[61] involved in the Treaty to compare their perspectives on the state of implementation of the Treaty and documenting these diverse perspectives constitutes an important contribution to the formal negotiating process, which has fed the second meeting of the WG-MLS.

However, the Informal Dialogue was not pursued further in parallel to the work of the WG-MLS. The information document summarizing the reflections expressed during the two-day workshop contains important aspects, which would have benefited from further explorations by the same participants in later gatherings, until the overall review of the MLS and Funding Strategy of the Treaty are completed. Such informal dialogue constitutes a good place to tackle highly technical issues (and take these issues far from the political sphere for a while), by those very persons who deal with these aspects in their everyday work (contrary to negotiators in official Governing Body meetings who are generally representatives of ministries and not direct users of PGRFA). Such open, de-politicized dialogue enables the development of ideas that might not persist if they were to be first introduced in the formal setting (whether at an Ad Hoc Working Group or Governing Body meeting). It seems that this initiative, which has been pushed by some stakeholders to create the space for 'out of the box' thinking, has been hesitantly adopted by the Governing Body through Resolution 2/2013 and has subsequently been given very limited space and scope, in order, perhaps not to override the official formal negotiations of the Ad Hoc Working Group. It is a pity that the complementary role of the initiative to the formal negotiations has not been better understood and that more resources and space have not been devoted to this Informal Dialogue for it to provide significant and useful inputs to the formal process. This is especially true taking into account the success in unlocking difficult negotiations between 'opponent' stakeholders that had occurred thanks to such informal process at the time of the Keystone Dialogues.

In this section, information on the rules and procedures for the governance of the Treaty showed that there is little space for other actors than states to govern the PGRFA issues at stake. However, History has shown with the Crucible Group and Keystone Dialogues the utility of involving all other stakeholders outside of formal negotiation processes in discussing problems and imagining solutions that would contribute to collectively face major future challenges. Moreover, one may wonder how come farmers' expertise, needs and solutions are not officially included in the governance system of the Treaty, when they are the ones dealing with seeds in their everyday practice to feed the world.

Conclusion

Regarding the Treaty's overall goals of food security and sustainable agriculture, the above study reveals that the way the MLS is designed and implemented does not seem to contribute significantly to reaching food security and sustainable agriculture. New ways of envisaging and implementing the MLS are necessary to achieve these overall goals. This could be done by attracting more concrete political attention and stronger legal levers for their realization *inter alia* by entrenching these goals into existing related international fora, conventions and instruments on related topics such as biodiversity in general (CBD and Nagoya Protocol), climate change (United Nations Framework Convention on Climate Change), nutritional health (World Health Organization), or (farm) labour (International Labour Organization). The attempt to relate the work being done by the GB to the 2030 Agenda on Sustainable Development shows that the Treaty is making such a move to open up outside of the food and agriculture forum.

On the matter of the difference in scope of the Treaty (all PGRFA) and of the MLS (only Annex-I PGRFA), the legal examination highlights the complications in the implementation process resulting from these discrepancies in scope. It also shows that Treaty stakeholders are more open than a few years ago in addressing this matter, *inter alia* by envisaging an enlargement of the annex I list of crops. However, this matter should be tackled together with the access and the Farmers' Rights issues, as a harmonization of scope will not happen as long as facilitated access constraints persist.

To conclude on the analysis of the implementation of farmers' rights, it is clear that in the consecutive Governing Body meetings the different Resolutions regarding FRs have become more substantial, more specific, and therefore more easily implementable. Even though there is still no agreement on one definition as to what is covered under the concept of FRs, this flexibility might be useful to adapt it to local specificities. However, the lack of formal recognition of FRs at the international level, coupled with a formal recognition of strong IPRs, has created a serious imbalance of rights, which prevents contracting parties from reaching the Treaty's objectives.

Regarding facilitated access, the analysis shows several things. First, that accessing seed by all PGRFA beneficiaries is not straightforward, i.e. the primary beneficiaries (smallholder farmers) do not benefit from Article 13.1). Second, that there remains a discomfort and distrust from a majority of contracting parties as to the potential enlargement of the Annex I list. This distrust is explained by the fact that many stakeholders do not benefit from the system. Consequently, reviewing the existing system will necessitate to leave some space to all stakeholders to participate in the process. Finally, it also reveals that further work is needed to design and promote sui generis PVP systems as well as adapt seed certification and commercialization regulations to recreate an effective farmers' exemption.

On benefit-sharing, the study shows that the system is very ineffective. The lack of funding is crucial, but more so is the position in which the Treaty places 'beneficiaries' of the Benefit-sharing Fund, i.e. farmers. This position of 'passive receiver' of the system contrasts with the central role and position of farmers in our world food chain as food producers, biodiversity stewards and crop variety innovators. This contrast creates structural dysfunction in the Treaty's stakeholder's relationship that hinders reaching the Treaties objectives. This central role should be reflected in the Treaty.

Addressing the information and knowledge topic, the Treaty analysis demonstrates that the Global Information System developed in response to the Treaty's obligations does not meet the needs of all stakeholders, but rather focuses on breeders and researchers. This exacerbates the imbalance of rights mentioned above. Furthermore, on the question of IP over knowledge and information related to PGRFA, the international institutions addressing these matters seem to be rather distant from the Treaty, and no concrete collaboration is foreseen in the future agendas of these international bodies to address this issue. Finally, the DSI bomb has shaken up the whole Treaty review process at GB7 and is likely to freeze any progress in the negotiation, should the seed industry remain on its position regarding free and open patentability of all subsequent innovation obtained thereof.

On the seventh Treaty topic, the various rules and procedures regarding monitoring, sanctions, dispute settlement and amendments to the Treaty were detailed. The assessment stresses the creativity of the Third Party Beneficiary instrument as an innovative international law instrument. However, this tool could have a much greater potential in facilitating the implementation of the Treaty and suffers from a lack of transparency. Besides, the compliance mechanism seems to be kept voluntarily void, thereby being inefficient in helping contracting parties complying with all Treaty obligations. Since the entry into force of the Treaty, no dispute has arisen between contracting parties, i.e. at the state level. However, a dispute has arisen in 2012–2013 through the Third Party Beneficiary procedure – regarding two transfers of PGRFA without the required SMTA – and was resolved fairly quickly. Finally, regarding the amendment provision, contracting parties seem to direct the Treaty review process towards a formal amendment of the SMTA and Annex I, excluding the choice of negotiating a protocol to the Treaty.

Last, regarding participation and governance issues, the legal analysis stresses that there is little space for other actors than states, as the only contracting parties to the Treaty, to formally play a role in the collective governance of PGRFA. However, history has shown – with the Crucible Group and Keystone Dialogues – the utility of involving stakeholders in discussing problems and imagining solutions to address and collectively face major future challenges. Added to this is the fact that specific stakeholders are particularly left aside, while taking into account their expertise, needs and solutions is vital for designing as sustainable and effective Treaty governance system.

Notes

1 Parts of the following description are drawn from Garforth and Frison (2007), and Frison et al. (2011: chapter 1).

2 This quote expresses very well the narrative that was dominant during the negotiation of the Treaty and the state of mind of the different parties at stake, i.e. the fact that seeds used to be freely accessible to all, that their appropriation has led to the genetic erosion of agrobiodiversity and that a solution to reopen access to conserved seeds was to be created. While I share the point of view that excessive appropriation has had serious negative consequences on biodiversity, farmers' social networks and livelihoods and even on innovation, I contest the idea that seeds have always been 'in open access' and exchanged under no rules. Rather, I claim that seeds are exchanged under rules defined by communities within seed networks (Coomes et al., 2015; see also Girard and Frison, 2018) and that international regulations have to take this fact into account when designing and implementing systems managing the conservation and sustainable use of seeds (Frison, 2018). Girard (2018: 123) explains that 'relying on archaeological evidence and inferences drawn from ethnographic data, it is well-established that seeds were domesticated and selected in "a decentralized manner, following largely informal but nonetheless binding rules and norms like reciprocity" (Schmietow, 2012: 74; also see Brush, 2004)'.

3 International Treaty on Plant Genetic Resources for Food and Agriculture, Food and Agriculture Organization, 2400 UNTS 303. The Treaty was registered with the Secretariat of the United Nations on 13 December 2006 under No. 43345.

4 Number of States listed as Contracting Parties on the Plant Treaty website on 1 March 2018. In comparison, the CBD has 196 Contracting Parties and the Nagoya Protocol has 104.

5 On 1 March 2018, signatory members to the Treaty were Colombia, Cabo Verde, Dominican Republic, Haiti, the Former Yugoslav Republic of Macedonia, Nigeria and Thailand. According to Article 18 of the Vienna Convention on the Law of Treaties, these signatory States have the obligation to refrain from acts which would defeat the object and purpose of the Treaty.

6 Vienna Convention on the Law of Treaties, 23 May 1969, 8 I.L.M. 679.

7 Book proposal submitted to Cambridge University Press.

8 An important exception to this general rule for underuse of PGRFA relates to the exploitation of in-situ wild relatives collected for direct use, i.e. plants that are collected from the wild and which are not cultivated. For these in-situ wild relatives, the 'tragedy of the commons' over-exploitation dilemma may exist.

9 The Nairobi Final Act stresses the importance of the agreements reached within FAO and called for the IU to be revised in harmony with the CBD. Resolution 3 from the

Nairobi Final Act (the relationship between the Convention on Biological Diversity and the promotion of sustainable agriculture) was adopted 22 May 1992 in Nairobi. Available at www.cbd.int/doc/handbook/cbd-hb-09-en.pdf (last accessed December 2010).

10 WTO was an observer to Treaty meetings at the Second Session of the Governing Body in 2007.

11 Adopted during the International Technical Conference on Plant Genetic Resources in 1996 (Fourth International Technical Conference on Plant Genetic Resources, Leipzig Germany, June 1996). The GPA lists twenty agreed priority activity areas organized into four groups: In-situ conservation, ex situ conservation, utilization of PGR and institutions & capacity building. The GPA was intended to be monitored, reviewed and updated by the FAO CGRFA as a rolling plan of action. It was endorsed by the FAO Conference at its 29th session in 1997.

12 National policies and structures for ex situ conservation are in place in most countries since FAO Member States have developed and implemented international policies which have been translated at the national levels. There is no need for more structure, bodies or policies (excepting the need for further funding, as in every field).

13 The GPA is included in the Treaty as a 'supporting component' in its Article 14.

14 FAO Member States had to submit their country reports for Second States of the World's PGRFA report. The country reports include a number of aspects of direct relevance to the implementation of Article 6, namely: information systems; PGRFA-related policies and regulations; and utilization of PGRFA. These reports also fulfil the countries' obligations under Article 17.3 of the Treaty to cooperate with the Commission in periodically assessing the state of the world's PGRFA in order to facilitate the updating of the rolling Global Plan of Action. See FAO, 'Second Report on the State of the World's Plant Genetic Resources for Food and Agriculture', 2010 See also the First report FAO, 'The State of the World's Plant Genetic Resources for Food and Agriculture', 1996 Available at www.fao.org/agriculture/crops/thematic-sitemap/theme/seeds-pgr/sow/sow2/en/.

15 The GCDT is considered a component of the Plant Treaty through the financial implementation of ex situ conservation.

16 For example, in comparing two funding from the Treaty for conservation activities, as for ex situ conservation activities the GCDT endowment fund received US$ 170 million, while the Benefit-sharing Fund of the Treaty dedicated around US$ 10 to 15 million for in-situ and on-farm conservation projects.

17 The committee met twice: in 2012 and in 2015. see www.planttreaty.org/content/ad-hoc-technical-committee-sustainable-use

18 The Stakeholder Consultation is accessible on the Treaty Website as an online survey; available at www.planttreaty.org/sites/default/files/files/ITPGRFA_SU_toolbox_stakeholder_survey_FINAL_EN.pdf.

19 For a synthetic view of these obligations, see the Resolution 7/2013 scheme on the Treaty website, available at www.planttreaty.org/sites/default/files/Res7_2013_schema_en.pdf.

20 At GB 6, Australia, Canada, Ecuador, Egypt, El Salvador, Germany, Italy, Kenya, Lesotho, Madagascar, Mali, Niger, Pakistan, Seychelles, Syria, Sweden, Tanzania, Uruguay and Zambia have submitted a report (with sometimes very limited content). Available at www.planttreaty.org/content/sustainable-use-submissions.

21 Cooperazione Rurale in Africa e America Latina (ACRA) and consortium partners, the GREEN Foundation, Programa Colaborativo de Fitomejoramiento Participativo en Mesoamérica (FMPA), the Consultative Group for International Agricultural Research (CGIAR) and The Southeast Asia Regional Initiatives for Community Empowerment (SEARICE). Available at www.planttreaty.org/content/sustainable-use-submissions.

22 There is a recognition that there is a multiplicity of terms that have been used in this area (including, *inter alia*, 'genetic sequence data', 'genetic sequence information', 'genetic information', 'dematerialized genetic resources', 'in silico utilization', etc.) and that further

consideration is needed regarding the appropriate term or terms to be used (Resolution 9/2017: § 12 note 47; see also Resolution 5/2017: § 9).

23 It should be noted that the text in the Treaty differs significantly from the 'Jurisdictional Scope' of Article 4 of the CBD.

24 I thank Esther van Zimmeren for sharing this idea with me.

25 See Treaty website for the list of material available through the MLS www.fao.org/plant-treaty/areas-of-work/the-multilateral-system/collections/en/.

26 The eleven Centres holding such collections in trust signed such agreements in 2006. The agreements are available on the Treaty website at www.planttreaty.org/content/agreements-concluded-under-article-15.

27 An example of such copyleft license in the plant field exists with the Open Source Seed Initiative (OSSI). Van Overwalle (2018: 95–6) clearly explains this initiative: 'Inspired by the free and open source software movement that has provided alternatives to proprietary software, OSSI was created to "free the seed" and "to make sure that the genes in at least some seed can never be locked away from use by IPRs". OSSI sets forth that it is dedicated to maintaining fair and open access to plant genetic resources worldwide in order to ensure the availability of germplasm to farmers, gardeners, breeders and communities of this and future generations. The core strategy for achieving these goals is the dissemination and propagation of the OSSI pledge and of OSSI-pledged varieties. The OSSI pledge claims to ensure the four Open Source Seed Freedoms for this and future generations: (1) the freedom to save or grow seed for replanting or for any other purpose; (2) the freedom to share, trade, or sell seed to others; (3) the freedom to trial and study seed and to share or publish information about it; (4) the freedom to select or adapt the seed, make crosses with it, or use it to breed new lines and varieties. The "copyleft" commitment ensures that the OSSI pledge is transmitted with any further distribution of the seed or the seed of any new varieties bred from it. In this way, OSSI aims to preserve the unencumbered exchange of plant germplasm for breeding purposes and guarantee the rights of farmers and gardeners to save and replant seed'.

28 Brazil has completely reversed its position on this issue, advocating since 2012/13 the need for expanding the MLS to all PGRFA, while it was initially one of the strongest countries opposed to this option. Personal communication.

29 For example, an 'African Model Legislation for the Protection of the Rights of Local Communities, Farmers and Breeders, and for the Regulation of Access to Biological Resources' was designed in 2000 where several provisions deal with the matter. However, according to the 'Farmers' Rights Legislation & Policy Database', in Africa only Ethiopia has adopted a legislation dealing with FRs. According to this database, one African country, eight Asian countries, one country in the Americas (Costa Rica), and no countries in Oceania or Europe have adopted legislations dealing with FRs. Available at www.farmersrights.org/database/index.html (Accessed on December 23, 2015).

30 The Benefit-sharing Fund website (changed in 2017) is available at www.fao.org/plant-treaty/areas-of-work/benefit-sharing-fund/overview/en/

31 Submissions of information on interrelations with UPOV and WIPO have been presented at Governing Body 6 by seven civil society organizations and one by the European Seed Association. Available at www.planttreaty.org/content/farmers-rights-submissions

32 With the notable exception of the only submission handed by an international organization, which was sent by the European Seed Association (ESA) on 27.01.2015, and which identifies no actual or potential difficulties in implementing both the Treaty's FRs and UPOV obligations. The ESA submission states that '[w]ith regard to Article 9, the UPOV Convention clearly should not be scrutinized on how it supports the various elements of Farmers' Rights (such as for example protection of traditional knowledge or the participation of farmers in decision-making on matters concerning the conservation and sustainable use of PGRFA) for the simple reason that it is not a task for UPOV to deliver on such goals; the joint exercise should nevertheless reflect on areas where there are some

clear interrelations'. This rather laconic declaration by ESA, which is supported by UPOV, ignores the arguments presented by the other seven submissions. Moreover, maintaining this issue as a question to be dealt with at the national level, denotes a clear will to avoid any in-depth collaborative work on the subject at the international level, denying the clear international aspects of the recognition and implementation of both rights systems. Available at www.planttreaty.org/sites/default/files/ESA_15.0015.1.pdf

33 Until 2013, the accessions provided by the CGIAR constituted more than half of the total MLS material available.

34 It is only on *10 March 2015* that the Secretariat of the International Treaty made available online the statistics on the Multilateral System in order to increase the understanding of germplasm flows, as part of the Global Information System on plant genetic resources for food and agriculture.

35 On September 2014, this number had already raised to 450, thereby confirming that 2013 was a shift year in the implementation process.

36 The Treaty Data Store contains SMTAs reported by genebanks of Contracting Parties and of five International institutions related to the Treaty through its Article 15 (CIMMYT, IRRI, Bioversity International, CIP and ICARDA). The Data Store functions since 2012. Available at https://mls.planttreaty.org/itt/index.php.

37 Research method implemented for this section was 'lighter' than for Phase One, that is to say that the official documents have not all been studied in-depth and in a systematic manner, as these documents only cover proposals not yet adopted by Contracting Parties. A thorough legal analysis will be conducted at the end of the MLS review process, once the revised SMTA will be adopted.

38 The Global Crop Diversity Trust is considered an essential element of the funding strategy. It is an endowment fund being set up by FAO and Bioversity International (then IPGRI) under the policy guidance of the Governing Body of the Treaty to provide funds in perpetuity for ex situ collections of PGRFA.

39 In October 2013, the Secretariat issued a figure identifying the main donors of the BSF. These are: Norway, the European Commission, Italy, Spain, IFAD (International Food and Agriculture Programme), Australia, Ireland, Germany and others.

40 Global Crop Diversity Trust website, available at www.croptrust.org/

41 In October 2013, the main donors of the BSF were: Norway, the European Commission, Italy, Spain, IFAD, Australia, Ireland, Germany and others.

42 In a Treaty Secretariat study, a model shows that with the current membership and availability in the MLS, it will be 38 years before the current fundraising target is reached (Stannard et al., 2012: 156).

43 The Brazilian Agricultural Research Corporation (EMBRAPA) and the Indonesian Agency for Agricultural Research and Development (IAARD) hosted a workshop in Brasilia, Brazil (June 2012) and a second workshop in Bandung (July 2013), with a number of stakeholders to discuss the establishment of the Platform for the Co-Development and Transfer of Technologies developed as a major capacity building instrument.

44 The Clearing House Mechanism (Article 18.3 of the CBD) is coordinated by the executive secretary and aims at the promotion and facilitation of technical and scientific cooperation; of information exchange among Parties, other Governments and stakeholders; and creating a fully operational mechanism with the participation of all CBD Parties and an expanded network of partners. www.cbd.int/convention/articles.shtml?a=cbd-18.

45 DOI webpage of the Global Information System of the Treaty, available at www.fao.org/plant-treaty/areas-of-work/global-information-system/doi/en/; www.planttreaty.org/news/updated-version-multi-crop-passport-descriptors.

46 The International Rice Research Institute (IRRI) has made the first contribution to the gateway by placing with it the genome sequences of more than 3,000 rice varieties. See www.fao.org/news/story/en/item/335405/icode/.

47 More information can be found on the GLIS portal. www.fao.org/plant-treaty/areas-of-work/global-information-system/en/.

48 The Global Sustainable Development Report was launched in June 2015 during the 2015 session of the High Level Political Forum (HLPF) on Sustainable Development. See https://sustainabledevelopment.un.org/globalsdreport/2015.

49 There is a lot to say about traditional knowledge related to genetic resources. This specific issue falls outside the scope of this research but would definitely require further work. This is one of the major limits of the present book. For information on the intersection of traditional knowledge and IPRs see *inter alia* Posey and Dutfield (1998); Grosheide and Brinkhof (2002); De Carvalho (2005); Srinivas (2008); Von Lewinski (2008); Kamau and Winter (2009); Dutfield (2004).

50 'These measures include the amendment of WIPO-administered patent systems (the International Patent Classification system and the Patent Cooperation Treaty Minimum Documentation). [. . .] WIPO has developed a toolkit to provide practical assistance to TK holders on documenting TK'. See www.wipo.int/tk/en/tk/

51 WIPO database on Traditional Knowledge, Traditional Cultural Expressions & Genetic Resources Laws, available at www.wipo.int/tk/en/databases/tklaws/

52 I was part of the Belgian delegation for two Governing Body meetings (the First Session in 2006 and Second Session in 2007), as legal advisor.

53 However, I have heard of such difficulties for a representative of a farmers' organization in other fora, for which it took four years to acquire the status of observer within the UPOV and WIPO, where the procedure might be different. Personal communication.

54 I have experienced this observer status under my University flag for four Governing Body sessions (2009, 2011, 2015 and 2017).

55 I have experienced this status within two other negotiating fora: the CGRFA and the WIPO-IGC. I was part of the delegation representing the CGIAR at the 13th meeting of the CGRFA in 2011.

56 It is important to note that Shakeel Bhatti was the former Head of the Genetic Resources, Biotechnology and Associated Traditional Knowledge Section at WIPO, where he was responsible for WIPO's work on intellectual property law in relation to genetic resources, biodiversity, traditional knowledge and biotechnology. Shakeel Bhatti has taught international patent law and genetic resource policy at several universities in India, Japan and Sweden. His background rooted in the IP field may explain some directions that have been taken in Governing Body governance and decisions, *inter alia* in relation to the Div-Seek initiative.

57 The Big Five are BASF, Dupont, Dow Chemical Company, Monsanto-Bayer, and Syngenta.

58 See same sub-section, point (4) below. The Informal Multi-stakeholder Dialogue initiative was endorsed by the Treaty Secretariat. See 'Rio Six-Point Action Plan for the International Treaty on Plant Genetic Resources for Food and Agriculture', Second High-level Roundtable on the International Treaty on Plant Genetic Resources for Food and Agriculture, United Nations Conference on Sustainable Development Rio de Janeiro, Brazil, 21 June 2012. Available at www.planttreaty.org/sites/default/files/ME253e(Rio_action_plan)01.pdf.

59 The composition of the WG-MLS is as follow: up to five members for Africa, Asia, Europe and Latin America and the Caribbean; up to 3 members for Near East; up to two members for North America and South West Pacific. Two co-chairs are elected from these regional representatives (Resolution 2/2013, point 23).

60 Prior preparatory meetings and/or conference calls took place between December 2013 and September 2014, but no official documents are openly accessible. Treaty Secretariat, 'Facilitator's Summary: Informal Stakeholder Workshop on Multilateral System of the ITPGRFA', Submissions Received from Stakeholders Groups and International Organizations: The Meridian Institute, document IT/OWG-EFMLS-2/14/Inf.4.1, Second

Meeting of the Ad Hoc Open-ended Working Group to Enhance the Functioning of the Multilateral System, Geneva, Switzerland, 9–11 December 2014.

61 Twenty-four participants from farmers' organizations, civil society organizations, the seed industry and research institutes as well as five staff members were present at workshop. The list of participants can be found in the meeting document IT/OWG-EFMLS-2/14/Inf.4.1

5 Seeds and people

A stakeholder's analysis of the Treaty

With this citation below, Jack R. Harlan, a noted American agronomist of the twentieth century, begins one of his most famous books, where he develops a philosophy of the evolution of crop plants and civilization.

> In this book, we shall be dealing with evolution. [. . .] We shall deal with the activities of man that have shaped the evolution of crops and with the influences of crops in shaping the evolution of human societies. Crops are artifacts made and molded by man as much as flint arrowhead, a stone ax-head, or a clay pot. On the other hand, man has become so utterly dependent on the plants he grows for food that, in a sense, the plants have 'domesticated' him. A fully domesticated plant cannot survive without the aid of man, but only a minute fraction of the human population could survive without cultivated plants. Crops and man are mutually dependent and we shall attempt to describe how this intimate symbiosis evolved.
>
> (Harlan, 1975: 3)

Through these first lines of *Crops and Man*, Harlan stresses that crops were developed harvest after harvest, in the hands of many different farmers, all over the world, generation after generation, following social, cultural and economic trends (Bragdon, 2004: 12–13). Domesticated crops are dependent on farmers and breeders, in the same way that farmers and breeders are dependent on these crops. As a consequence, a crop variety bred nowadays is the result of this past collaboration and interdependence (Palacios, 1997). Similarly, the world's population relies on farmers and breeders for producing the necessary food to survive, thereby widening the interrelation and interdependence circle between crops and men.

Does the current international setting for the management of PGRFA (the above identified 'regime complex', see Chapter 2) apply this philosophy? Is there another choice than respecting (*inter alia*) both principles of collaboration and interdependence in setting up a sustainable PGRFA regime complex, responding to human needs (and to the planet's needs too). Following the analysis conducted under chapters 2 and 3 of the present dissertation, it seems that the equilibrium in the interactions between plants and men has not been respected over the last decades. Rather, the 'hyperownership' trend (Safrin, 2004) seems to have led to the predominance of human activities over the evolution of crops, dismissing some of the plants' vital and intrinsic characteristics: i.e. evolution as a universal, dynamic, cooperative, diverse and interdependent system. Chapter 4 has highlighted how Treaty stakeholders have attempted to legally re-establish some sort of equilibrium by creating the Treaty and its Multilateral System of access and benefit-sharing. However, the legal analysis of the Treaty has shown that its implementation is difficult and that stakeholders are currently unable to reach their goal due to intrinsic conceptual constraints.

To further assess (and cross-check with the legal analysis) whether the current regime of the Plant Treaty allows to reshape an equilibrium by reaching the set objectives of conservation, sustainable use and exchange of PGRFA, a Treaty stakeholders' analysis (Freeman, 2010; Friedman and Miles, 2002) was carried out through the publication of an edited book: *Plant Genetic Resources and Food Security: Stakeholder Perspectives on the International Treaty on Plant Genetic Resources for Food and Agriculture* (Frison et al., 2011a). By briefly describing who these actors[1] are and by summarizing the findings of the stakeholders' analysis, the present chapter gives a voice to PGRFA actors that have participated in the negotiation and current implementation of the Treaty. The objective is to identify the various needs and interests of these stakeholders, their expectations regarding the Treaty and assess whether the Treaty satisfies their needs. The overall goal is to allow stakeholders to identify constraints they are faced with when implementing the Treaty. These constraints are then used in Chapter 6, as a basis for suggestions to mitigate the dys-functioning of the Treaty. As the reader will see, the constraints identified by stakeholders below confirm and complement the results of the legal analysis of the Treaty in the preceding Chapter 4.

Several remarks are to be noted regarding the edited book. While the effort to have Treaty stakeholders' views expressed in a publication constitutes an original contribution in itself, it has some limitations. First and foremost, the book's sole ambition is to provide a compilation of stakeholders' views on the Treaty and a preliminary analysis of these views.[2] The book is not a rigorous social science study complying with all the requirements and methods of stakeholder analysis (Reed et al., 2009); rather, the purpose was to gather information directly from Treaty stakeholders in support of the legal assessment of the Treaty. Second, deriving from this first limitation, the stakeholders who participated in the book are those people already present and active within the FAO/Treaty forum.[3] A classical definition of stakeholders is 'any group of individuals who can affect or is affected by the achievement of the organization objectives' (Freeman, 2010: 46). Stakeholders may be natural persons, groups or legal entities; they are not limited to 'insiders' within the organization (for a narrow definition see Bowie, 1988; for a wider definition see Starik, 1995; Hubacek and Mauerhofer, 2008). The participating authors of the book were the people who identified themselves as Treaty stakeholders and which were (relatively easily) accessible within the limited means of this research (i.e. the 'usual' accredited representatives present at FAO Treaty meetings, well-known in the field by the other stakeholders). It is acknowledged that this constitutes a bias, in that the voices of other people active in PGRFA management, but not present in the Treaty forum, are not represented in the book. However, an attempt was made to mitigate partly this bias with the inclusion of a chapter representing views of consumers. Several attempts to obtain the participation of a representative of the food processing industry failed, resulting in no chapter from that category of actors. Third, the content of the book remains cautious. Although it is the first time that Treaty stakeholders have spoken relatively openly on the subject in a collective book, the very fact that the means chosen to express their voice is a written publication has led stakeholders to write cautiously. Views, which might have been expressed quite openly during conversations, have necessarily been translated into a 'politically' correct language, without retrieving on the basis for their argumentation. Finally, the book was published in 2011. Although seven years have passed, the up-to-date legal assessment of the Treaty in Chapter 4 shows that the Treaty evolution concur with the results of the stakeholders' analysis. It also further confirms the suggestions made by the book editors in 2011 on ways and options to reach the Treaty's objectives and to improve its implementation.

Despite the above mentioned biases, the edited volume fills in an information gap and provides a published and freely accessible source of information on the subject.[4] It constitutes a useful contribution for researchers wishing to better understand the role of stakeholders and of their relationship during the negotiations of the Treaty, especially when taking into account the fact that there exists no verbatim (written or oral recording) of negotiation meetings expressing stakeholders' positions at the time of the negotiations.[5]

This chapter is divided into two sections. First, major stakeholders who participated in the Treaty negotiations are identified and their specific interests and concerns are explained. This picture is followed by a table recapping (in an overly simplified manner) stakeholders' various (and contradicting) views. Second, based on the content of the 18 book chapters, the challenges identified by the authors are identified in a table listing 17 'specific implementation challenges and constraints' (Frison et al., 2011b: 276–277).

A description of Treaty stakeholders

In theory, all Treaty stakeholders have an overall common interest, i.e. to conserve seeds to be able to use them to produce food. Notwithstanding this self-evident statement, stakeholders have very different agendas and objectives, specific institutional structures, diverse economic means to reach their goals, etc. Hence, there is a diversity of direct short- and medium-term objectives, which may be competing and which may hinder reaching the identified common overall interest. The historical account of PGRFA management (Chapter 2) has revealed the complexity of the connections between the fields of agriculture, environment, scientific advances and international regulatory developments, especially regarding intellectual property rights. This complexity is reflected in the relationships between the stakeholders involved in the international management of PGRFA, which include states, public and private institutions, individuals, communities as well as commercial and non-profit actors. This section aims at describing these actors. It is essential to comprehend how they function, what their claims and interests are, and how they participate in the PGRFA regime complex in order to understand the roots of the current setting; to foster ways of bridging stakeholders' diverse interests; and to propose solutions that are likely to be accepted and implemented by these stakeholders. Indeed, as Reed et al. (2009: 1935) put it '[o]nly by understanding who has a stake in an initiative, and through understanding the nature of their claims and inter-relationships with each other, can the appropriate stakeholders be effectively involved in environmental decision-making'. The actors who contributed to the edited volume are the following: states (contracting parties); the CGIAR; the Global Crop Diversity Trust; genebanks and collections; the seed industry; farmers and farmers' organizations; and NGOs.

States

States as sole recognized decision-making actors within FAO

States are the primary actors in intergovernmental organizations such as FAO. There are 186 nations that are member of FAO and 145 are contracting parties to the Treaty. In international law, all states are equal, as every state has a legal personality and there is no hierarchy among them (Cragg et al., 2012). Within international fora such as UN agencies, one state equals one vote according to the sovereign right of states to negotiate international treaties (FAO Constitution, Article 3.1; Vienna Convention on the Law of Treaties, Article 6 stating that every state possesses capacity to conclude treaties). State sovereignty can be seen as the warranty for state compliance with the designed international regulatory instruments (Chayes and Chayes, 1998). Decisions are generally taken by a two-thirds majority, although unanimity or systems with weighted voting are also frequent for specific types of decision (Cragg et al., 2012; Vienna Convention on the Law of Treaties, Article 9). Notwithstanding this equality of status, it is a fact that states with greater economic power and political clout use their influence to impose their views.

States' diversity of objectives in creating and complying with regulatory instruments

While each state has its own national policy regarding PGRFA management, it is impossible for any state – no matter how powerful it is – to impose its own economic, social and environmental political agenda on all other FAO members. While states share a common concern for the conservation of biological diversity, for their economic growth and health, or for their social development, it is clear that they all have divergent agendas and objectives according to their specificities, wealth, or stage of development, etc. Furthermore, states' negotiating behaviour will reflect the concomitant constraints of both a domestic political game and an international game (Putnam, 1988; Evans et al., 1993). During the Treaty negotiations for example, the seed industry strongly lobbied within developed countries' delegations such as the Netherlands or the U.S. to develop access rules that do not affect their businesses too much, i.e. that do not interfere with their IPR policy; while conservationists or farmers' organizations tried to weigh-in on the voting power of other states (with more or less success).

States' negotiating practice within FAO

Within FAO's CGRFA, the usual practice is that countries from the same region meet in regional groups to develop consensus positions and therefore have stronger influence on the negotiation outputs. CGRFA member nations[6] are subdivided into seven geographic regions (CGRFA Rules of Procedure, Rule III): Africa, Asia, Europe, Latin America and the Caribbean, Near East, Northern America and South West Pacific. Gerbasi (2011: 29) recognizes that '[w]hile the existence of these regions responds to technical needs, it is also true that this has political implications', thereby increasing states' political and economic weight during the negotiations. Other groupings have emerged, such as the European Union (EU),[7] the G-77,[8] the OECD group[9] or the mega-diverse countries,[10] each with the aim of giving more weight to states with overall common objectives during negotiations and decision-making processes. However, none of these groups really align with the political divide that occurred during the Treaty negotiations between gene-poor developed countries and gene-rich developing countries (Frison et al., chapters 1 to 8, in particular Gerbasi, 2011).

As for all international negotiations, the resulting instrument reflects the balancing powers and interests of most parties. Although in theory, states have equivalent powers and one state equals one vote, in reality not all states can exert the same level of influence. In explaining why Africa did not obtain as much as initially hoped, representatives from the African group contend to have 'given in' too quickly and hence to have lost their major bargaining power (Egziabher et al., 2011: 52).

The Consultative Group on International Agricultural Research (CGIAR)

A brief history of the CGIAR

The CGIAR grew out of the international response to widespread concern in the 1950s, 1960s and early 1970s that many developing countries would succumb to hunger.[11] Experts predicted prevalent and devastating famine between 1970 and 1985, with hundreds of millions starving to death. The roots of the CGIAR go back almost three decades before its formal inauguration in 1971, beginning with a collaborative programme between Mexico and the Rockefeller Foundation (Byerlee and Dubin, 2010). The CGIAR is a strategic partnership of countries, international and regional organizations and private foundations supporting the work of 15 International Agricultural Research Centres united in a Consortium (CGIAR, 2012; Moore and Frison, 2011). The CGIAR has strong links with university and private research, *inter alia* through public-public or public-private partnerships. Its activities are mainly financed with public funds from donor countries, developing countries, regional organizations, international organizations and foundations.

The CGIAR's objectives

According to their website,[12] the CGIAR is dedicated to the conservation and use of agricultural biodiversity to improve the livelihoods of poor people. It fosters sustainable agricultural growth through high-quality science aimed at benefiting the poor through e.g. stronger food security, better human nutrition and health, higher incomes and improved management of natural resources. The CGIAR research centres do so in collaboration with national agricultural research systems, civil society and the private sector, as complementary stakeholders to states' actions in international agricultural policies. The CGIAR Constitution does not explicitly mention that one of its objectives is to provide international agricultural goods and services which states fail to provide on their own (i.e. such as food security and sustainable agriculture). However, several studies have shown that the CGIAR has de facto acted as an early provider of such goods (Sagasti and Timmer, 2008; Shands et al., 2008; Anderson, 1998) either by providing global knowledge or specific products and services in agricultural research (Sagasti and Timmer, 2008: 27; see also the GPG Project of the genebanks of the CGIAR Centres).[13]

The role of the CGIAR and of Bioversity International in the Treaty negotiation

The CGIAR has developed many policies, programmes and tools to reach the above mentioned objectives.[14] Indeed, by collecting, characterizing and conserving PGRFA, and by training, transferring knowledge, technology and material in an 'open access' manner, the CGIAR has significantly contributed to designing a new international regime for germplasm development and transfer (Falcon and Fowler, 2002). The research, conservation and training activities are mainly focused on developing countries. Since its inception, the CGIAR has contributed to facilitate the access to as many PGRFA as possible, first by creating a global network of gene banks and collections, then by holding PGRFA in trust for the global community,[15] and by making information available to the community (*inter alia* through databases such as SINGER or Genesys).

Bioversity International is one of 15 centres supported by the CGIAR.[16] A study was published in 2003 on the role and political influence that Bioversity (at that time named IPGRI) exerted between 1996 and 2001 in the international negotiations revising the IU. The study shows that

> the provision of timely and relevant technical inputs directly linked to IPGRI's area of expertise was the most successful means of influencing the negotiations. [. . .] [P]olitical neutrality and reliability were seen as factors that enhanced IPGRI's ability to influence [negotiations]. (Sauvé and Watts, 2003: 307)

Bioversity International and the CGIAR are accredited to participate in Treaty meetings as 'observers from intergovernmental organizations'. Albeit being a key stakeholder in the Treaty negotiation and implementation, the CGIAR lacks the resources and formal rights that endow states, and this fact ultimately limited IPGRI's ability to directly influence the negotiations (Sauvé and Watts, 2003).

Outcomes of CGIAR's participation in the Treaty negotiating process

The CGIAR has actively contributed to the creation of the Treaty's innovative MLS (Cooper et al., 1994; Falcon and Fowler, 2002). As a matter of fact, during the Treaty negotiations, the CGIAR and especially Bioversity International, has played a crucial role in unblocking negotiations on technical matters, in particular regarding the access to and exchange of PGRFA, as well as the SMTA (Moore and Frison, 2011: 156–159). As the historical accounts have shown, the CGIAR objectives and policy instruments have functioned as a driver to design and adopt the policy instruments and objectives for the international management of PGRFA under the Treaty.

During the implementation process, the CGIAR continues to be a central actor of the PGRFA regime. Indeed, its role has been recognized and integrated within the Treaty system under Article 15. The CGIAR Centres and other international institutions holding PGRFA collections in trust are the main distributors of Annex I (and non-Annex I) PGRFA following the MLS provisions (i.e. using the SMTA). Even more, the CGIAR is now subject to policy guidance of the Governing Body for the ex situ collections held by them (Plant Treaty, Article 15.1(c)).

The Global Crop Diversity Trust

The mission and objectives of the Global Crop Diversity Trust

The mission of the Global Crop Diversity Trust (GCDT) is to ensure the conservation and availability of crop diversity for food security worldwide. It contributes to sustaining PGRFA collections that are critical for food security and sustainable development and to secure PGRFA in the long-term.

The objectives of the GCDT are to safeguard collections of unique and valuable PGRFA held ex situ, with priority being given to those that are included in Annex I to the Treaty or referred to in Article 15.1(b) of the Treaty. This includes the funding of projects for the regeneration, characterization, documentation and evaluation of PGRFA and the exchange of related information, as well as securing the availability of PGRFA. The GCDT also aims at supporting the implementation of the Global Plan of Action for the Conservation and Sustainable Utilization of PGRFA and at participating in national and regional capacity building initiatives (Hawtin and Fowler, 2011: 215–216).

Legal status of the GCDT

The GCDT is an independent organization under international law, officially established in October 2004,[17] although discussions on the need for such an institution had been ongoing for several years already. The adoption of the Treaty unblocked the discussions and allowed for its swift creation.

A 'Relationship Agreement between the Governing Body of the International Treaty and the Trust' was formally approved during the first meeting of the Governing Body of the Treaty in June 2006 (IT/GB-1/06/Report at §§ 35–40). Although the agreement recognizes the GCDT as an 'essential element' of the funding strategy of the Treaty and provides for the Governing Body to give policy guidance to the Trust and to appoint four members of its Executive Board,[18] many developing countries were unhappy with the establishment of the Trust as an independent organization. Besides the lack of real control of the Governing Body over the Trust, one of the reasons of this resentment was the fact that they saw significant funding going to the Trust for ex situ conservation, while they would have liked to see more investment in in situ conservation.

Strategies and policy results of the GCDT

The GCDT uses a constellation of strategies and formal policies to reach its objectives (Hawtin and Fowler, 2011). It finances ex situ conservation projects all over the world and reports on its activities to the Treaty's Governing Body.

Since its inception, the GCDT was able to provide long-term maintenance grants to many collections (aroids, banana, barley, bean, cassava, fava bean, forages, grass pea, pearl millet, rice, sorghum, wheat and yam). In addition, the GCDT has funded projects around the world in partnership with a large number of other institutions (including private corporations). As of December 2015, the GCDT had received from donor countries, foundations, the private sector and international organizations total pledges of support amounting to over US$444 million and of this more than US$309 million had already been received.[19]

Genebanks and plant genetic resources collections

The predominance of ex situ conservation of PGRFA

Genebanks are the facilities where major crop plants and their wild relatives are stored for their conservation and use. They provide the raw material for the improvement of crops by professional plant breeders. 90 per cent of the six million accessions conserved in ex situ collections worldwide are stored as seed because it is practical and relatively cheap. Other conservation methods include freezing plant tissues or pollen (i.e. cryopreservation), tissue cultures grown in test tubes (in vitro conservation), live plants in greenhouses and live plants grown in the field (field collections). Various types of genebanks exist, i.e. crop-based collections which focus on one crop (e.g. rice, wheat, potato), or national (e.g. US National Plant Germplasm System), regional or international collections, which store a variety of crops. N.I. Vavilov was the first to create a genebank in Saint-Petersburg (Leningrad) at the beginning of the twentieth century; his activities where focused on crop improvement. Pistorius identifies the US National Seed Storage Laboratory – created in 1958 – as one of the first real ex situ collection facilities to conserve threatened landraces and wild species (Pistorius, 1997: 7). In 2010, FAO registered some 1750 genebanks around the world (FAO, 2010: 55).

Objectives of genebanks

Genebanks have various functions. Article 2.1 of the IU distinguished between a 'base collection of plant genetic resources' and an 'active collection'. 'Base collections' hold plant genetic resources for long-term security to preserve the genetic variation for scientific purposes and as a basis for plant breeding. They constitute a safe long-term duplicate of PGRFA that are also conserved in 'active collections'. The Svalbard Global Seed Vault,[20] which is probably the most famous genebank, provides for an additional safety back-up. 'Active collections' complement base collections in that they maintain seed samples that are drawn on for use. They acquire (through exchanges or exploration and collection trips), conserve, multiply, regenerate, study (*inter alia* through characterization and evaluation) and distribute genetic resources to users such as breeders or researchers in various fields, including genetics, plant and seed physiology, in vitro culture, cryopreservation or information technology. In the Treaty, a general inclusive definition of 'ex situ collection' is provided in its Article 2 as a collection of PGRFA maintained outside their natural habitat.

Conservation strategy at the international level

According to some experts,

> maintenance of seed viability and genetic integrity remain the cornerstones of gene bank management. The quality and sustainability of any genetic conservation effort depends on how seeds are processed and conserved. Inappropriate procedures for seed handling lead to accelerated deterioration, making conservation more expensive.
>
> (Rao et al., 2006: 2)

While ex situ conservation has invaluably contributed to the mitigation of diversity losses and has constituted the core 'raw' material for major public research on food security issues, genebank curators have pointed to the lack of sufficient financial, human and technological means to efficiently conserve and safeguard PGRFA (Cubero, 2011). Therefore, there is a need to develop the coherent genebank management strategies and policies required to respond to the increasing pressure to improve cost efficiency and effectiveness. Such efforts are made by international actors such as FAO or Bioversity International through the development of guidelines for breeders and researchers, a common framework for seed policies, information sharing mechanisms,[21] or standardization of material transfer agreements (SGRP, 2003). During the Treaty negotiations, genebanks have mainly been involved indirectly through either the CGIAR or through national delegations. The policy implemented through the supporting components of the Treaty (the GPA, the role of the CGIAR, the international PGRFA networks and the GLIS) reinforces the global strategy for ex situ conservation. A remark is made on the fact that this strategy clearly focuses on providing material mainly to one type of users, i.e. breeders and researchers.

Plant breeding and the seed industry

Companies within the breeding and seed industry

While seed companies constituted a very fragmented market until the 1960s (van den Hurk, 2011), a massive industrialization of agriculture has led to a ruling by a few oligopolistic companies (Kloppenburg, 2004: 11–18). The global seed market increased from US$ 12 billion in 1975, to around US$ 20 billion in 1985 and was estimated at US$ 36.5 billion in 2007 (van den Hurk, 2011: 164). The rapid progress of technology, especially in molecular biology and genetic engineering, has introduced new players into the field (Howard, 2009; Footer, 2000: 51). The wave of acquisitions that took place in the 1980s has significantly concentrated the seed market (LeBuanec, 2007: 10), where mega-seed companies (Herdt, 1999: 8–9) expanding on the agrochemical market aim at controlling everything from genetic engineering of seeds to the selling of seeds to farmers, to marketing plant-grown drugs, modified foods and industrial products. Aoki (2008: 113) cites Janet Hope saying that 'the merger-mania was driven primarily by the need to avoid high transaction costs associated with clearing multiple IPR, [. . .] and that most key enabling technologies are now in the hands of only a handful of firms'. According to Laird and Wynberg (2005: 18), '[t]he seed industry is characterized by three levels of companies: life science giants, large multinational firms, and small and medium-sized enterprises'. By reminding that in 2005, ten companies controlled 49 per cent of the global seed market, with an increased trend towards acquisitions and mergers between seed and agrichemical companies Laird and Wynberg (ibid) confirm that '[m]ost of the larger companies also have active interests in agro-chemicals and pharmaceuticals. An intensifying trend over the past decade has been the continued consolidation of the seed, crop protection and plant biotechnology industries'. Today, the agro-chemical-seed market is dominated worldwide by the Big-Six that is to say BASF, Bayer, Dupont, Dow Chemical Company, Monsanto and

Syngenta, which is becoming the Big Five, with the ongoing Bayer-Monsato merger. In 1996, Monsanto was ranked number 12 in the overall ranking of seed company revenues with 170 million Euros, whereas it reached the first place in 2006, with a company turnover exceeding 4 billion Euros (LeBuanec, 2007: 10). The concentration tendency is continuing, as shown by the 2015 failed tentative US$46.5 billion takeover bid of Monsanto over Syngenta[22] and is now also shifting towards farm machinery industry, thereby potentially creating a 'one-stop shop for farm inputs' for all farm machinery, seeds, fertilizers and chemicals (ETC Group, 2015: 3).

Besides, in specific countries, such as the Netherlands where plant breeding has a strong historical trade background, or in developing countries where the breeding industry is emerging, small and medium-size seed enterprises manage to continue to evolve in the international seed market.[23] Laird and Wynberg (2005: 18) confirm that 'small and medium-sized seed companies, of which there are several thousand, are also significant and occupy different market niches, [such as] vegetables, grasses and more marginal crops'. These small and medium-sized companies hold different interests and positions than the mega-companies.

Activities of the private seed industry

Seed companies are market entities, whose objectives are primarily profit-oriented. They exploit the potential of isolating and manipulating specific genetic characteristics in seed varieties (Footer, 2000: 51). Over the years, plant breeding has largely concentrated on the following aspects of major crops: increased yields, earliness, resistance to biotic stresses and tolerance to abiotic stresses; these characteristics are incorporated in new varieties. Some qualitative characteristics like taste or nutritional value were also improved for some crops (Bruins, 2009: 23). According to the first report of the state of the World PGRFA,[24] the private sector

> tends to focus on crops that either cover large areas (maize, soybean, wheat, rice) or that generate high per hectare income (tomatoes, sugar beet, etc.). It also tends to concentrate on those crops that offer the strongest protection, either through IPR legislations or more often through the technical or physical characteristics of the seed such as hybrid seed or the terminator technology. This indicates a continuing need for public investment in plant breeding to cover the gaps in private-sector efforts.
>
> (FAO, 1996a: 165)

The seed industry's strategy in the international PGRFA forum

To represent the interests of the mainstream seed industry at a global level, seed companies have grouped into the International Seed Federation (ISF, formerly ASSINSEL). The mission of ISF is to facilitate the international seed flow, related know-how and technology; to mobilize and represent the seed industry at a global level; to inform its members; and to promote the interests and the image of the seed industry.[25] ISF has been very active during the negotiations on the Treaty and its SMTA (van der Hurk, 2011). Two aspects are crucial for the seed industry: IPRs and access to material.

Regarding the first aspect, companies insist on the fact that the cost of R&D is substantial. Therefore, they perceive intellectual property protection on the varieties they produce as the most appropriate manner to ensure a return on R&D investments (Footer, 2000: 51). The possibilities offered by intellectual property protection of plant varieties and biotechnological inventions have encouraged companies to increase their spending on R&D: the plant breeding industry spends on average 10 to 15 per cent of its annual turnover (Bruins, 2009: 22). Therefore, during the Treaty negotiations, ISF has strongly supported the protection of crop improvements based on plant breeders' rights (UPOV 1991 rather than the 1978 Convention) and on patents. ISF argues that it provides a stronger protection of plant varieties against inappropriate exploitation by others (van den Hurk, 2011: 172), which therefore allows for a fair return on investments. Notwithstanding this position, the European Seed Association (ESA, 2014) has shown support to the Treaty and to the MLS, which is 'seen as the more suitable system by the plant breeding sector'. Furthermore, ESA recently contributed to the Benefit-sharing Fund for 300.000 EUR, thereby confirming its willingness to have a functional MLS.[26]

Regarding the second aspect, i.e. access to seeds, Ten Kate and Laird highlight that 'shared access and reciprocity (applied to genetic resources and information on research results) constitute the norm between breeders'. As one private breeder explained (Ten Kate and Laird, 2002: 148), '[i]t is an unwritten rule of ethics for breeders that when someone provides genetic resources, breeders will send them information relating to the research done'. Van den Hurk confirms that it is important for the seed industry to have

> sufficient freedom to operate to carry out their breeding activities and have access to the necessary plant genetic resources. This means that a flow of genetic resources should continue to take place. It is important to realize that access is required both in developed and developing countries. In the latter it may become even more important as the plant breeding sector is expanding. Moreover, genetic resources should be available for all type of users, be they small, medium-sized or large enterprises.
>
> (Van den Hurk, 2011: 173)

ESA members use the SMTA when accessing MLS seed, because the SMTA recognizes existing IPR systems over seeds. This recognition of IPRs is the cornerstone of ESA's position (ESA, 2012): the MLS shall respect their need for a 'suitable framework that ensures adequate return on investment to the plant breeder for his accomplishment', such suitable framework being the plant breeder's rights system (PBR) as provided for within the framework of the UPOV Convention. On the contrary, the Big Five, which are more focused on high-tech biotechnological products, favour a protection based on patents.

Succeeding in imposing their voice

In 2006, the global seed market amounted to US$ 30 billion and 49 per cent of its value was owned by the ten major agro-biotech life science companies (Laird and Wynber, 2008: 15, table 3). In 2008 they controlled 55 per cent of the commercial seed market and 64 per cent of the patented seed market (Chiarolla, 2011: 49). During the Treaty negotiations, seed companies were active as observers and as experts being part of national delegations (e.g. the US or the Netherlands). Moreover, their voice was heard during the negotiations on the SMTA between 2004 and 2006 and recently during the review process of the MLS (ESA, 2011), especially regarding the rate adopted for the benefit-sharing payment schemes.

Farmers and farmers' organizations

The many faces of farmers

The first farmers were also the first breeders (Cubero, 2011: 197): they domesticated wild species to transform them into crops. Farmers are the ones selecting, sparing and sowing seeds, which is a method of crop improvement called 'bulk selection'. Bulk selection consists in choosing the seeds of the best individuals of a harvest, and to mix them to form the sowing bulk for the next season. Today, farmers have many different faces: from the prosperous entrepreneur managing highly industrialized large-scale farms in Brazil or the US, to the small-scale traditional subsistence farmer in Africa (Batta-Bjørnstad, 2004: 30–31). Within the scope of this research, farmers refer primarily to smallholder farmers (or subsistence farmer) and rural people who produce their food for survival. The 'Rural Poverty Report' published by the International Fund for Agricultural Development[27] (IFAD, 2011: 16 and 46) shows that 55 per cent of the total population in developing countries are rural and that despite the current shift towards urbanization, poverty remains largely a rural problem.

Regarding the relations between farmers and seeds, Louwaars (2008: chapter 3; see also De Boef et al., 1996) describes two independent yet complementary systems of crop development, with two types of farmers acting within two types of seed systems. On the one hand, farmers from industrialized countries mainly get propagating material from the formal sector. Formal seed systems are constituted by an organized and often regulated chain that includes genebanks, breeders, seed producers and seed marketing and distribution organizations. They provide tested, certified seed to farmers in an organized manner. On the other hand, small-scale subsistence farmers usually get seeds from their own holding, from neighbouring farmers and from local markets (McGuire and Sperling, 2016). These informal seed systems are

> by far the most important suppliers of seed, and are particularly important for resource-poor farmers. [. . .] In practice, these different systems operate side by side to serve the needs of different types of farmers for different types of crops.
> (Louwaars, 2008: 29)

And indeed, Halewood (Halewood et al., 2006: 179) states that 'farmers' systems of informal exchange are crucial to: (i) their ability to constantly introduce new material into their cropping systems; (ii) maintaining high levels of diversity; (iii) maintaining relatively stable yields'.

Farmers' movement defending their rights to save, use, exchange and sell seeds

While farmers have always exchanged seeds freely and traditional agriculture depended on the constant exchange and movement of PGR, current regulations tend to hinder this practice more and more, leading to the privatization of resources. This trend limits the exchange of seeds, whereas the right of farmers to save, use, exchange and sell seeds is seen by farmers as one of the most basic foundations of the farmers' system of PGR management, as this is how PGRFA diversity has been created and maintained (Pelegrina and Salazar, 2011: 178).

Farmers' movements at the international level have emerged in the 1970s in reaction to the corporate driven intensification of agriculture. The most active one within the FAO fora is La Via Campesina,[28] which defines itself as an international movement bringing together peasants, small and medium-size farmers, landless people, women farmers, indigenous people, migrants and agricultural workers from around the world to defend small-scale sustainable agriculture as a way to promote social justice and dignity (Claeys, 2014a). Other non-profit international organization supporting small farmers include GRAIN or the International Planning Committee for Food Sovereignty (IPC).

La Via Campesina, GRAIN and IPC were very active during the Treaty negotiations, although only participating as accredited observers, without any decision-making power. They fought most intensively for Article 9 on Farmers' Rights in order to safeguard their practice of saving, using, exchanging and selling seeds freely, thereby allowing farmers to grow the diversity needed for their own food, provide for the different other needs of farming communities, deal with climate changes and other biotic and abiotic stresses.[29] Besides, Pelegrina and Salazar note that

> while there are farmers and farmer groups who have started expounding on Farmers' Rights, a large number of farmers and their organizations have yet to identify themselves with this "social construct". No one can teach farmers about Farmers' Rights because it is imbedded in them and it is the role of governments and other stakeholders to ensure that farmers can continue with what they have been doing or strengthen their knowledge and skills for global public good.
>
> (Pelegrina and Salazar, 2011: 177)

Farmers' strategy

As the small-scale farmers' priority strategy is to meet their immediate needs related to cultivation, storage, processing and consumption (Berg et al., 1991: 16), being able to maintain several varieties is essential to select those varieties best adapted to different fields or for different uses. This results in specific adaptation to micro-level agroecological niches and to cultural, economic and social needs.[30] To maintain this practice, farmers' organizations demand to fully recognize Farmers' Rights at the international and national levels. La Via Campesina calls for an increased awareness of the importance of farmers' practices and for the participation of farmers and farmer communities in decision-making processes that have an impact on their lives. Pelegrina and Salazar (2011: 177–181) recall that farmers are primarily concerned about their livelihood, [i.e.] the return on their inputs and hard labour in the form of sufficient (preferably with surplus) food supply and income. It is natural for farmers to test and innovate as part of risk management measures to ensure their livelihoods. They argue that it is not just about protecting the germplasm materials, but using it to satisfy their needs. Therefore, farmers organizations stress that they need to be able to access seeds, to use and exchange them freely, including modern cultivars, to be able to feed the world's population, to continue participating in the development and conservation of PGRFA diversity and to face unpredictable pressure such as climate change. Jarvis et al. (2011: 126) argue that enabling farmers to take a greater role in the management of their resources could increase the conservation of PGRFA diversity and therefore contribute to achieve global food security.

Farmers' rights: a claim partially heard

It is clear that farmers' communities as such did not have much influence during the Treaty negotiations. Many farmers and farmers' communities around the world had no clue about the Treaty negotiations happening in Rome. Those who were informed and registered as observers were divided into several groups, with little training, information and coordination. Although developing countries and some developed countries (notably Norway) have fought for their cause, the result is limited. The Treaty under its Article 9 recognizes Farmers' Rights, but they are subject to national legislation (Article 9.2) and therefore to the goodwill of national legislators and politicians. Therefore, most farmers are reluctant to engage in the Treaty discussions, as they have yet to see concrete results out of the Treaty, as translated into national policies (e.g. seed regulations) and programmes on PGR.

Farmers' rights are a crucial aspect of the Treaty, and I believe that the Treaty's implementation can only be successful if the farmers' rights provisions are fully operating.

Non-governmental organizations

NGOs at FAO

NGOs are local, national or international not-for-profit groups or associations acting outside political and governmental institutions. They perform a variety of functions e.g. bring citizens' concerns to governments, advocate and monitor policies or encourage political participation through the provision of information. An attempt to define NGOs is provided by Peter Willetts (2011: chapter 2), where he contends that it is easier to agree on what NGOs are not than on a widely accepted definition of NGOs. Depending on the room left to NGOs to operate within specific international fora, they will influence the debate with more or less success.

Within the CBD forum, 'non-governmental organizations, including environmental not-for-profit organizations, played a leading role in the initial conception, negotiation and adoption of the Convention on Biological Diversity, and continue to shape policy development' (Djoghlaf, 2006). Indeed, the CBD has created an Alliance of civil society organizations[31] aimed at facilitating more diverse, coordinated and effective civil society input into CBD policy-making. FAO[32] has a more restrictive definition in which an NGO is 'an organization which seeks funding, hires staff, and undertakes programmes, but does not realize a profit'.[33] Within the limits of this research, the focus will be directed on NGOs which have been active in the agricultural field and which have the status of observers at FAO. They are less numerous than in the wider biodiversity field. Within FAO, there is no permanent fixed location for the NGO delegate community to meet and coordinate, contrary to WTO and WIPO fora or now even within the CBD (with the Alliance).

ETC Group (Action Group on Erosion, Technology and Concentration, formerly RAFI) was the first civil society organization to draw attention to the socio-economic and scientific issues related to the conservation and use of PGRFA, IPRs and biotechnology. Within the PGRFA fora, although civil society organizations were very dynamic in the beginning of the CGRFA era (during the 1980s) (Mooney, 2011: 143), few organizations have had a strong influence on the Treaty negotiations from the 1990s onwards. Mooney states that

> [i]f not sooner, the 1991 Commission meeting was certainly the last that was dominated by civil society. By the time governments met again in 1993, the Commission was thoroughly institutionalized and government delegations coming to Rome had marching orders from their capitals that demanded obedience. We could still cajole and tease but we could not decide.
>
> (ibid: 145)

Some NGOs have been more active over the last few years, such as: the Berne Declaration, the Third World Network, or the International Planning Committee for Food Sovereignty (IPC). Some of these organizations have grouped under the initiative 'No patents on Seeds',[34] which calls to rethink European patent law in biotechnology and plant breeding and to support clear regulations that exclude patentability processes for breeding, genetic material, plants and animals and food derived thereof.

The role of NGOs in the Treaty forum

NGOs are information carriers making the link between people and the politics in negotiating fora. National, regional and international accredited NGOs may participate in the international PGRFA meetings as observers, but they cannot participate in the decision-making process. For example, the Southeast Asia Regional Initiatives for Community Empowerment (SEARICE, established on June 1977) is a regional non-government development organization that promotes and implements community based conservation, development and sustainable use of plant genetic resources in partnership with civil society organizations, government agencies, academic research institutions and local government units in Bhutan, Lao PDR, the Philippines, Thailand and Vietnam.[35]

A note is made regarding the way institutions are listed as observers in Governing Body reports. Under the category 'observers from non-governmental organizations', all sorts of entities are listed, which do not fall under the above mentioned definition of an NGO, including companies from the seed industry. Under the same category one can find universities, 'true' NGOs or civil society organizations, research centres, foundations, national programmes, the International Seed Federation, etc. Therefore, one should be cautious when looking at which NGO had the status of observer for Governing Body meetings.

According to Matthews (2007: 1371), NGOs' objectives are mainly to provide information and support to country delegates, to enhance the negotiating capacity of delegates, particularly from developing countries (i.e. raising awareness, providing advice and technical expertise, keep delegates informed), and to achieve coherence on policy positions in different multilateral for a (ibid: 1372). However, when looking at ETC group's website for example, the NGO clearly goes beyond that and claims that it influences public policy and institutional change. An example is given with the seed wars in the mid-1980s and 1990s, when ETC Group called attention on what they called the 'terminator technology' (Kloppenburg, 2004: 319), or on the then new concept of biopiracy (Lenné and Wood; Brush, 2004: 154). ETC claims that their action has led to revoke patents on crop species and human tissues.[36]

NGOs' strategy and mixed influential outcome

Agenda 21 (chapter 27)[37] recognizes that NGOs hold well-established and diverse experience, expertise and capacity, and that they offer a global network that should be exploited, further enabled and strengthened in order to support efforts in achieving environmentally sound and socially responsible sustainable development. NGOs interact in formal and informal settings where they exchange information and share knowledge, ideas and expertise. They act as facilitators by building trust. They try to increase coordination and interaction between NGOs and developing countries (Matthews, 2007: 1372–1373). Indeed, during the Treaty negotiations, the organization of a series of informal meetings – called the Keystone International Dialogue Series on Plant Genetic

Resources and the Crucible Group – allowed NGOs and many other actors to meet in a neutral setting in order to openly discuss their views and interests. These discussions were effective in allowing these actors to seek consensual solutions to a range of critical issues that were then transmitted to the negotiating fora. The process created bonds of cooperation and sometimes, comradeship that have held up over the years. It did not really cause people to change positions so much, but to at least be able to understand one another's positions and find common ground where common ground was occasionally useful (Mooney, 2011: 142; Frison et al., 2011a: 8). However, unlike during the IU period, it has to be recognized that during the first decade of implementation of the Treaty, NGOs lacked influential participation and negotiating power within Governing Body meetings (Mooney, 2011).

This state of affairs contrasts with the commonly admitted view that today a larger participation of civil society in international policy negotiations has the capacity to increase efficiency, impartiality, transparency and democracy in the global policy-making process (Bevilacqua and Duncan, 2010). Indeed, within FAO, this progressive position has been officially acknowledged through the reform of the Committee on World Food Security (CFS), qualified by Olivier De Schutter (2013: 4, also see 2009) as being the 'single most significant development in the area of global food security in recent years'. The new CFS aims at being

> the foremost inclusive international and intergovernmental platform for a broad range of committed stakeholders to work together in a coordinated manner and in support of country-led processes towards the elimination of hunger and ensuring food security and nutrition for all human beings.
> (CFS, 2009)

The CFS does so by creating a forum where different international stakeholders including governments, but also international organizations and transnational networks of civil society organizations 'can work together to ensure that their policies converge, rather than undermining each other's efforts' (De Schutter, 2013: 16). Perhaps this recent progress in FAO governance will little by little 'contaminate' the Treaty forum too and enable NGOs' voice to be better heard, for the benefit of all.

The Keystone International Dialogue Series and the Crucible Group

An ephemeral but important informal setting

Following the adoption of the CBD and before the conclusion of the Uruguay round of multilateral trade negotiations in 1993–1994, 28 individuals active in the field of PGRFA met three times in an informal way to discuss contentious issues related to the management of PGRFA: 'policy concerns related to intellectual property over biomaterials' (The Crucible Group, 1994: xi). This group, hosted by IDRC and coordinated by Bioversity International (at that time

IPGRI) was called the Crucible Group (I and II). The group was composed of people from public and private institutions, developed and developing countries, civil society organizations and the seed industry.[38] They met to continue the work done by the Keystone International Dialogue on PGR, and in particular finish the 'unfinished business' regarding the IP agenda (ibid). They held very different views on many contentious issues regarding the international agenda for genetic resources. However, they all agreed that decisions taken in international fora could imperil the availability of PGRFA for world food security and agricultural development. They believed that it was imperative to inform decision-makers on the reality of scientists, researchers, breeders, collection holders: i.e. all PGRFA users. The Crucible Group did not last after the adoption of the Treaty. However, a short explanation of its output is provided below because it played an important role during the Treaty negotiations as information provider.

A fruitful and useful open multi-stakeholder dialogue without consensus

Crucible Group members could not reach agreement on many issues (and indeed, the objective was not to reach agreement, but rather to express the differing views and positions to enlighten people on each other's interests). After several meetings and in-depth discussions, they published a book (the Crucible group, 1994; Halewood, 2000) where they identified 28 recommendations for policymakers, many of which related to IPR issues. Five years later, while it was not the initial intention to have a second round of discussions, most of the same persons and many others (up to 45 persons participated) decided to organize a Crucible II cycle because the developments in biotechnology and genetic resources-related policies called for further solutions. Two books where published as a result of Crucible II (The Crucible Group II, 2001).

What is important to note from this initiative, is its spontaneous, informal, multi-stakeholder and non-consensus modality (Halewood, 2000: 1). Based on the beliefs and goodwill of individuals, discussions were initiated, points of views were openly exchanged, opposing parties met in an attempt to understand each other and move forward for a common goal: safeguarding access to and conservation of PGRFA for world food security and poverty alleviation. This second informal dialogue setting (the first one being the Keystone Dialogues) has accompanied and fed the formal negotiation of PGRFA management.

A failed revival

In 2013, some stakeholders wanted to recreate a 'Crucible Group' type of dialogue through the organization of an informal multi-stakeholder dialogue, in order to progress on the difficult review process of the Treaty MLS and funding strategy (see above Chapter 4). While the initiative was officially welcomed at the Fifth Session of the Governing Body (Resolution 2/2013: point 7), there was little concrete subsequent support neither from the Treaty Secretariat nor from contracting parties, and this despite the strong interest expressed by the different stakeholders participating in a two-day workshop in September 2014 (IT/OWG-EFMLS-2/14/Inf.4.1).[39] The workshop was jointly organized by Bioversity International and the Meridian Institute (under the guidance of Michael Lesnick and Timothy Mealey who were both facilitators during the Keystone dialogue between 1988 and 1991). Unfortunately, not much has happened following this workshop and no other meeting is scheduled. Regrettably, stakeholders were not able this time to mobilize the resources to take benefit of such informal dialogue-type initiative.

Table 5.1 Synopsis of stakeholders' objectives, positions and strategies

Stakeholders	*Primary objective*	*Secondary Objective*	*Position on IPRs*	*Conservation strategy*	*Main technology*	*Exchange Networks*
Gene-rich & technology poor countries	Benefit-sharing	Access to seed & technology	No IPRs	*In situ* & *ex situ*	Conventional plant breeding	Mainly informal networks
Gene-poor & technology rich countries	Access	Protection of IPRs	IPRs	*Ex situ*	Conventional plant breeding & biotechnology	Formal networks
Seed industry big	Protection of IPRs	Access	Strong IPR (patents)	*Ex situ*	Biotechnology breeding	Own collections & Formal networks
Seed industry small & medium	Protection of IPRs	Access	Flexible IPR (PBRs)	*Ex situ*	Conventional plant breeding	Formal networks
Genebanks & GCDT	Conservation	Facilitated access	Flexible IPR + non IPR	*Ex situ*	Conservation technology	Formal networks
CGIAR	Facilitated access	Conservation & Sustainable Use	Flexible IPR + non IPR	*Ex situ* & *in situ*	Conventional plant breeding; biotechnology; various conservation technologies & participatory plant breeding	Formal networks & sharing with farmers
Farmers & farmers' organizations	FRs	Access & Sustainable Use	No IPRs	*In situ*	Farmers' selection & participatory plant breeding	Informal networks

The above description of Treaty stakeholders (see Table 5.1) provides a concise view of who these actors are, as well as how and why they are involved in the Treaty. To clarify at a glance their divergent interests, a summary table is provided below, stating stakeholders' objectives with the Treaty, their positions on IPRs, their conservation strategy and the main technology they apply when using PGRFA, as well as their favourite exchange networks to access seeds. The objective of this table is to show (in a rather synthetic manner) the divergences in positions. This oversimplification is necessary to help understand why stakeholders have difficulties in implementing the Treaty and reaching its objectives in a speedy and collaborative way. It is contended that the reality in their positions is much more nuanced.

List of Treaty constraints identified by stakeholders

Following the presentation of Treaty stakeholders and of their various interests and positions, I will now summarize the challenges and constrains faced by Treaty stakeholders in their implementation of the Treaty. The list of 17 constraints identified by stakeholders originates from the edited book, where a full analysis can be found (Frison et al., 2011b: 257–280). To provide a rapid and concise summary of the views expressed, the identified constraints have been listed in a simple manner under the table 5.2 below. It is noteworthy to mention that most of these issues have been confirmed as needing solutions in the Treaty review process which began in 2013 at the Fifth Session of the Governing Body. Furthermore, while recognizing that many actions could be taken at the local, national or regional level, the list of actions and their related needs remain at the Governing Body level, as the scope of the present work is limited to the international law level.

The above list of constraints is based on the stakeholders' experience with the implementation of the Treaty. They are very diverse and deal with scientific, technical, legal, political and/or economic aspects of the Treaty implementation. Many constraints cover several of these aspects at the same time. This makes it even harder to tackle them. In an attempt to facilitate the analysis of all the constraints identified, they have been categorized into four types of associated needs: the need for more clarity; the need for review and update; the need for further development; and the need for more coherence and coordination. This classification should help in identifying what type of legal procedures could respond to these needs (i.e. amending the Treaty, adopting agreed interpretations, adopting codes of conducts, etc.). This categorization is not exhaustive but allows to raise the following four questions (Frison et al., 2011b: 268–270).

Is there a need for clarification of Treaty provisions?

Many stakeholders have indicated the need to clarify various Treaty provisions to guide its implementation, in particular, of Article 11.2 and Article 12.3(d).

Table 5.2 List of constraints identified by Treaty stakeholders

Treaty Part		*Specific implementation challenges and constraints identified by stakeholders*	*Needs related to the constraint*			
			Clarification	*Further Development*	*Review/ Update*	*Coordination/ Coherence*
Part I ***General Constraints***	1	Policy coherence between the ITPGRFA & other conventions (CBD, TRIPS, UPOV)		X	X	X
	2	Public awareness & capacities at the national level		X	X	X
	3	Trust between contracting parties	X	X		X
	4	Clarity of Treaty provisions (e.g. Articles 6, 9, 11.2 and 12.3(d))	X	X		X
Part II ***Conservation & Sustainable Use of PGRFA***	5	Financial, technical & scientific limits		X		X
	6	Weak implementation of *in-situ* conservation obligation		X		
Part III ***Farmers' Rights***	7	Recognition & National implementation	X	X		X
	8	Participation of Farmers' organizations		X		X
Part IV ***The Multilateral System of Access and Benefit-sharing***	9	Modification of Annex I list to face new challenges		X	X	
	10	Limitations in access to PGRFA	X	X	X	
	11	Notification of PGRFA Inclusion in MLS	X			X
	12	Limited realization of benefit-sharing		X	X	X
	13	Limited realization of non-monetary benefit-sharing		X		X
Part V ***Supporting Components***	14	Limited implementation of the GPA		X	X	X
	15	Little use of existing formal & informal networks		X		X
	16	Limited implementation of the GLIS		X		X
Part VI ***Financial Provisions***	17	Limited and unpredictable funding		X	X	X

A clear example of ambiguity concerns the scope of the following expression 'PGRFA [. . .] that are under the management and control of the contracting parties and in the public domain' (Article 11.2). The need to provide guidance in the interpretation of this and other ambiguities related to the MLS has led to the establishment by the Governing Body of an Ad Hoc Technical Advisory Committee. It aims to provide *inter alia* some guidance regarding the identification of plant genetic resources for food and agriculture under the control and management of contracting parties and in the public domain (IT/AC-SMTA-MLS1/10/4; see also Chapter 4 above). Another example relates to Article 12.3(d), where different interpretations can be given to the terms 'parts and components' and 'in the form received' and therefore to the definition of the material that can be protected by IPRs or not.

Is there a need for further development of Treaty mechanisms and strategies?

Stakeholders have pleaded for rapid action by the Governing Body to develop further mechanisms and strategies in various aspects. A major example concerns the non-monetary benefit-sharing obligations (Article 13.2(a), (b) and (c)), which is poorly implemented according to many authors. A second example relates to the need to further develop financial mechanisms helping countries to implement the GPA priority activities, and especially the priorities that are not directly covered by funds already established (the GCDT or the Benefit-sharing Fund). Indeed, many priorities of the GPA do not foresee an appropriate and specific financing mechanism to implement them yet. This could perhaps be done by taking advantage of the experience of other existing fundraising mechanisms and funding organizations active in the agricultural sector (such as the Global Environmental Facility (GEF), the United Nations Development Programme (UNDP) or the International Fund for Agricultural Development (IFAD)). A third example where further development is needed relates to the Global Information System. At the last Governing Body meeting, progress was made on this question (see Chapter 4 above).

Is there a need for review or update of Treaty mechanisms and strategies?

The text of the Treaty and its implementation mechanisms and strategies request, in certain cases, such review and update processes. Several examples can be mentioned, such as the review by the Governing Body of the levels of payment in the SMTA (Article 13.2(d)(ii)); or the periodic establishment of a funding target (Article 18.3). Another example concerns Article 17.3 of the Treaty, which requires contracting parties to collaborate with the CGRFA to periodically reassess the state of the World's plant genetic resources for food and agriculture in order to update the GPA. A last example is constituted by the priorities set for the Benefit-sharing Fund, where Annex 1 of the Funding Strategy sets out eligibility, selection criteria and additional requirements that can be updated regularly by the Governing Body. In addition, the editors further consider the possibility to modify and review Treaty mechanisms and strategies in reaction to external circumstances, which were not foreseen at the moment of the Treaty negotiations and which may have a substantial impact on its implementation. A good example would be the updating of Annex I list as a consequence of external factors. Indeed, the list of crops and forages was negotiated according to the double criteria of interdependency and food security, which are currently being affected by climate change and technological developments.

Is there a need for a stronger coordination and coherence to facilitate the Treaty implementation?

Many stakeholders have stressed the limited coordination and coherence at three levels, resulting sometimes in numerous competing and/or conflicting international obligations: (1) between governing bodies and secretariats of international institutions; (2) between national representatives attending different but related international fora such as the WTO, the CBD and the ITPGRFA; and (3) between different sectors and people at the national level responsible for the implementation of these different international obligations. (1) At the secretariat and governing body level, periodic meetings between Secretaries and joint meetings between Governing Bodies of different international organizations could be two options leading to the development of common programmes and activities, mitigating the limited coordination and coherence problems. At GB 5 and 6, such impulse to collaborate further with the CBD, UPOV and WIPO has been provided by the Governing Body and the Treaty secretariat. An example of inter-sectorial cooperation in the negotiating process is provided by the mutual recognition and support between the Treaty and the Nagoya Protocol on Access and Benefit-Sharing. The Protocol expressly refers to the Treaty as a complementary instrument of the international ABS regime (CBD, COP Decision X/1: §§ 6 and 11 of the preamble). (2) At the national delegation level, common preparatory meetings and inter-sectorial composition of delegations could be envisaged to prepare for international meetings. (3) At the national level, coordination by national inter-sectorial committees could contribute to improve coherence and coordination when implementing international obligations at the national level.

Conclusion

Already in 2011, stakeholders had pointed at the risk that the lack of appropriate and quick decisions and actions to speed up the implementation process might lead to a 'decreased level of confidence in the general framework set up by the Treaty' (Frison et al., 2011b: 257). The legal analysis of the Treaty in Chapter 4 has confirmed this tendency and shown that the implementation process requires a review of the MLS and of the Funding Strategy, which is being undertaken by the Governing Body. Specific conceptual constraints have been highlighted throughout the legal analysis and the Stakeholders' analysis comes to the same results with the list of 17 identified Treaty constraints and limitations.

The further analysis of these constraints, as well as the guidance of Treaty stakeholders – some of which had named the MLS a 'global crop commons' – have led to the recognition of a link between these identified limitations and the concepts and principles deriving from the theory of the commons: stakeholder participation in governing collectively a common resource; rights of access to, use and management of a common resource; sustainability; equitable use of the resources and benefits deriving from their collective management; the importance of boundaries and of the community; etc. Therefore, Chapter 6 will explore the application of the commons theory to the PGRFA field. The aim is to provide recommendations on what could be done to mitigate the identified constraints and allow stakeholders to reach the Treaty's objectives and how it could be done, using the theory of the commons as guidance.

Notes

1 In the present work, actors and stakeholders are used interchangeably.

2 Authors had a great liberty in the content of their chapter. Identical very general guidelines were provided to all, without further requirements to keep their voice as autonomous, free and objective as possible. The guidelines requested the authors to talk about the past, present and future of the Treaty, and to identify what difficulties and constraints they were facing in their experience with the implementation of the Treaty.

3 The food processing industry was also approached to provide input, but the editors did not receive a positive response from the several persons approached.

4 The book can be downloaded freely at www.bioversityinternational.org/e-library/publications/detail/plant-genetic-resources-and-food-security/

5 Though the 'Earth Negotiation Bulletin' does retrace the day by day development of the negotiation for each GB meeting, and often refers to positions of X or Y. However, it is not an official and public verbatim tracking. See http://enb.iisd.org/enb/

6 As of June 2011, 173 countries and the European Union are Members of the Commission. Membership is open to all FAO Members and Associate Members, upon request.

7 The EU group has a specific position as it is the only regional group which is also a party to the Treaty.

8 This group was founded on 15 June 1964 by the 'Joint Declaration of the Seventy-Seven Countries' issued at the United Nations Conference on Trade and Development (UNCTAD). It integrates 131 developing countries. In the G-77 there are countries from the following regions: Latin America and Caribbean, Africa, Asia, Near East and Pacific.

9 The OECD was created in 1960 with 31 developed countries. Country members of OECD are from North America, Europe, Latin America and Asia and the Pacific.

10 This group of countries is active within the CBD forum. The initial member countries of the Like-Minded Group of Mega-diverse Countries were Brazil, China, Colombia, Costa Rica, Ecuador, India, Indonesia, Kenya, Mexico, Peru, South Africa and Venezuela. Since then, Bolivia, the Democratic Republic of the Congo, Madagascar, Malaysia and the Philippines have also joined.
11 Founding Resolution of the Consultative Group for International Agricultural Research, Washington DC, May 1971, available at www.worldbank.org/html/cgiar/publications/founding.html.
12 www.cgiar.org/
13 GPG project webpage of the CGIAR (accessed May 2015), which does not seem to be accessible anymore in July 2018, www.sgrp.cgiar.org/?q=node/158. See also its continuation project: the Genebank Knowledge Base, an initiative of the System-wide Genetic Resources Programme (SGRP) of the Consultative Group on International Agricultural Research (CGIAR). It was developed as part of the World Bank funded project *Collective Action for the Rehabilitation of Global Public Goods in the CGIAR Genetic Resources System, Phase 2 (GPG 2)*, available at https://cropgenebank.sgrp.cgiar.org/index.php/home-mainmenu-70/about-this-site (accessed August 2018).
14 These include *inter alia* the System Wide Programme on Property Rights and Collective Action (CAPRi), which is one of several intercenter initiatives of the Consultative Group on International Agricultural Research (CGIAR) created to foster research and collaboration among the CGIAR and national agricultural research institutes on the institutional aspects of natural resource management. CAPRi contributes to policies and practices that reduce rural poverty by analysing and disseminating knowledge on the ways that collective action and property rights institutions influence the efficiency, equity and sustainability of natural resource use. More information at www.capri.cgiar.org.
15 Signed in October 1994, the in-trust agreement state that the CGIAR agree to hold designated germplasm in trust for the international community under the auspices of FAO, thereby making this material freely available for research and crop improvement purposes.
16 Bioversity International is based in Rome, Italy. It was formerly named the International Plant Genetic Resource Institute (IPGRI) up to 2004 and the International Board for Plant Genetic Resources (IBPGR) from its creation in 1974 up to 1992. www.bioversity international.org/
17 Establishment Agreement and Constitution of the GCDT. Available at www.croptrust.org/main/governance.php?itemid=5.
18 For more information, see Executive Board of the Global Crop Diversity Trust: Policy and Governance, www.croptrust.org/about-us/governance-policy/executive-board/.
19 Global Crop Diversity Trust Pledges available at www.croptrust.org/.
20 The Svalbard Global Seed Vault was built as a last resort safety back-up repository, which is now freely available to the world community for the long-term storage of duplicate seed samples. More information at http://nordgen.org/sgsv/ and www.croptrust.org/main/arcticseedvault.php?itemid=211.
21 See World Information and Early Warning System (WIEWS) on Plant Genetic Resources for Food and Agriculture, website at http://apps3.fao.org/wiews/wiews.jsp.
22 With Monsanto being the world leader in seeds and genetically engineered traits and Syngenta in insecticides, fungicides and herbicides, the merger would have created an agricultural behemoth with the largest market share in the world in both seeds and agricultural chemicals. Available at www.nytimes.com/2015/08/27/business/dealbook/monsanto-abandons-47-billion-takeover-bid-for-syngenta.html?_r=2.
23 According to Plantum, the Dutch association for the plant reproduction material sector, Dutch plant reproductive materials sector in 2014 consists of around 300 specialized breeding and propagation companies. See its website available at www.plantum.nl/321519619/Basis-for-the-Green-Economy.

24 For an extensive historical and socio-technological account of plant improvement mechanisms and innovation contexts, and an illustration of the impact of enclosure mechanisms on plant improvement stakeholders, see Batur (2014).
25 More information at www.worldseed.org/isf/home.html, International Seed Federation website – Representing the Seed Industry at a Global Level.
26 ESA sends strong signal of support to the ITPGRFA, www.euroseeds.eu/esa-sends-strong-signal-support-itpgrfa; www.euroseeds.eu/esa-hands-300000%E2%82%AC-voluntary-financial-contribution-fao-it.
27 IFAD is a specialized agency of the United Nations. It was established as an international financial institution in 1977 as one of the major outcomes of the 1974 World Food Conference. The Conference was organized in response to the food crises of the early 1970s that primarily affected the Sahelian countries of Africa. For more information, see www.ifad.org/governance/index.htm.
28 The movement was born in 1993 in Mons, Belgium, from a group of farmers' representatives from the four continents. La Via Campesina comprises about 150 local and national organizations in 70 countries from Africa, Asia, Europe and the Americas, representing altogether about 200 million farmers. It defines itself as an autonomous, pluralist and multicultural movement, independent from any political, economic or other type of affiliation. More information at http://viacampesina.org/en/.
29 For an account on their strategy of using Human Rights to support their claim, see Claeys (2012, 2014b).
30 Batta-Bjørnstad (2004: 30–31) argues that '[i]n formal plant breeding, the strategy is the opposite. The breeding-and seed industry cannot economically handle a great number of varieties. The breeders therefore have to opt for stability in order to produce varieties, which can be used by as many farmers as possible. The strive for stability is also based on the desire for plant variety protection, which requires stability as one of the conditions for receiving such protection. Thus, the informal system of crop development produces genetically diverse farmers' varieties (traditional varieties), while the formal system contributes genetically homogeneous cultivars (high-yielding varieties)'.
31 The CBD Alliance is a loose network of activists and representatives from NGOs, community-based organizations, social movements and Indigenous Peoples organizations advocating for improved and informed participation in CBD processes. The Alliance is premised on the belief that global policy-making should be a transparent and democratic undertaking. It aims to increase the informed and effective participation of Southern NGOs, Indigenous Peoples, Community Based Organizations and social movements. The Alliance exists to help them be more effective in their CBD-related advocacy by facilitating communication among Civil Society representatives and other organizations, Parties to the Convention, media and the CBD Secretariat – to change and ultimately improve biodiversity-related policy at international, national and community levels. More information at www.cbdalliance.org/.
32 FAO Constitution Article 3.5 provides that the 'Conference may invite any public international organization which has responsibilities related to those of the Organization to appoint a representative who shall participate in its meetings on the conditions prescribed by the Conference. No such representative shall have the right to vote'. Available at www.fao.org/docrep/x5584e/x5584e0i.htm. The basic texts of FAO also rules on the matter in its sections L, M and N, Resolution. 39/57 and 44/57; available at www.fao.org/3/a-mp046e.pdf.
33 FAO Forestry Department, Glossary and Acronyms, FAO corporate document repository, available at www.fao.org/docrep/X5327e/x5327e03.htm.
34 No patents on seeds website, https://no-patents-on-seeds.org/en/about-us/home
35 More information at SEARICE website, www.searice.org.ph/
36 See 'ETC Group: A Brief History', www.etcgroup.org/content/etc-group-brief-history

37 Agenda 21, Chapter 27 'Strengthening The Role of Non-governmental Organizations: Partners for Sustainable Development', www.un.org/esa/dsd/agenda21/

38 In September 2014, a similar attempt to form a 'Crucible Group type' of meeting took place through the Informal Multi-stakeholder Dialogue workshop jointly convened by Bioversity International and the Meridian Institute. See Chapter 3.

39 Prior preparatory meetings and/or conference calls took place between December 2013 and September 2014, but no official documents are openly accessible.

6 Inspiring an effective Plant Treaty with the 'theory of the commons'

The twentieth century was marked by worldwide genetic resource erosion, in reaction to which the international community (in particular countries from the North) developed large ex situ conservation policies. These policies culminated with the creation of significant national and international genebanks and research centres (such as those of the Consultative Group on International Agricultural Research – CGIAR), as observed in Chapter 2 of the present book. At that time (between 1900 and 1970/80s), the state and public institutions were urged to take action to conserve biodiversity by scientists that were alarmed by the state of biodiversity loss worldwide. Since the 1960s and mostly from the 1990s onwards, another actor was perceived to be the most apt to regulate biodiversity management: the market. The science (modern breeding and biotechnology) and the law (IPRs and international trade laws) were used as enabling tools for the market to dominate biodiversity management. While some actors – mainly FAO, Bioversity International and the CGIAR – advocated the need to conserve genetic diversity in trust for the benefit of humanity,[1] the general commodification of genetic resources trend was already well settled, and all the international agreements adopted to protect biodiversity were entangled within these market beliefs (see chapters 2 and 3). Indeed, the Convention on Biological Diversity formalized the objectification of biodiversity as mere economic resources, the use of which could generate benefits, but also consecrated the market as the most appropriate regulating instrument for reaching biodiversity conservation and sustainable use objectives.

While in the 1980s, the FAO Commission on Plant Genetic Resources had attempted to establish an 'in trust for humanity' status for the most important plants feeding the world,[2] the negotiations of the International Undertaking on Plant Genetic Resources soon turned out to a non-legally binding instrument, unable to counterbalance the general appropriation of biodiversity trend. The idea that seeds should be governed differently through an international legally binding instrument, and most of all collectively, germinated in some minds (Cooper et al., 1994). As a reaction to the commodification scheme and to resolve the important tensions explained in Chapter 3 of the present work, the Multilateral System of access and benefit-sharing of the Plant Treaty was developed, thereby definitely recognizing the special status of the most important plants feeding the world (Falcon and Fowler, 2002). Since then, the MLS is considered to function as a Global Commons (Halewood et al., 2013; Dedeurwaerdere, 2013; Halewood, 2013), thereby implicitly endorsing a collective governance policy for PGRFA. Yet, Chapters 4 and 5 of the present book demonstrate that, because of the hyperownership views that has dominated the negotiation of the Treaty, as of 2016, the Treaty implementation process is entangled in significant problems. These constraints clearly show that many contracting parties have difficulties implementing the Treaty and that specific stakeholders' voices and needs (i.e. farmers) are neither heard nor met.

The hypothesis underlying the present chapter postulates that picturing the MLS as a 'true' global commons, with intrinsic consequences on its governance, may contribute to a more efficient implementation of the Treaty and to better reaching the Treaty's overall goals of food security and sustainable agriculture, similarly as what authors have postulated for governance mechanisms of microbial resources (Dedeurwaerdere et al., 2009; Halewood, 2010). Hence, considering seeds as common goods constitutes an alternative way to overcome the problematic dichotomy that appeared above in the Treaty analysis between seeds defined exclusively as private goods and seeds characterized as public goods. According to Ostrom (1990), common-pool resources institutions are an answer to manage resources in a sustainable manner, to avoid depletion and to allow communities to live perpetually from a resource (fishing; irrigation, etc.). Local and collective self-organization by community users is seen as a third and often overlooked possibility for effectively managing a resource sustainably, compared to market regulation or state intervention (Ostrom et al., 2007).

Considering this hypothesis, there are many questions that arise in order to build further the last step of the inductive research carried out in the present work: how would the concept of commons better allow for reaching the Treaty's direct objectives and overall goals? Can the current review process of the Treaty resolve all the PGRFA management constraints listed at the end of Chapter 6? Can one instrument answer all the different needs and expectations of all the different stakeholders? Would the theory of the commons enable the Treaty to address the central issue of heterogeneity of resources (landraces/breeding lines/genomic resources), heterogeneity of stakeholders (small farmers/breeders/Big Five/consumers/ funders), heterogeneity of uses (self-consumption/local market/global trade market/speculation and financing-investment) and heterogeneity of contexts (various development perspectives) (Batur, 2014; Batur and Dedeurwaerdere, 2014)? How could 'the law' and its necessary transformation solve the identified issues? Would the commons theory allow the Treaty to intrinsically respect the interdependence criterion, which is fundamental for the seed governing system to realize the necessary balance, equilibrium and resilience between the above-mentioned aspects? These are the questions to be explored when relating the research results of the legal and stakeholders' analysis to the theoretical framework of the commons.

This final chapter is divided into four sections, two of which have already been published in a previous book[3] (Girard and Frison, 2018a). First, I briefly explain the theory of the commons, from Hardin to Ostrom including the 'new vogue' with Mattei, Dardot and Laval, as well as others. Second, I show how this theory is a pertinent theoretical framework in resolving the Treaty's dysfunction. Third, I do this by focusing on six invariable principles pertinent both for the Treaty and the theory of the commons. Finally, I analyse the eight identified Treaty topics in light of the theory of the commons and explore how a deeper 'commonization' might contribute to enhance the Treaty's implementation. This final section aims at transforming the current MLS into a 'real global seed commons', by making normative proposals for possible solutions to boost reaching the Treaty's objectives.

A brief history of the theory of the commons

The concept of *Commons* is not very well defined and has grown fuzzier with globalization and the complexification of wider resource governing systems (Berge and van Laerhoven, 2011). To provide some clarity on the concept as addressed in the present work, this section describes the evolution of the theory of the commons from the 1960s to nowadays. As mentioned above, this is not a political sciences research; that is to say that the use of the theory of the commons will be selective and partial for serving specific purposes. Therefore, this section will be limited in two ways. First, explanations will focus on the recent developments of the theory, from the 1960s onwards and will not cover historical commons since the middle ages (Bravo and De Moor, 2008; De Moor, 2008, 2009, 2011). Second, regarding the scope and definition of the commons, the focus will expand on what has been described by Ostrom as common-pool resources (at any scale: local, regional or global, see *inter alia* Blomquist and Ostrom, 1985: 383). It will not cover the debate on what part of the international law literature calls 'global commons'[4] and which relates to the concept of 'common heritage of (hu)mankind', i.e. the high seas and deep seabed, Antarctica, outer space and the global atmosphere (Buck, 1998; Vogler, 2012; Ashley Roach, 2012), but remain attached to the notion of resource (whether physical i.e. the seed, or informational, i.e. the knowledge attached to the seed).

The purpose here is not to be exhaustive on the prolific literature on the commons and on the so many definitions and strings of the commons theory, but rather to highlight some key phases in the recent history of the theory. After briefly explaining the revival of the concept of commons following Hardin's publication in the 1960s, the important contribution of Ostrom will be summarized and complemented with recent developments, what I call the 'new vogue' of commons' scholars.

Hardin's 'tragedy of the commons'

In 1968, Garrett Hardin published an over-exploited allegory named the 'Tragedy of the Commons', where he analysed problems related to the over-exploitation of finite resources under unlimited and free access conditions to all, in the context of an ever-growing world population. He took the example of grazing and posed the precondition that rational people would always try to get the maximum and immediate profit from their individual use of what he calls a 'common resource' (Hardin confuses common goods with good in open access),[5] and therefore lead to overgrazing and the destruction of the common pasture. Hardin states that

> [e]ach man is locked into a system that compels him to increase his herd without limit – in a world that is limited. Ruin is the destination toward which all men rush, each pursuing his own best interest in a society that believes in the freedom of the commons. Freedom of the commons brings the ruin to all.
>
> (Hardin, 1968: 1244)

This view is supported by Mancur Olson (1965: 2) in his work on the logic of collective action, who states that 'unless the number of individuals is quite small, or unless there is coercion or some other special device to make individuals act in their common interest, *rational, self-interested individuals will not act to achieve their common or group interest*' (emphasis in original text). Although Olson was much more cautious than Hardin in the proposed solution to the 'tragedy', leaving the question of common management open, he suggests that common goods – i.e. goods that are not under exclusive property rights according to him – are due to be over-consumed, for men are unable to rationally exploit such common resource.

Hardin proposed three options to mitigate his tragedy: reducing world population to avoid overconsumption (which, as a Malthusian, is the solution he is strongly pleading for in his article, see Locher, 2013, 2016); or establishing an external institution to regulate the use of the resource, whether through public management (state) or through the market (i.e. enclose the commons using property rights) (Hardin, 1968: 1245). At that time, the dominant liberal economic trend was strongly pushing for private property rights to be established as the warrant for an optimal free market, avoiding free-riding and internalizing the 'social cost' (Coase, 1960; Demsetz, 1967; Alchion and Demsetz, 1973). Hardin's paper was then used as justification to promote the reinforcement of exclusive property rights and to develop intellectual property rights in many new fields, including over living organisms, as initiated by the US Supreme Court *Diamond v. Chakrabarty* decision in 1980a.[6]

While Hardin's work has received significant diffusion (his paper has been cited almost 30,000 times according to Google Scholar), his work is now viewed quite differently by many academics from different disciplines. Criticism has bloomed on different aspects of Hardin's tale, notably on his 'explanation for the need to enclose the commons [which] confounded the resource with its governance regime' (Berge and van Laerhoven, 2011: 161), or on the fact that Hardin mixed up open access with community management, as well as common property rights with absence of property rights (Ciriacy-Wantrup and Bishop, 1975; Bollier, 2007; Bollier and Helfrich, 2014).

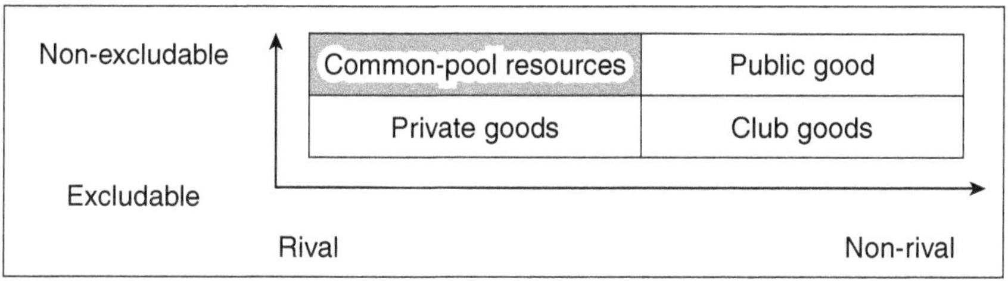

Figure 6.1 Quadrant of the economic classification of goods

To understand this, one need to go back to the economic classification of goods (see Figure 6.1), according to which goods are subject to two characteristics – rivalry and excludability – as shown in the following quadrant. Under this definition, common-pool resources are goods that are rival but hardly excludable, such as the grass in the grazing 'commons'.

In his article, Hardin confounded the good (public good versus CPR), the property rights attached to the good (private, public or collective rights) and its governing regime (open access versus commons) (see the definitions of these terms in Cornu et al., 2017). As a matter of fact, in reality, it is not the good and its rival and/or excludable characteristics that determine its mode of governance but rather the institutional setting designed by the user-community, as a collective, long-during, and evolving action in response to specific needs.

In addition, Hardin considers human being as objectively rational people, i.e. excluding the influence of other values and ethical aspects in community behaviours. Hence, according to Sen, who worked on welfare economics, people's values and commitments influence economic policies in terms of their effects on the well-being of the community (Sen, 2003: 15, 40; see also Sen, 1981). Other criticism can be formulated against Hardin's views, including the fact that in real life, people communicate and are rarely put in a situation where a common resource is used by different person who do not talk to each other and discuss how to manage the resource commonly (Ostrom, 1990: 7).

Ostrom's institutional analysis of common-pool resources

In response to the presumed supremacy of property rights (whether by the state or the market) as being the optimal system to manage resources, Elinor Ostrom spent her career studying the collective management of common resources (Blomquist and Ostrom, 1985; Ostrom, 1987, 1990, 2005, 2007, 2010a; Schlager and Ostrom, 1992; Ostrom et al., 1999; Hess and Ostrom, 2003, 2006; Poteete et al., 2010; Vollan and Ostrom, 2010).[7] To get a deeper comprehension of the conditions for sustainable resource use and governance regimes, she analysed Common-Pool Resource (CPR) institutional arrangements based on extensive field studies. In her famous book *Governing the Commons* Ostrom conducted wide meta-analysis of existing common-pool resources case studies mainly in agricultural production systems, e.g. irrigation, forestry, fishery management systems, etc. Building on the dominant liberal system and from the usual economic classification of goods (based on the notions of rivalry and excludability), in her understanding a commons is any natural or man-made resource that is or could be held and used in common. By focusing on the institutionalization of managing regimes, Ostrom showed that stakeholders can effectively set up rules together (i.e. self-organization) to manage sustainably and efficiently resources established in a local common-pool for their own use, and outside of the market or governmental intervention (i.e. self-governance). Relating practical problems to theoretical thinking following an inductive research path, Ostrom (1990: 90 table 3.1) used these data and her observations to identify recurring principles useful to govern long-term efficient CPR systems:

1 Clearly defined boundaries (i.e. effective exclusion of external unentitled parties);
2 Congruence between appropriation and provision rules and local conditions;
3 Collective-choice arrangements (i.e. allow most resource appropriators to participate in and modify the operational rules);
4 Effective monitoring (by monitors who are part of or accountable to the appropriators);
5 Graduated sanctions (scale of sanctions for appropriators violating community rules);
6 Conflict-resolution mechanisms (cheap and of easy access);
7 Minimal recognition of rights to organize (the self-determination of the community is recognized by higher-level/governmental authorities);

Plus, for CPRs that are parts of larger systems

8 Nested enterprises (organization in the form of multiple layers of nested enterprises, with small local CPRs at the base level).

These design principles are very helpful (but not compulsory; see Ostrom and Delville, 2009: 8, 13). Actually, Ostrom leaves much space for heterogeneity and diversity in systems and places, insisting on the fact that the institutional arrangement should always be adapted to local needs and conditions in order to be efficient (which implies that other design principles may be better adapted to different situations) (Ostrom, 2009; also see Cox et al., 2010).

Following her seminal book, Ostrom refined her approach further by defining the term common property resource in a paper written with her colleague Schlager, in which they integrated a 'bundle of rights' approach.[8] Schlager and Ostrom (1992: 259) propose a scale of property rights, in which authorized users, claimants, proprietor and owners exercise different types of de jure or de facto rights (right to access and withdraw; right to manage; right to exclude and right to alienate a resource). They distinguish between operational-level property rights and collective-choice property rights (see below in the 'participation and governance' paragraph at the end of this chapter). According to them:

> [o]perational activities are constrained and made predictable by operational-level rules regardless of the source of these rules. By the term "rules" we refer to generally agreed upon and enforced prescriptions that require, forbid, or permit specific actions for more than a single individual. [. . .] Operational rules are changed by collective-choice actions. Such actions are undertaken within a set of collective-choice rules that specify who may participate in changing operational rules and the level of agreement required for their change. In regard to common-pool resources, the most relevant operational-level property rights are "access" and "withdrawal" rights and the collective-choice property rights include management, exclusion, and alienation.
>
> (Schlager and Ostrom, 1992: 250–251)

This distinction is crucial, as it expresses the 'difference between exercising a right and participating in the definition of future rights to be exercised' (ibid). Hence, according to Schlager and Ostrom (1992: 259), '[a]ssigning full ownership rights does not guarantee an avoidance of resource degradation and overinvestment' (see also Larson and Bromley, 1990; Clark, 1973, 1974; Ginkel, 1989). What is essential is the institutionalizing process between users of a common resource to manage collectively and sustainably their commons. In concluding their publication, Schlager and Ostrom (1992: 260) call for further research to be conducted in order to understand the functioning of various efficient property rights regimes and address the following three issues: 1) the conditions to enhance or detract from the emergence of more efficient property rights regimes; 2) the stability or instability of these systems when challenged by various exogenous or endogenous changes; and 3) the costs of enforcing regulations that are not agreed upon by those involved.

Many academics have followed Ostrom's path since then. The International Association for the Study of the Commons was created,[9] along with the International Journal of the Commons,[10] thereby contributing to the dissemination of Ostrom's innovative and interdisciplinary thinking. The commons' research has spread outside the thematic scope of physical agricultural resources, as an answer to what Boyle (2003b) has identified as the 'second enclosure movement'. Works in other fields have developed such as the study of complex socio-ecological systems (Berkes et al., 2008), the internet and knowledge in general (Hess and Ostrom, 2003, 2007), microbial commons (Hess and Ostrom, 2006; Dawyndt et al., 2006; Dedeurwaerdere, 2006, 2010; Dedeurwaerdere et al. 2009), the science commons (Cook-Deegan, 2006; Dedeurwaerdere, 2008), global environmental commons (Brousseau et al., 2012), micro-psychological foundations for governance regimes (Kahneman and Tversky, 2000; Gintis et al., 2005), the role of trust in communications between members of the governing community

(Six et al., 2015) in particular in a computer-mediated forms of communication setting (Kiesler et al., 1984; Bente et al., 2008), etc. (the list here could be much longer).

After 2008: the 'new vogue' of the commons theory

Since Ostrom, and in particular since the global financial crisis of 2007–2008 (Quinn, 2009; Stiglitz, 2010),[11] the theory of the commons has bloomed into prolific theoretical and practical developments, as a response to major difficulties in managing sustainably our societies and ecosystems.

Regarding the theoretical moves, scholars from various disciplines have built on Ostrom's vision to provide answers to current societal challenges in many different fields: economics (Coriat, 2013, 2015), law (Capra and Mattei, 2015; Mattei, 2011a, 2011b; Lucarelli, 2011a, 2011b, 2013; Dani, 2014), socio-philosophy (Dardot and Laval, 2014; Gutwirth and Stengers, 2016). They respond to specific issues which Ostrom's work has only partially covered. These scholars form what one may call a 'New Vogue'[12] and go beyond Ostrom's legacy by questioning governing systems for specific resources/regimes which are conceived by some as being 'essential resources' for humanity (Pistor and De Schutter, 2015). Building on Ostrom's solid conceptual basis, academics have pointed to several issues to be further developed or reconsidered, in particular following a 'political constitution of the common as an alternative generalizable rationality' (Dardot and Laval, 2014: 156, personal translation).

Going beyond Ostrom's influence of the economic approach to the 'good/resource'

Dardot and Laval pose the fact that, notwithstanding the crucial advancement and evolution of Ostrom's thinking over several decades, Ostrom remains within the ambit of the public/private good economic theory (Ostrom and Ostrom, 1977: 12, table of goods). They see this as a limit to the further development of 'a political constitution of the common' as 'alternative generalizable rationality' (Dardot and Laval, 2014: 156–157). Indeed, for Ostrom, the institutionalization of governing systems remains based on pooled physical or informational resources, i.e. the characteristic of the resource explains why people get together to institutionalize its governance regime (Verhaegen, 2015: 116–117; Dardot and Laval, 2014: 157). It also works within the boundaries of the established concept of property over a resource (ibid: 137–148). According to Capra and Mattei (2015: 12, 169–188), doing so may hinder the development of different types of institutional arrangements with a more 'holistic view', i.e. an 'ecological understanding of life' and a 'new kind of systemic thinking' towards reaching an 'ecolegal revolution' (ibid: 176). Prolonging Ostrom's thought, Bollier, considers that a commons is not just about the resource that is governed. Commons are

> paradigms that combine a distinct community with a set of social practices, values and norms that are used to manage a resource'; it is '*a resource* +

> *a community + a set of social protocols*. The three are an integrated, interdependent whole.
>
> (original emphasis, Bollier, 2014: 15)

A note is made regarding the fact that contrary to Dardot and Laval who see 'le commun' with a political/ideological perspective, Bollier (2014: 34) rather thinks of 'commoning' as a 'vernacular movement', that is to say 'shared spaces of a community in which people assert their collective moral values and political interests, over and above, those of the state, the corporation and other institutional powers'. This understanding sheds the light not on the resource or the role of individuals forming the community in designing rules, but on the relation, the interactions between 'resources' and people. As recognized by Hess and Ostrom (2007: 13) and developed by the literature on the knowledge commons (Madison et al., 2010), the focus now lays at the interface between resources and community 'if we want to capture the full potentiality of the commons' (Girard and Frison, 2018b: 8). Rival (2018: 153) points that other aspects characterizing communities should indeed be taken into account such as the 'internal politics of attachment, membership, sharing, redistribution, obligation, gift, reciprocity'. Her reading of Gudeman's anthropological investigation of property and access demonstrates that Ostrom's model overemphasizes market ends and the achievement of efficiency (hence the focus on the 'right to exclude'; see Girard and Frison, 2018b: 8). Going even further, Capra and Mattei challenge Ostrom's thinking who addresses the problem with a 'subject-object' position (i.e. individuals interact to govern an object/resource). They consider that it excludes other possible narratives where plants are not considered as objects/resources but as other living parts of ecosystems constituting the 'web of life' (Capra, 1996; Moore, 2015) and interacting in constant relationship with all other elements, i.e. the 'object' is in relation with the individual, at the same level and their relationship results in a governing ecosystem (Capra and Mattei, 2015: 176–177). According to them, the law may serve as a 'commoning instrument' if its ecological dimension is recognized.

> [A]n ecological vision of law does not reduce law to a professionalized, preexisting, objective framework 'out there', separate from the behavior it regulates and tries to determine. Instead, law is always a process of 'commoning', a long-term collective action in which communities, sharing a common purpose and culture, institutionalize their collective will to maintain order and stability in the pursuit of social reproduction. Thus the commons – an open network of relationships – rather than the individual, is the building block of the ecology of law and what we call an 'ecolegal' order. Such an ecolegal order is built on the recognition that human survival on this planet is not guaranteed by the destruction of life and by the domination of nature in search of growth. Rather, it seeks a quality of economic life aimed at nurturing our living planet and focusing on generative, complex patterns of relationships.
>
> (Capra and Mattei, 2015: 14–15)

The role of power relations and of rationality by members of a community

For Dardot and Laval (2014: 156), another limit in Ostrom's work lies in the limited study of the impact of power relations between users within a community (also see Bechtold, 2015: 3–4; Weinstein, 2013: 7). When studying a CPR, Ostrom will focus her work on studying the rules governing the system and on how the rules emerge or change (Ostrom and Basurto, 2011). In her earlier work, Ostrom further assumes that individuals are 'broadly rational' (1990: 33). She later widens her understanding (1998: 1) by stating that 'individuals achieve results that are "better than rational" by building conditions where reciprocity, reputation and trust can help to overcome the strong temptations of short-run self-interest' (also see Ostrom, 2005: 101, 130–131; Ostrom, 2007). Nonetheless, Ostrom will not necessarily analyse whether and how, within the community, the rules have been equitably decided amongst members of the community. Dardot and Laval contend that she argues that social conditions will favour or inhibit the establishment of the rules, which is an important step outside of the dominant economic 'rational' thinking. However, they perceive (2014: 156) that more study could be focused on the power relations inside the community and the effects of a systemic domination over behaviours of one individual or group of individual within the community (also see Weinstein, 2013: 31–32). Recent publications argue that Ostrom's bundle of rights approach was recuperated by the 'core of property' school of thought (Girard, 2018). This might have reinforced the 'sovereignty of the proprietor', viewed by the neo-liberal Chicago School as the only way to reach economic efficiency (Ost et al., 2015; also see Orsi, 2013), thereby encapsulating the analysis of the relationships into an economic efficiency dynamic and partially abiding the intra-inequalities within a community. This point is confirmed by Agrawal (2001), who argues that better off group members are likely to gain a larger share of benefits from a resource. He later confirms (2002: 60) that 'inequalities within a group are not necessarily reduced because group members are willing to cooperate in the accomplishment of a collective goal'.

This is an important point when addressing the objective of 'fair and equitable' sharing of the benefits deriving from the use of the resource (Tsioumani, 2018). If one takes the example of an Indian community managing the irrigation system of their village, it is not difficult to imagine that the voice of a member from the Brahmins' cast could have more weight than the voices of the Dalit members that are part of that community. One could legitimately wonder whether the rules established by that community really answer the needs of all its members, including the Dalits. Ostrom remains attached to the idea that people are rational and that this rationality will lead the community to take decisions in its collective interest, without analysing deeply existing social, political or economic inequalities that the CPR system may reproduce, i.e. privileges, casts, etc. While 'there are pragmatic advantages to rational choice-based approach approaches, they miss much individual and social behaviour' (Singleton, 2017: 2; also see

Saunders, 2014; Peterson and Isenhour, 2014). However, this is an important question to dig, as it entails consequences on the members of the community, *inter alia* on the fair and equitable distribution of revenues, resources or other advantages resulting from the collective management system (Weinstein, 2013: 18–19; see also Baland and Platteau, 1997, 1998b). This is important when addressing the issue of imbalance of rights (IPRs vs. FRs) between different users of the 'Treaty community' and the question of stakeholders' participation in the Treaty decision-making process.

Beyond the boundaries of the CPR: the question of scale and of internal vs. external environment

To fil a gap in her analysis, which focused mainly on the micro-institutional level (Verhaegen, 2015: 117), Ostrom was influenced by institutional theories of polycentricity, i.e. 'the relationships among multiple authorities with overlapping jurisdictions' (Andersson and Ostrom, 2008: 71; see also Ostrom, 2010b). In her later writings, she developed a framework to enable

> scholars to organize analyses of how attributes of (*i*) a resource system (e.g., fishery, lake, grazing area), (*ii*) the resource units generated by that system (e.g., fish, water, fodder), (*iii*) the users of that system, and (*iv*) the governance system jointly affect and are indirectly affected by interactions and resulting outcomes achieved at a particular time and place.
>
> (Ostrom, 2007: 15181)

The framework aims at identifying 'how these attributes may affect and be affected by larger socio-economic, political and ecological settings in which they are embedded, as well as smaller ones' (ibid; also see Ostrom et al., 2007).

However, as she contends herself, the

> framework presented [. . .] will obviously need further development. [. . .] Policy analysts need to study and record the unintended effects of particular policy interventions, so that dangerous combinations of policies devised at diverse tiers or attributable to particular aspects of a resource system and resource units can be avoided.
>
> (Ostrom, 2007: 15186)

Weinstein (2013: 19) points to the need to take into account the network of complementary institutions within which every commons' system is integrated, through market and non-market relations of the commons and of the different individuals and groups which constitute it, with the rest of society (see also Coriat, 2013: 14; Coriat, 2015). This highlights the difficulty of studying a commons, which may be very diverse in scale and heterogeneous in its composition, in relation to the context (historical, social, political, etc.) in which it is embedded. A final interesting issue is raised by McCarthy (2005: 20) when he affirms that '[t]o assert a commons at one scale is almost necessarily to deny claims at another', thereby exacerbating the tension that might exist between coexisting systems and communities managing a same resource-type (i.e. the sum of heterogeneous collective management systems governing a same type of resource: PGRFA). This is of course directly relevant for a global seed commons, where multiple stakeholders and multiple layers relate and overlap within the system itself and within the network or context in which the global seed commons is embedded. Already in 1986, Rose was wondering whether a commons 'inside' would not function as private property 'outside', excluding people outside the community from accessing the resources managed inside the community (Rose, 1986; also see Bailey and Mattei, 2013: 993).

Towards 'inappropriability'

In reaction to the lack of protection and promotion of the collective interest through the market or the state institutions, authors have questioned the role of private property in managing resources. In line with Ostrom's views, for Bollier (2014: 102) a commons can coexist with private property; they can be 'mutually compatible and even work hand in glove'. However, he recognizes that the character of the commons is quite different from that of property; commons is not a variant of property as 'commons is less about *ownership* as we usually understand it than about *stewardship*' (ibid: 102, emphasis in original text). Dardot and Laval go much further in contesting the current property rights regime attached to commons. They argue (2014: 233 et seq) that some resources/regimes should be institutionalized as non-appropriable by society. The fundamental institution of property is refuted for resources that are considered to be managed by the community for the collective interest, in perpetuity, as a 'political constitution of the common' (ibid: 156). This theoretical step allows them to pass from the concept of 'the commons' (i.e. individual interests joining in a common objective) to that of 'a common' without an 's' (i.e. *the* collective interest, superseding individual interests), thereby widening significantly the definition of what a 'common' is. This conceptual move implies, according to them, to reject the current system, and rebuild a new mode of governance. They propose a 10 point 'memo' on ways to 'institutionalize the inappropriable' ('instituter l'inappropriable', 2014: 578–583). While this idea can seduce a utopian believer and defender of the collective interest, the political and economic context dominating our Anthropocene makes it certainly difficult to put it in practice easily today (in the review process of the Plant Treaty for example). However, one could argue that the 'common' would touch upon what Allaire (2013: 18) calls 'ideal goods', and which Ostrom classifies as public goods (e.g. peace and security of a community, national defence, knowledge, fire protection, weather forecasts, etc.); the difference being that it would not be the state as provider of that 'public good'/'ideal good'/'commons', but the members of the community at large.

Commons and human rights

A wide and brief overview of some important outputs of the new commons vogue has been summarized above. There is another interesting move to be mentioned, which examines the relationship between the commons and human rights, i.e. how can the commons serve human rights and vice-versa? Although this Human Rights (to food/to seeds) approach falls outside the scope of this research, it is interesting to point a couple of recent developments that will not be analysed in much detail. First, mention is made of the Rodotà Commission which, in Italy, has defined the commons as goods *essential* to the satisfaction of *fundamental human rights* (emphasis added, Rodotà, 2013; also see Dani, 2014; Lucarelli, 2011a, 2011b). Following this thought, commons must be accessible for all present and future generations. The property rights over these goods must be exercised in a way that their collective function is warranted. When the state holds this function, the common goods are placed outside of the commercial market. The state is reduced to a simple administrator. Citizens then benefit from a 'diffuse ownership', which allows to control the use of the good and to warrant access (Sezioni Unite della Cassazione, numero 3813, 16 February 2011). Following this idea of the state as mere administrator enabling access to and use of common goods by citizens, Pistor and De Schutter (2015) propose the recognition of a new category of goods – 'essential resources' – which access would necessitate to rethink governance regimes, where people could express their 'voice' and 'reflexivity'. They define essential resources (2015: 3) as 'resources that are either absolutely necessary for the survival or every human being', i.e. water, basic food, clothing and shelter, or resources that are 'indispensable for minimum existence in a given society', i.e. land, electricity, etc. In their book, they develop the idea that a shift needs to occur from 'the tragedy of the commons to the tragedy of exclusion', that is to say that there need to be a 'critical reassessment of existing governance regimes and their distributional effects', towards a shift in scarce resource management from efficiency to essentiality (ibid). An area where progress has taken place is in the field of water management and the right to water, where the European Water Movement declares that water is a common good and a universal fundamental right.[13] As Verhaegen (2015: 120) puts it when relating human rights and the commons, 'the purpose of the "commons" is highlighted, a purpose of *global justice in the right to access and use essential resources*' (emphasis added).[14]

Second, in the specific field of PGRFA, literature has acknowledged 20 years ago already the possible contradiction between Human Rights and the implementation of UPOV and TRIPS, which jeopardizes access to seeds and the right to food (Correa and Yusuf, 1998; UNDP, 2000). Since then, the implementation of instruments such as the CBD or the ITPGRFA promoting benefit-sharing mechanism to fairly and equitably compensate for the use of resources has not convincingly resolved this contradiction. Strangely enough 20 years later, Morgera (2016: 3) notes that the intersection of the 'variety of international human rights bodies and processes with the concept of benefit-sharing from the use or conservation of traditionally owned or used natural resources remains understudied'. As highlighted by Tsioumani (2018: 52), the debate remained focused on the impact of IPRs over Human Rights and in particular on the unrecognized right to seeds following the restriction of farmers' privilege in implementing UPOV 1991 (mainly with the work of former Rapporteur on the Right to Food Olivier De Schutter between 2008 and 2014, see in particular De Schutter, 2009b). The ongoing negotiation of a UN Declaration on the Rights of Peasants and Other People Working in Rural Areas demonstrates an international will to recognize specific rights to specific seed stakeholders (Claeys, 2014a, 2014b). However, these rights remain attached to an individual/universal character without formally recognizing the role of communities in their realization (Bessa, 2015). To my view, there would be an interesting path to explore with the recent formal recognition of the role of communities in conservation and sustainable use of biodiversity as a group with Article 6 (6.2 and 6.3.f) of the Nagoya Protocol. Morgera et al. (2014: 146) explain that the 'Nagoya Protocol goes beyond existing international human rights standards by extending the community (prior informed consent) requirements to local communities' (also see Morgera, 2017). The recognition of local communities' role as stewards of biodiversity formalizes a collective right, which level is missing in Human Rights international law. For example, the Right to Food is both an individual and a universal rights, but it is hard to implement (Claeys, 2015; De Schutter, 2014) and former rapporteur on the Right to Food recommends states to legally recognize customary systems of tenure for example, and encourages customary ownership systems rather than individual titling (De Schutter, 2010a). I hypothesize that formally recognizing the collective group level would allow to make the link between the individual and the universal level of recognition, collective level which I believe to be crucial for an effective recognition and implementation of these rights. However interesting this research trend might be, it will not be pursued in the present work.[15]

Commons and social movements

As for the practical developments, the theory of the commons has served several strong citizens' movements to protest against the hyperownership domination of the economy. The purpose is not to provide extensive details here, but simply to mention that people have appropriated themselves the concept of commons to defend their causes. As expressed by Capra and Mattei,

> [i]n the current state of affairs, recovering the commons is not the business of lawyers or politicians, whose intellectual and institutional landscape is framed by ideology of modernity. This process belongs now to plain people who, from choice or necessity, participate in caring for something that they recognize as a common.
>
> (Capra and Mattei, 2015: 153)

Before the law is effectively mobilized to transform society, people have not waited to take action, *inter alia* in the field of food and agriculture (Claeys, 2014a).

Recent famous movements are the 'Nuits Debout' in France,[16] Italy's various movements on the Commons (Teatro della Valle,[17] Naples water management),[18] Podemos and the15M Indignados in Spain,[19] Gezi Park in Istanbul etc. These movements show the role that citizens currently play as growing actors (Pleyers, 2017; Glasius and Pleyers, 2013; Pleyers and Capitaine, 2016) in co-creating collective institutions to manage resources, which the state or the market fail to manage sustainably in the collective interest. In addition, the global wave of 'cities in transition'[20] that is currently sweeping a huge number of cities worldwide demonstrates the capacity and role that citizens can play (i.e. what Capra and Mattei refer to as 'making community sovereign', 2015: 136–142).

These various elements of the commons' new vogue build on Ostrom's seminal work. One of the trend developed in particular by Capra and Mattei focuses on the important role of law – and its potential – in the institutional management of any system. They advance that

> [t]he legal order is the most important vehicle through which a worldview is enforced and transformed into social action, and thus human law is also the agency through which we may implement new ideas and values. We must rethink our human laws and their relationship with the laws governing the ecology of a living planet. Such a rethinking, a kind of Copernican revolution in the law, must use nature as a mentor and model, putting the commons and a long-term vision at centre stage. We must move from thinking of a 'mechanism of law' and move toward an 'ecology of law'.
>
> (Capra and Mattei, 2015: 12)

In an effort to propose ways forward for an effective Global Seed Commons (at the end of the present chapter), one needs first to identify which underlying principles of the theory of the commons (i.e. Capra and Mattei's 'laws governing the ecology of the living planet') respond to invariable principles relevant for stakeholders to manage seeds according to the Treaty's direct objectives and overall goals. These invariable principles are identified below as potential principles to transform the current legal system into an ecolegal order, i.e. principles that may transform existing rules into a 'generative ecological law', where the law would be disconnected from power and violence, where it would render to the community level their sovereignty by recognizing their stewardship role, and where ownership would no longer be extractive but generative/inclusive (Capra and Mattei, 2015: 131–146). These invariable principles are then proposed as guiding compass to mitigate the conceptual constraints identified in the legal and stakeholders' analyses.

The commons and the International Treaty on Plant Genetic Resources for Food and Agriculture

The above-mentioned fruitful literature on the Commons allowed differentiating collective action issues and various ways to address them, in particular when the collective action problem lays in the underuse of the resource rather than its overuse. Below, I explain the link between the Treaty and the theory; the justification why the commons is a relevant theory to mitigate the Treaty's dysfunction; and how to bring solutions to an existing international law instrument using tools and concepts from a governance theory. Again, a caveat is made as to the non-exhaustive character of the analysis of the commons; this work is not a study of the theory of the commons but a study of the Plant Treaty in light of the theory of the commons.

Certainly, one specific area of interest for the international management of seeds is the work on 'anticommons' (Fennell, 2010; Aoki, 1998; Heller, 1998; Heller and Eisenberg, 1998). Referring to literature developing the 'antitragedy' of the commons, di Robilant explains that

> [t]he debate is multi-faceted. It provides explanations for the frequent reversal of the Demsetzian path from open access to private property rights. It asks which values and goals a commons regime ought to promote and facilitate. Finally, it considers which legal rules or design principles would best accomplish these goals.
>
> (di Robilant, 2012)

With her work, she addresses the challenges involved in the design of common ownership institutions, and proposes to refocus the design work on promoting greater 'equality of autonomy' (ibid). Of course, this path bounds to conceive seeds as a commodity over which property rights ought to be acquired. It excludes other apprehension of the seed (e.g., its cultural, spiritual, biological, social values), in particular as a relational component of a global ecosystem between man/woman and nature (as opposed to the current restrictive economic vision of good out of which a commercial value has to be extracted). However, I believe that in the current state of the debate, such discussion on property rights over seeds remains crucial to attempt englobing these other perceptions of the many values attached to seeds.

Little legal scientific literature was published on the Plant Treaty during its first years of implementation.[21] Since then, authors have gained interest in the topic and the Multilateral System has been assimilated to a 'commons-type' regime (Dedeurwaerdere, 2012), i.e. a global crop commons (Halewood et al., 2013) or PGRFA commons (Halewood and Nnadozie, 2008), a global genetic commons (Safrin, 2004: 644; Falcon and Fowler, 2002: 200; also see Helfer, 2005: 219–220), or semicommons (Bertacchini, 2007). However, analysing the Plant Treaty and its MLS as a global seed commons is not an easy task. In 2013, Michael Halewood published a paper where he 'identifies which subsets of PGRFA are (or could be) included in an evolving global plant genetic resources commons, [as well as] options for policy reforms to provide better tailored institutional support for the plant genetic resources commons' (Halewood, 2013: 278). He clearly explains the particularity of the seed commons, in that 'PGRFA do not fit neatly within the institutional frameworks of analysis that have been developed for natural resources commons on one hand, and constructed cultural commons on the other' (ibid). He analysed the Treaty system according to Ostrom's eight design principles and identifies loopholes in their realization for an effective common-pool resource for seeds (ibid).

One of the identified difficulties relates to the global dimension of the MLS, as opposed to the generally small and local character of communities studied by Ostrom (Dedeurwaerdere, 2012; Brousseau et al., 2012). As confirmed by Henry and Dietz (2011) or by Stern (2011), a transposition of the design principles from the local to a global setting is not self-evident. Ostrom's work is inspiring and should be seen as a complementary conceptual input to adapt the Treaty's governing regime under international law, but it is certainly not sufficient to be transposed as such, given the very different situations and conditions of governance regimes. Another problem lies in the fact that contracting parties (i.e. states) have designed the existing institutional arrangement (even if it is based on prior existing practices by specific PGRFA stakeholders), and are managing it, with no formal space for all (non-state) stakeholders to participate in the management of the MLS. Other challenges are related to the little trust among stakeholders in the Treaty's Governing Body forum (Six et al., 2015: 164–167), or to the complex technicalities of the implementation tools and instruments developed by the governing body. How can Ostrom's theoretical approach, enriched by the developments undertaken by authors from the new vogue, contribute to an efficient Global Seed Commons? To explore this path, potential useful commons principles are explored.[22]

Identifying invariable principles for the global seed commons

In an attempt to build on Halewood's work and examine these (and other) loopholes in the MLS (Frison, 2016), I focus on different principles to be taken into account in the design of an effective global seed commons. They all play a role in the property/ownership debate over seeds and must be further examined in any attempt to resolve the tensions around 'hyperownership' (Safrin, 2004) over seeds. The purpose of this Section is to identify what important principles of the theory of the commons could constitute the laws governing the ecology of the global seed commons, i.e. what I call invariable principles. These invariable principles could be useful in the review process of the Treaty in order to move towards a truly effective Global Seed Commons, thereby participating to the transformation of the law (i.e. the Treaty) into a generative ecological law. Each of the six invariable principles identified are expounded below: 1. sustainability; 2. interdependence; 3. the anticommons dilemma; 4. the material and informational dual character of PGRFA; 5. the global seed community; and 6. diversity, heterogeneity and complexity.

Invariable Principle 1: Sustainability

One of the shared characteristics of the CPRs studied by Ostrom and her followers is sustainability: sustainability of the resources and of the governing institutions (Ostrom, 1990; Agrawal, 2002: 44). These two levels of sustainability are interrelated and interdependent. Berge and van Laerhoven remind us that CPR theory is 'key to understand[ing] under what conditions it can be expected that resource governance regimes *may* result in *more sustainable forms of resource use*' (Berge and van Laerhoven, 2011: 161, our emphasis). This idea fits with the conservation and sustainable use objectives of the Plant Treaty and with its long-term overall goals of food security and sustainable agriculture for the benefit of the whole community (on this term, see below). Therefore, for the MLS to function sustainably, PGRFA have to be conserved and use sustainably. Capra and Mattei strongly argue that to 'be sustainable, human laws should serve, rather than exploit and plunder, the web of life' (Capra and Mattei, 2015: 29). For his part, Agrawal argues that 'sustainable resource management can never be independent of sustainability of the collective human institution that frames resource governance, and that local users are often the ones with the greatest stakes in sustainability of resources and institutions' (Agrawal, 2002: 41). Keeping in mind and implementing sustainability as a firm objective is a prerequisite for any institutional management system to function. While the conservation and sustainable use objectives of the Plant Treaty undeniably fit with this invariable principle, the tools and instruments designed by the Governing Body to implement the Treaty and make it function do not sufficiently take this objective into account. Long-term objectives for the benefit of the global community should be translated into concrete rules within the global seed commons, which integrate 'more interactive and participatory processes between scientist, policy makers and stakeholders' (Dedeurwaerdere, 2014: 24).

It should be further noted that sustainability is seen as a 'dynamically maintained system condition rather than a static equilibrium' (Agrawal, 2002: 59), i.e. users of the community manage a resource with the perspective of duration and renewal in an adaptive relationship with each other and with the resources (see the importance of this aspect in Bollier, 2007). This dynamic aspect should also be expressed/allowed in the governing system at the global level (see below principle 6 for explanation of the concept of 'dynamism').

Finally, for sustainability to be expressed, another principle has to be realized, i.e. interdependence. As confirmed by Agrawal (who refers to Wade, 1988 and to Baland and Platteau, 1996), the 'importance of greater interdependence among group members as a basis for building institutions that would *promote sustainable resources management*' (Agrawal, 2002: 60, our emphasis). This statement highlights the very close relationship between sustainability and interdependence and leads us to the following invariable principle.

Invariable Principle 2: Interdependence

Interdependence goes hand in hand with sustainability. As expressed by Haas, '[i] nterdependence, far from being the description of a condition, becomes something to be realized – a purpose' (Haas, 1975: 839). He argues that actors of many entrenched networks feel 'enveloped in a massive "collective situation" to which there can only be a "collective response" if anyone is to attain his objectives' (ibid: 868). He adds that 'to study how actors learn to cope, [. . .] we must keep in mind why actors may wish to do better in managing resources of common concern. This brings us back to the issue of political purpose' (ibid), i.e. the political construction of a common in the line with Dardot and Laval (2014). Haas therefore suggests the need for a coordinated and collective response to a collective action problem, highlighting the importance of 'interdependence' both as a precondition and above all as a purpose to be realized.

Moving to a wider level of 'interdependence', Capra and Mattei state that 'recogniz[ing] the interconnectedness of our global problems [would] enable us to find appropriate, mutually supportive solutions that [. . .] would mirror the interdependence of the problems they address' (Capra and Mattei, 2015: 159). This is particularly well adapted to the food and agriculture field, where interdependence is the result of long run human cooperation and collaboration in the exchange of seeds across the world. Farmers and breeders have selected, exchanged and bred seeds to develop such or such characteristic over millennia that respond to specific needs and to local conditions.[23] There is therefore an ongoing need to exchange plant genetic resources from countries all over the world.[24] It is this human-level sense of the word interdependence that constitutes the foundation of the concept of benefit-sharing.

It is argued that PGRFA interdependence contains a dual social and economic dimension underpinning the concept of benefit-sharing. The social dimension is understood as encompassing the formal and informal (Kalaugher et al., 2002) networks involved in governing the flows of PGRFA (Frankel, 1970: 469–489). These human networks of farmers, breeders and scientists have therefore a crucial role in safeguarding the availability of and accessibility to PGRFA diversity.[25] As for the economic dimension of countries' interdependence, it is a consequence of the rapid globalization and economic integration, and of growing cross-boundary flows of trade, financial capital, technology and know-how. More specifically, interdependence between supply and use of genetic resources is much higher for the agricultural sector compared to other sectors using genetic resource such as pharmaceuticals or bio-engineering industries. Both social and economic interdependencies between stakeholders and states are intensifying and it has been argued that these trends limit states' leeway to deal with these challenges autonomously because 'internal dynamics are to an increasing extent determined by external processes' (Oosterveer, 2005: 32). This judgement may partly explain why the benefit-sharing concept is so well entrenched in the management of PGRFA and so intrinsically integrated into the Plant Treaty specifically.[26]

When applying this to PGRFA management, one can argue that awareness and solid understanding of the factual interdependencies of crops (as a scientific characteristic) and of countries (as a socio-economic result of globalization) are preconditions to put into place the multilateral access and benefit-sharing mechanism. But for the MLS to reach its overall goal of food security and sustainable agriculture, it is argued that a more profound degree of mutualism in the understanding of the criterion of interdependence is needed (Mattei, 2011a: 101–102). Interdependence is therefore an existing factual element, but is should also become a philosophical and political goal to be attained by all countries in the implementation of the international regulations for them to reach global food security. PGRFA actors (countries, gene banks, researchers, farmers, NGOs, etc.) and their interactions within PGRFA networks therefore create a complex relationship of various interdependencies, at various levels, but without being able to translate this principle into the international governance regime. Mol says that more and more 'nation-states and national political actors are embedded in broader frameworks of governance and politics, consisting of multiple layers, from local to global, and multiple actors from private firms to nongovernmental interest groups' (Mol, 2001: 219). Commenting on Mol's citation above, Oosterveer adds that

> [c]onsequently, the resulting changes in the interaction patterns between different states and between different state and non-state actors lead to a variety of innovative forms of governance involving diverse social groups at different spatial and sectoral scales. This is generally referred to as "multi-sector and multi-level governance" or "network-based governance".[27]
>
> (Oosterveer, 2005: 23, internal references omitted)

Therefore, it is argued that interdependence should be reflected in the international governance regime as both an intrinsic element necessary for the viability of the regime, and also as a political and philosophical objective to be reached and maintained, for a sustainable management of the resource in the collective interest. Digging into Burris, Drahos and Shearing's work on nodal governance could be one way to search how this principle could be better expressed in the complex network(s) of seeds worldwide, from the local to the global level (Burris et al., 2005; Holley and Shearing, 2017).

Invariable Principle 3: Anticommons Dilemma – Underuse of Seeds as Main Erosion Risk

Related to the sustainability and interdependence invariable principles, another key issue for the conservation and management of seeds is the 'anticommons dilemma' (Heller, 1998; also see Fennell, 2010; Aoki, 1998). Heller defines the anticommons dilemma as occurring when 'there are too many owners holding rights of exclusion, [then] the resource is prone to underuse' (Heller, 1998: 624). Regarding seeds, the dilemma is not that overconsumption leads to depletion of the resource, but rather that underuse leads to erosion. Through 'the process of domestication and co-evolution with humans, crops have become dependent on human beings for their continued existence; they cannot exist on their own in the wild' (Halewood, 2013: 291 – referring to Wilkes, 1988). This is why over the last decades, as access to seeds has become more and more restricted, erosion of seed diversity has never been so wide. Halewood emphasizes that this 'aspect of PGRFA informs the need for collective action institutions that are necessary to support their continual creation/ evolution as well as ensuring that they are conserved and available for use' (Halewood, 2013: 291). Therefore, the only sustainable way of managing seeds and avoiding the anticommons dilemma to erode PGRFA diversity is to facilitate their exchange, access and use by all users,[28] not only breeders and researchers, as provided for by the Treaty, but at the global level for every farmer feeding the world, including by supporting and protecting existing seed networks. Indeed, farmers constitute the large majority of day-to-day users of seeds. Hence, they are the first stewards of PGRFA conservation and sustainable use. When associating these invariable principles – sustainability, interdependence, anticommons – it is easy to conclude that promoting the widest use of and access to PGRFA leads to a 'comedy of the commons', i.e. enhancement of use and value of the resource, as framed by Rose (Rose, 1986; also see Verhaegen, 2015: 116). Like a virtuous circle, enhancing PGRFA use increases *inter alia* benefits, sustainability, interdependence, which in turn enhances PGRFA diversity and conservation for the global community.

Invariable Principle 4: Physical and Informational Components Inextricably Bound to the Use of Seeds

Following Ostrom's lead on studying governing regimes for natural resource commons, academics have expanded the research field to information (Benkler, 1997: 2003), knowledge (Hess and Ostrom, 2003, 2006) and science commons (Reichman and Uhlir, 2003; Cook-Deegan; Dedeurwaerdere, 2006). Hess and Ostrom contend that advances in law and technology 'have generated greater access to important information about history, science, art, literature and current events, while at the same time enabling profit-oriented firms to extract value from resources previously held in common and to establish property rights' (Hess and Ostrom, 2003: 112). Ostrom reminds us that

> [f]or most of human history, the [global commons] remained unclaimed due to a lack of technology for extracting their value and for establishing and sustaining property rights. To our peril, the technology to extracting value from [the global commons] has developed more rapidly than have the appropriate legal mechanism for establishing an effective property regime. The treasured resources for all mankind are threatened by the very technological abilities that we have mastered during recent eras.
>
> (Foreword by Ostrom in Buck, 1998: xiii)

With the advent of the Internet, CPR studies have spread to intangible material, pointing to 'the second enclosure movement' phrased by Boyle in 2003 (Boyle, 2003b), i.e property rights over intangible material rather than physical goods. Hess and Ostrom explain that '[t]he enclosure is caused by the conflicts and contradictions between intellectual property laws and the expanded capacities of new technologies', as an 'outcome of new technologies and global markets' (Hess and Ostrom, 2003: 112; also see Benkler, 1997; Boyle, 1996: 6–7). In the field of seed innovation, IPR and new technologies go hand in hand in enclosing information, technologies and access to knowledge and material traditionally available, thereby facilitating the commodification of a more abstract innovation, for which a new economic value is created on the global market. See for example Monstato's patent claim on the genetic structure of the neem tree (Marden, 1999).

As regards PGRFA, the physical and informational components are inextricably bound to the use of seeds.[29] Dedeurwaerdere confirms that PGRFA are somewhere in between the exclusive 'natural resource commons' and the exclusive 'knowledge commons', containing both a physical component and an informational component (Dedeurwaerdere, 2012, 2013). This dual component as physical and informational asset should be taken into account when considering the institutionalization of a global seed commons. It would require specific governing rules which change and adapt with the evolution of the IP protection scheme, for instance including the concept of 'reflexivity', i.e. 'the ability to assess the actual consequences of existing practices in order to reform them if needed' (Brousseau et al., 2012: 350). However, one must bear in mind that reclaiming an 'intellectual public domain' (Hess and Ostrom, 2003; also see in plant breeding: 2014) might not be sufficient to dis-enclose the 'knowledge commons', particularly when the intellectual public domain deals with high-tech information and knowledge. Public domain

> is generally defined in the negative, as that which copyright does not (or no longer) protects: it is composed of elements that are by themselves unprotected (Dusollier, 2013). In patent law, elements may fall into the public domain as a result of public ordering: non-patentable subject matter, expiry of a patent; or they may be the result of private ordering: the decision of an inventor to publish, or the decision not to patent.
>
> (Van Overwalle, 2018: 94; see also Dusollier, 2007)

As Van Overwalle (ibid, emphasis added) demonstrates, material included in the public domain is not protected from appropriation: 'dedicating an invention to the public may stop short of establishing and strengthening access. When renouncing a (potential) patent right and dedicating the invention to the public domain, *the inventor can no longer safeguard his invention against capture*'. This means that the public domain, as it stands today, is not a means to protect from misappropriation.

Information on seeds is arguably enclosed because of IPRs of course; but also because of the high-tech nature of the information. Since, for this information to de facto be accessible to people through a reclaimed 'intellectual public domain' would necessarily imply transfer of the related technology and training, allowing users to understand and use the technology and information. This entails that facilitating access to material (whether physical, informational or both) without creating a protective scheme from misappropriation will likely continue to favour the same users (industry and research institutions from the Global North, which have the financial and technical means to use and develop the material into new products).

Examples of such attempt in plant breeding can be found in the launch of the DivSeek initiative and the Global Open Genome Sequence Data Framework (Piotrowski, 2015; Warthmann and Chiarolla, 2015). These initiatives to reopen access to breeding information are interesting in that they put forward an inclusive and sharing characteristic in order to promote research in the field of food and agriculture. However, besides the fact that these initiatives will only reach specific categories of seed users (biotech industry and researchers), de facto excluding users who do not have the adequate training and technology to benefit from them, they contribute to pushing further the erasing of the material/physical component of the seed, making it more and more intangible and therefore more and more disconnected from the environment and ecosystem in which varieties live and evolve. These various elements can be seen by some stakeholders as constitutive blocks to recreate an open arena for the private sector to use resources without being obliged by international benefit-sharing obligations from the Treaty or the Nagoya Protocol.

Furthermore, other attempts to 'reopen' access to seed varieties through 'open source' systems arise in several countries, such as the Open Source Seed Initiative (OSSI) in the USA,[30] the Cambia BiOS (Tsioumani et al., 2016), but still need to demonstrate their effectiveness (see Diebel, 2018; Van Overwalle, 2018 for further information on these open source initiatives). In addition, these initiatives using copyleft licensing tools are based on IPRs (Santilli, 2012: 257–270). As stressed by Van Overwalle (2018: 97), 'this is by far the most striking – and unexpected – feature of the open source model: open source is based on IP' (also see Benkler, 2002; Rai, 2005).

Invariable Principle 5: Global Seed Community

Whether studying the meadows and forest communal tenures of Törbel in Switzerland or of Hirano in Japan, or the Huerta irrigation institutions in Valencia and Alicante in Spain, Ostrom has systematically analysed a collective management system from specific and relatively clearly defined small/local communities (Ostrom, 1990). Bollier insists that a community is one of the three constitutive elements of a commons, along with a resource and a set of social protocols (Bollier, 2014: 15).

A commons becomes a *commons* only when *commoners* decide collectively to *commoning* resources, i.e. manage in a fair and equitable way the access to and use of a resource in the collective interest (ibid: 15). But how can a community be identified? What/who constitutes 'a community'? Who are 'the commoners' (ibid; also see: Bollier and Helfrich, 2014)? Is there a 'Plant Treaty community' or several complementary communities involved in the Plant Treaty system? This paragraph will bring more questions than answers.

Baland and Platteau define a community as an arena where 'all members of a social group have an access to the local resources' (Baland and Platteau, 1998a: 644). This definition implies belonging to an identified social group, and a notion of (local) scale. The question of scale has focused attention of many commons scholars, in particular in trying to identify whether group size was necessarily a factor of success (or failure) of a long-enduring commons. Agrawal summarizes these studies and states that the impact of group size on effectiveness of collective action is mediated by many different factors: production technology of the collective good, its degree of excludability, jointness of supply and the level of heterogeneity in the group (Agrawal, 2002: 59–60). He stresses that more research is needed on the relation between group size and success of collective action. In the Plant Treaty arena, scale of the community is certainly a crucial issue. Authors have referred to the 'Global Seed Commons', setting the scale of reference at the global level. Are we to understand that a global community of seed users exists? Would this global community be constituted by all the sub-communities coexisting and co-managing PGRFA? For now, the Treaty answers fairly well the needs of the researcher-breeder sub-community and not of the farmers. However, seeds are used by many other stakeholders, the largest group being farmers.

This raises questions of legitimacy in the recognition of de jure and de facto holders, and leads to the second aspect, i.e. the notion of '(Global) Community' and therefore of heterogeneity of members. Rights of de facto holders, i.e. farmers, should be taken into account and heard to create and develop the global seed community apt to sustainably manage the global seed commons. But how to define farmers' communities? Is it simply 'all farmers'? Smallholder and subsistence farmers? What about local communities (as referred to in the Treaty), and indigenous communities (as enshrined in the CBD)? The Multilateral System of the Treaty facilitates access to Annex I PGRFA for 'research, breeding and training' purposes (Article 12.3(a)). Does this imply that the MLS community is restricted to breeders, researchers and trainers? If so, this is clearly inconsistent with other invariable principles (sustainability, interdependence, anticommons dilemma) which call for a widest access and use of seeds as possible for reaching the Treaty's overall goals of food security and sustainable agriculture. Restricting the community to breeders and researchers could also be seen as contradicting Articles 5 and 6 (on the conservation and sustainable use of PGRFA) and 18 (on the funding strategy) of the Treaty which rather call for including farmers in the Treaty's 'global community'.

Defining who is part of the community is crucial as it sets the legitimacy for rights claims for managing the resources. In other words, only those members that are part of the community may collectively manage (and benefit from) the resource. In the Treaty, farmers are clearly identified as the target group for benefit-sharing (the first benefit of which being facilitated access to seeds). Should this then automatically equate to a right to collectively manage the resource[31] at that global level (and not relegate this right subject to national legislation – and therefore recognition). To be congruent with the objectives of the Treaty, recognizing the fundamental role of farmers in the sustainable use and conservation of PGRFA and in their key role as food producer, as well as identifying the community of farmers as a targeted beneficiary should automatically integrate this category of 'users/commoners' as part of the 'management team' of the Treaty, i.e. the Governing Body and its resulting tools and instruments.

A final note is to be made on the concept of community and its related implication with the notion of exclusion. Community necessarily implies members being part of the group, and people being out of the group, i.e. excluded. This clearly points to the notion of boundaries and scope of a management system. But can we talk about a community and therefore exclude people from the MLS when we are dealing with accessing and producing food, which is a universal human need? Is the community, and its attendant exclusive dimension, a relevant concept to tackle a subject for which interdependence of plants, people and institutions are so deeply intermingled into one another that inclusion should be the rule rather than exclusion?

What role would states have in contributing to define and implement the rules for this global community as a fundamental norm as well as a limitation to the imbalance of powers expressed in international law instruments? Can/should states participate in representing the interests of farmers and their communities, thereby actively trying to include weaker stakeholders in the negotiation and implementation forum?

Invariable Principle 6: Diversity, Heterogeneity and Complexity

Ostrom's eighth design principle on 'nested enterprises' (Ostrom, 2009) is premised on the assumption that larger commons might be more complex to govern than smaller ones. Later, further studies developed this intuition, showing that heterogeneity, diversity and complexity in CPRs (see *inter alia* Berkes et al., 2008; Ostrom, 2005; Jungcurt, 2007; Cox et al., 2010) were important aspects to take into account. In studying the character of an adaptive system to a changing context, Dedeurwaerdere pointed to the importance of the modular character of organizational architecture (Dedeurwaerdere, 2012). This modular character of organizational architecture has to be recognized and facilitated in the design of the institutional managing systems. I understand modularity as allowing adapted 'governance nodes' (i.e. communities at various levels and scales), both on horizontal and vertical scales, to govern, react and evolve according to external and internal factors. This embeddedness of systems enables to express diversity, heterogeneity and complexity at all levels and scales. As Ostrom and Basurto put it

> [w]e do not need to be complex, however, just for the sake of being complex, but we do need to get over our simplicity hang-ups. Obviously, our theories will always be simpler than the worlds we study, or we are trying to reproduce these worlds rather than a theory of these worlds. Given the complex, nested systems of the biophysical world, however, we need to develop a social science of complex, nested systems.
>
> (Ostrom and Basurto, 2011: 337)

This is particularly true in today's context where climate change and other hazards impose quick, reactive and adaptive responses.

Unfortunately, globalization and the homogenization of biodiversity governing regimes hinder the emergence of institutional diversity. Roa-Rodríguez and Van Dooren stress that

> [t]he dynamics unleashed by IP and sovereign regimes are transforming the varied common spaces, with their multiple modalities of access, use and alienation of resources, into a de facto homogeneous commons space where the negative and exclusive characteristics are predominant. This is a highly undesirable outcome if our true goal is the conservation and sustainable use of [plant genetic resources] for the well-being of society at large.
>
> (Roa-Rodríguez and Van Dooren, 2008: 193–194)

Indeed, it is crucial to allow local practices and traditional knowledge related to seeds to continue being used and exchanged by farmers. Imposing a single (market-driven) path for seed conservation and use governance regime is not an option, as demonstrated by the drastic erosion of agrobiodiversity over the last decades.

In my view, effectively reaching the Treaty's objectives of conservation, sustainable use and ABS is only possible when taking into account the need for flexibility, diversity and dynamism in the management of all PGRFA, of PGRFA networks and PGRFA conservation systems worldwide. This does not mean that the proposed governance mechanism has to be complex. Quite the contrary, its core elements should remain simple to be easily implemented in all different settings. But simple does not mean homogeneous. Indeed, setting a homogenous, one-size-fits-all solution (that of the commercial value of seeds as highly technologically improved material) whereby exclusion is the prevailing characteristic in the management rules, will not work out for seeds. Seeds can only survive and develop through diversity and heterogeneity, movement, adaptation and constant use and human interaction. However, Agrawal warns that '[h]eterogeneities of endowments have a positive effect on resource management while heterogeneities of identities and interests create obstacles to collective action' (Agrawal, 2002: 60 – citing Baland and Platteau, 1996; also see Libecap, 1994). Legal and stakeholders' analyses of the Treaty (Frison, 2016; Frison et al., 2011a,2011b) have revealed the heterogeneities of identities and of interests. In my view, this is a crucial element to take into consideration when investigating solutions to the MLS governing constraints.

Redesigning a global seed commons

Undeniably, the Treaty can be considered a fertile ground for institutional innovation: it invented the Multilateral System of access and benefit-sharing as a unique collective exchange mechanism; it created the Third Party Beneficiary concept to safeguard the collective interests of the MLS; it took first steps to formally recognize Farmers' Rights; and it established a Benefit-sharing Fund (BSF) as trust account for farmers; thereby de facto creating an instrument reflecting concepts inspired from the theory of the commons (for a more critical view see Thomas, 2017).

The legal and stakeholder analyses (in this book Chapters 4 and 5) have shown that what authors have call 'global crop commons' exists but that its functioning is not efficient in reaching the Treaty's objectives of food security and sustainable agriculture. The review process of the Multilateral System of the Treaty launched at the Fifth Session of the Governing Body in 2013 confirms this statement. The ongoing review shows that the Treaty forum is reactive and willing to function effectively, and that it remains a fertile ground for innovation in creating institutional settings for collective management of common resources.

Along this line of thought, it is hoped that the present work may modestly contribute to bringing new ideas and different perspectives to the Governing Body debates. To this end, the final section of this chapter aims at formulating proposals to mitigate the constraints identified within the eight topics covered in Chapter 4. The recommendations made are centred on the conceptual constraints identified (e.g. imbalance of power between FRs and IPRs), rather than on the concrete technical constraints (e.g. difficulties for contracting parties to 'designate' to the Treaty secretariat the PGRFA under their management and control). The reason for this choice is that making concrete proposals to address technical problems would require very specific and different expertise and competencies, which the author of the present work does not claim to have. In the meantime, it is hoped that proposing possible solutions to conceptual issues would constitute the first step in designing concrete answers to technical problems, later on in the Treaty review process. A clear caveat is therefore made, that the recommendations proposed below are not ready-to-implement solutions to the technical constraints identified during the Treaty analysis. Rather these recommendations nourish the conceptual apprehension of the issues at stake to guide towards concrete solutions that all Treaty stakeholders will have to find together. The objective is to open up the debate to new ideas and different ways of thinking to add to the Treaty's review process.

Similarly to Chapter 4, this section is divided into eight paragraphs (see Table 6.1): sustainable agriculture and food security; scope; Farmers' Rights; facilitated access; benefit-sharing and the Benefit-sharing Fund; information and

Table 6.1 Treaty topics, conceptual constraints and relevant commons invariable principles (original material for this book)

Treaty topics	*Conceptual constraints*	*Commons underlying principles*
1 ***Sustainable agriculture & food security***	Overall goals of Treaty not reached because not recognized as direct objectives	Sustainability (of resources and institutions) Interdependence Anticommons dilemma
2 ***Scope***	Difference between scope of Treaty and scope of MLS leading to dysfunction	Interdependence Anticommons dilemma Physical and informational components inextricably bound to the use of seeds
3 ***Farmers' Rights***	No real recognition at international level in the same terms as IPRs (recognition of their role but not their rights)	Anticommons dilemma Community Interdependence
4 ***Facilitated access***	Facilitated access is absent for the ultimate beneficiaries: farmers	Anticommons dilemma Community Interdependence
5 ***Benefit-sharing/ Benefit-sharing Fund***	Puts the farmer in a situation of receiver/beneficiary instead of actor/ user/ stakeholder	Community Interdependence Anticommons dilemma
6 ***Information/ knowledge***	Appropriation, Protection Availability mainly of one type of information	Physical and informational components inextricably bound to the use of seeds Interdependence
7 ***Third Party Beneficiary***	Preservation of MLS rights, but what about preservation of all stakeholders' rights? Lack of system to balance powers	Diversity, heterogeneity, complexity Interdependence Community
8 ***Participation/ governance***	Governance of MLS remains at state level Problem of trust Need to include all stakeholders at all levels	Community Diversity, heterogeneity, complexity

knowledge; Third Party Beneficiary; and participation and governance. Each paragraph is composed of a first part stating the conceptual constraint related to the Treaty Topic, and a second part proposing recommendations in light of the commons invariable principles detailed above in section 2. On a preliminary note, one cross-cutting aspect that appears in almost every topic is the lack of recognition – translated into concrete obligations, instruments or procedures in the Treaty implementation – of the role and rights of smallholder farmers. In order to provide a quick overview of this section, a table summarizes its content.

Sustainable agriculture and food security

For this first Treaty topic, the main conceptual constraint identified in the legal and stakeholder analyses and impeding the achievement of a truly effective global seed commons relates to the fact that food security and sustainable agriculture are two overall goals of the Treaty instead of being direct objectives. Indeed, having sustainable agriculture and food security as overall goals maintains some distance with the concrete implementation actions undertaken by contracting parties. Decisions and actions taken by the Governing Body are directed towards conservation, sustainable use and ABS purposes, not necessarily towards food security and sustainable agriculture as a whole.

Furthermore, the above analyses have shown that the implementation of the Treaty does not reach its direct objectives of conservation, sustainable use and ABS for all its stakeholders, and hence its overall objectives of food security and sustainable agriculture. These direct objectives will not be achieved as long as they are dissociated from the primary activity of farming and seed 'commoning' (Bollier, 2014), i.e. food production. Focusing more strongly and more directly on the overall goals of the Treaty could contribute to formally designate food security and sustainable agriculture as the main objectives of the Treaty. This would also be consistent with recognizing that the primary role of farmers is producing food.

Bringing back the sustainable agriculture as a central objective of the Treaty would respond to the important underlying principle of the commons theory: sustainability of resources and institutions are crucial for effective commons to last in time and adapt to changes. Pushing further this principle towards sustainability of ecosystems (i.e. integrating the commons 'resource-institution' pair into its wider ecosystem) might be even more effective.

Doing so can be justified in several ways. First, it would follow the idea expressed in the preamble of the Treaty whereby the Treaty is positioned within an already established set of international instruments and fora dealing with sustainability and food security. Indeed, paragraphs 4, 5 and 15 of the Treaty preamble place the negotiation of the Treaty process within the ambit of FAO, and the existing (non-binding) voluntary international instruments dealing with PGRFA conservation and sustainable use, i.e. the Rome Declaration on World Food Security, the World Food Summit Plan of Action and the Global Plan of Action for the Conservation and Sustainable Use of PGRFA.[32] Reinforcing this link would also fit in the current international community's discussion on Sustainable Development Goals adopted in 2015 and with the developments taking place in the UN Human Rights Council on the Rights of Peasants and Other People Working in Rural Areas. Second, this would also allow for a debate on whether the objective of the Treaty should be food security or its competing concept food sovereignty, thereby opening a discussion on the self-determination of farmers in a transparent manner.[33] This would contribute to opening the Treaty system to all stakeholders, as all stakeholders are concerned with food security and sustainable agriculture, while within the current terms of the MLS, not all stakeholders are concerned with ABS (i.e. only breeders, researchers and trainers). Doing so could contribute to widening the scope of the MLS. Finally, this would contribute to decreasing the hyperownership over seeds supremacy by mitigating a fundamental contradiction within the Treaty principles. Indeed, the Treaty provisions formally recognize international IP law in its body text. The recent general trend in IP law reinforces protection rights over plants, thereby limiting access to and use of seeds. Limited access and use of seeds are contrary to conservation objectives but also to innovation processes (Heller and Eisenberg, 1998; Heller, 2010). Recent research by Petra Moser (2016) finds out that narrow and short-lived IPRs benefit innovation but wide, strong and long-term protection rather has opposite results. The fact that the MLS integrates IPRs in its functioning de facto reinforces the commodification trend over biodiversity due to the 'second enclosure movement' (Boyle, 2003b). This is fundamentally in contradiction with the invariable principles of the Treaty (in particular interdependence and the anticommons dilemma) and its overall goals of food security and sustainable agriculture. Strengthening the recognition of food security and sustainable agriculture as direct objectives to the Treaty could therefore limit the negative impact of the hyperownership trend over seeds.

Recommendation 1: Formal recognition of food and nutrition security and sustainable agriculture as direct objectives of the Treaty.

Scope

Regarding the second Treaty topic, the conceptual constraint identified in the Treaty analysis lies in the difference in scope between the Treaty and the MLS. The problem of the boundaries of the global seed commons is a real problem and is twofold. First the Treaty applies to all PGRFA and rules mainly on conservation and sustainable use obligations while the MLS applies to the Annex I list of 64 crops and forages and rules on access and benefit-sharing obligations. Second, at the level of the contracting parties, the Treaty is not (yet) truly global, as there are 145 members, with important countries – in terms of genebanks and genetic diversity – remaining outside of the system (e.g. China, or Russia; the US having ratified in December 2016). This allows for stakeholders to continue free-riding the system.

Regarding the first issue, the difference in scope leads to an overall dysfunction in the implementation of Treaty and MLS obligations by contracting parties. Having a MLS functioning only for 64 crops and forages significantly complicates the governing rules of the system. Indeed, in order to design the tools and instruments to implement these obligations, contracting parties have had to create tracking and identification obligations (not foreseen in the Treaty) which constitute heavy administrative burdens, complicating the implementation of the system. Examples of such burden are the difficulty for many contracting parties to designate to the Treaty Secretariat what PGRFA are under their management and control and the resulting SMTA tracking obligation. Indeed, the Treaty requires for the access to seeds to be 'accorded expeditiously, without the need to track individual accessions and free of charge'. However, the SMTA creates a tracking obligation by the need to list the material provided in an annex to the SMTA and to report systematically to the Treaty Secretariat. Therefore, there is a contradiction between the initial obligation (expeditious access without tracking) and the instrument designed by the Governing Body to implement that obligation (which creates tracking). It is therefore crucial to resolve this conflict of obligations between provisions of the Treaty and the MLS and those of the SMTA.

The second issue related to the scope deals with the boundaries of the Treaty memberships. The fact that China, Russia, the US and other countries are not yet contracting parties allows for an easy free-riding of the system, i.e. PGRFA can be accessed in genebanks from non-member countries without using the SMTA. Indeed, the US for example detains the largest genebank in the world. China and Russia also have very large collections of PGRFA. If these three nations alone were contracting parties to the Treaty, the boundaries of the global seed commons would be geographically much closer to being truly global and free-riding would be much more difficult.

It was shown above that in order to have an effective global seed commons, the invariable principles of interdependence and anticommons dilemma have to be taken into account in the institutional design. Widening the scope of the MLS to all PGRFA and expanding the membership boundaries of the Treaty would allow for these invariable principles to be respected. Indeed: 1) interdependence requires collaboration and exchanges between all countries and for all PGRFA; 2) resolving the anticommons dilemma for seeds requires an easy access and use by as many users as possible. However, this solution alone would not be sufficient. It would need to be coupled with a review of the payment scheme funding the MLS, which is under review by the Governing Body. Indeed, access is currently related to the payment of monetary benefits to the BSF, under specific conditions. Access to PGRFA and payment should be dissociated in order for the proposal to include all PGRFA in the MLS to constitute an effective answer to the problem of dichotomy of scope and free-riding.

Recommendation 2: harmonize the scope of the MLS with that of the Treaty to include all PGRFA and expand the boundaries of the Treaty to make it truly global, while turning the MLS into a 'protected public domain' to avoid misappropriation.

Farmers' Rights

The conceptual constraint related to this third Treaty topic is certainly one of the most important identified in the overall analysis of the Treaty: there is no real recognition of FRs at the international level while there is a very strong recognition of IPRs. There is recognition of the role farmers played for millennia in conserving and developing PGRFA diversity, but this recognition is not accompanied by actual de jure rights protected at the level of international law. This creates a strong imbalance in rights penalizing farmers. This imbalance is entrenched in the Treaty's tools and instruments, i.e. the MLS and SMTA, which recognize innovation as being the role of breeders and not of farmers. This reflects the strong beliefs that technological innovation is the only relevant innovation apt to end hunger and biodiversity erosion, thereby ignoring the relevance of different, informal innovations by farmers (Coomes et al., 2015; Demeulenaere and Goulet, 2012; Pautasso et al., 2013).

The Treaty provides only a weak recognition of farmers' rights to be implemented at the national level. The Treaty does not provide for facilitated access to PGRFA for direct use by farmers and it does not facilitate access to improved seeds and technology particularly to the smallholder farmers of the South. Neither does it provide clear tools to promote and respect existing seed networks, which are core to farmers' resilience and food production. This represents a major contradiction with the fact that those farmers are feeding the world (FAO, 2010, 2014a; Graeub et al., 2016). With major climatic hazards likely to arise more often and more severely in the future, it is imperative that farmers be able to access seeds and relevant technology, enabling them to continue to maintain a wide agrobiodiversity, to innovate and to adapt in order to face these risks and to continue feeding the world population.

Furthermore, there is a major discrepancy between the intended beneficiaries of the MLS and those who really benefit. The benefit-sharing clause (Plant Treaty Article 13) focuses on farmers as main beneficiaries. But the first benefit, i.e. the facilitated access, is essentially directed towards breeders, while farmers clearly still do not benefit directly from this facilitated access, which is limited to research, breeding and training purposes.[34]

FRs should be formally recognized at the international level, at the same level as IPRs so as to rebalance rights and to stop penalizing farmers. One way of formalizing this recognition is to grant direct access to farmers to MLS seeds for their direct use, exchange, etc. Doing so would allow for a sort of 'renewed farmers' exemption' to exist for PGRFA covered by the MLS. A complementary way is to include farmers in the decision-making process, not only at the national level (and encourage this through sharing of experience, help in design of national legislations and policies etc.), but also at the international level in the management of the Treaty, the MLS and the Benefit-sharing Fund.

Doing so would resolve the anticommons dilemma by allowing for maximum users of PGRFA diversity to access, sustainably use and conserve seeds. It would also safeguard the interdependence link binding all PGRFA users, thereby increasing communities' self-determination in producing their own food (i.e. enhancing sustainability and reaching food security). Finally, it would expand the sense of belonging to one same global community that is composed of different heterogeneous communities, instead of ostracizing farmers from the community of Treaty stakeholders.

Besides, it is the right moment to recognize farmers' rights at the international level. Indeed, progress is made on the implementation of the right to food, on the 'Zero Hunger' sustainable development goal, and on the negotiation of the UN Declaration on the Rights of Peasants and other People Working in Rural Areas. Now is the moment to progress on this fundamental issue. The Global Consultation of farmers' rights which was requested by the Governing Body to follow-up the implementation of Resolution 5/2015, took place in 2016 in Bali.[35] The Treaty secretariat had prepared an electronic survey to gather views, perceptions, options and approaches and possible strategies and options for the implementation of farmers' rights. The survey was also aimed at gathering inputs for the preparation of a study on lessons learned. Following this impulse, the Seventh Governing Body of the Treaty adopted a Resolution creating of a working group on farmers' rights. Let us see whether this important step be accelerate the pace of their recognition both at national and international levels.

Recommendation 3: Formal recognition of farmers' rights at the international law level and commitment to implement these rights at the national level.

Facilitated access

One of the key innovations of the MLS is the design of a collective mechanism to facilitate access to PGRFA. It is the fourth Treaty topic covered in Chapter 5. The main conclusion resulting from this analysis is that facilitated access functions for specific Treaty users (i.e. breeders and researchers) but that it fails to benefit the supposedly primary beneficiaries of the MLS (i.e. farmers), thereby creating an unequal dual system. Two conceptual constraints explain this situation. First, farmers are relegated to a role of beneficiary of the MLS, but the first benefit (accessing seed) is not recognized for them. This is at odds with their fundamental role in feeding the world and conserving PGRFA sustainably. Second, hyperownership through expanded IPRs has almost suppressed farmers' exemption in accessing protected (improved) PGRFA.

The issue of PGRFA access by farmers for direct use

When integrating the commons invariable principles most relevant to the question of facilitated access to seeds (anticommons dilemma, community and interdependence), the need to recognize that access should first and foremost be facilitated for farmers (i.e. for all their needs, not just for breeding, research and training purposes) appears clearly as a necessity. Treaty provisions dealing with conservation and sustainable use activities create a first set of obligations in which farmers play a central role; the MLS, which aims at promoting breeding and research activities to develop improved varieties, constitutes a separate set of obligations, in which breeders and researchers are the main actors. These two sets of obligations stress the fact that different PGRFA communities are expected to play different roles. This leads to two main questions: first, what about the primary role of a farmer, i.e. producing food? Should this role not be recognized and integrated in the Treaty system? Second, why dissociate the roles and sets of obligations (farmers conserving and using sustainably/breeders improving PGRFA)? Doing so recognizes only modern technology as valuable innovation and negates informal innovation by farmers. However, these different sets of obligations are all interrelated. There may be no efficient conservation and sustainable use of PGRFA without access to as wide a diversity as possible of PGRFA varieties and without use of that diversity by as many different stakeholders as possible. It is unrealistic to imagine that conservation and sustainable use of PGRFA will be achieved while dissociated them from the end activity of producing food. This is directly linked to the issue of the objectives of the Treaty (above in § 1), where it is recommended that sustainable agriculture and food security be recognized as direct objectives. Furthermore, it is unrealistic to imagine that conservation and sustainable use of PGRFA will be achieved as long as these separate communities (farmers vs. breeders and researchers) are kept separate.

Acknowledging this state of facts, the CGIAR centres have always provided access to PGRFA for farmers, whether for research and breeding or for direct use for cultivation, and whether for unimproved or improved material. The material provided to farmers was transferred under favourable conditions, avoiding excessive cost and stringent IP conditions. However, this takes place for a relatively limited number of species and for pure varieties, while smallholder farmers are more interested in heterogeneous varieties.

Yet, this practice has not been recognized as official interpretation of the related Treaty obligations by contracting parties. Therefore in 2010, the Ad Hoc Technical Advisory Committee on the SMTA and the MLS was requested to examine the issue, which was addressed at its second and third meetings (IT/AC-SMTA-MLS 2/10/Report at §§ 52–60 and Appendix 7; IT/AC-SMTA-MLS 3/10/Report, Appendix 3; IT/AC-SMTA-MLS 2/10/7 and IT/AC-SMTA-MLS 3/12/3). The Committee contends that there is no problem with providing PGRFA for direct use by farmers when the material is not received under an SMTA (IT/AC-SMTA-MLS 2/10/7, at §§ 9–13). The main difficulty lies with PGRFA received under an SMTA,

> since the terms of the SMTA require that the use of the material be restricted to research, breeding and training. If material acquired from the Multilateral System under the SMTA is to be made available for direct use for cultivation, this would require the express permission of the provider that included the PGRFA in the Multilateral System.
>
> (IT/AC-SMTA-MLS 3/10/Report at § 23)

The Committee specifies that making PGRFA available for direct use for cultivation should not be made under the SMTA.

The question was first explicitly limited to the transfer for direct use by the CGIAR centres, but was then expanded to transfers for direct cultivation by contracting parties too (IT/AC-SMTA-MLS 3/12/3 § 33; IT/AC-SMTA-MLS

3/10/Report at §§ 22–25). While it is encouraging that this issue was raised explicitly during an inter-sessional process, shedding more attention on farmers' needs, a more progressive position could have been taken by contracting parties. Explicitly recognizing the legality of transferring Annex I material for direct use for cultivation would be a manner of recognizing and implementing part of the obligations covered under Article 9 on Farmers' Rights. Doing so without using the SMTA would not increase administrative burden.

Over-IPRization and the disappearance of the farmers' exemption

At the frontier of this issue of access to PGRFA by farmers, several problems remain, one of which relates to the intellectual protection of improved varieties (Girard and Noiville, 2014). Indeed, how to provide access to PGRFA when more and more improved varieties are protected by strong intellectual property rights with no or very limited farmers' and breeders' exemptions?[36] Correa has attempted to provide a solution by proposing the development of a sui generis system for plant variety protection (PVP) as an alternative to UPOV 1991 obligations. This sui generis system would be compatible with existing international obligations (WTO, TRIPS, but also CBD, Nagoya Protocol and the Treaty). It would counterbalance the 'over-IPRization' process occurring with the expansion of the boundaries of patents over plants. The general idea would be to create a system close to the UPOV 1978 convention (i.e. plant variety protection for new uniform plant varieties with clear breeders' and farmers' exemptions), complemented with a protection for new farmer and other heterogeneous varieties as well as a recognition for traditional farmers' varieties, in order to prevent misappropriation of varieties developed by farmers and farmers' communities. This is also a means to recognize farmers' role in the breeding innovation process. The purpose of the present work is not to dig into these technical issues but rather analyse their impact on the conceptual constraint related to access; further information on this interesting proposal can be found in Correa's publication (2015; also see Singh et al., 2011; Saez, 2016: 2).

At the conceptual level, would this sui generis PVP system mitigate whole or part of the access problem in the Treaty implementation? If the system enables farmers to be integrated in the breeder/research community at the same level (i.e. their PGRFA are protected, their rights are recognized and their role in the innovation process is recognized at the same level), then one can believe that such a system may significantly improve the Treaty implementation and contribute to reaching its objectives and overall goals. However, this means adapting the existing plant protection system by restricting the ever-wider patent scope, and perhaps, as Van der Kooij (2010) suggests, creating a breeders' exemption in patent law. Some European countries have recently attempted to supple their patent law in this regard. According to Prifti,

> [a] recent exception to patent law is the so-called breeder's exception or breeding exception. The breeding exception allows for the free use of

> patented biological material limited to the breeding or discovery and development of new varieties. Four European countries, France, Germany, the Netherlands and, Switzerland have already adopted this type of exception into their patent laws.
>
> (Prifti, 2015, 50)

Additionally, Article 27.c of the recent Agreement on a Unified Patent Court (AUPC) incorporates a provision on the breeding exception. So far, the AUPC has not yet entered into force (as of March 2018), thus, 'it remains to be seen whether EU countries will implement such provision into their national patent law' (Prifti, 2015: 50). Moreover, as Girard explains it, this solution might not be so straightforward.

> Even if the exception is sometimes interpreted broadly and even if there are countries where an exemption for breeding purposes has been introduced into patent law, breeders are allowed to use patented biological material only for breeding purpose, not for commercializing the variety developed. If they wish to exploit their work, they must request authorization from the patent holder, who is thus in a position to require payment of significant royalties, or to refuse. That is the reason why there is an ongoing debate in the Netherlands, initiated by Plantum (the Dutch association of plant breeders), around the opportunity to set up a comprehensive breeder's exemption (as opposed to the limited breeding exemptions that have just been mentioned), thereby vesting the breeders with the right to commercialize a plant variety regardless of whether it contains a patented technology.
>
> (Girard, 2015: 51, references omitted)

Some countries (e.g. Thailand, Malaysia or India) have decided to develop sui generis legislation but, as Girard further notes, these legislations have raised many concerns.

> [F] irst, these rights may have a disincentive effect on the flow and exchange of valuable genetic materials and a propensity to create tensions between communities; second, as the concept of property over germplasm is alien to some traditional cultures, their inception may adversely affect the cosmologies of certain communities; and actually serve unwillingly the expansion of the religion of property. Finally, farmers' varieties have more often than not incredibly complex pedigree and are part of populations of landraces, if not metapopulations, that requires the very difficult task of determining the contribution made by different landraces and communities (delineation of the boundaries of the innovation) before attributing rights over the variety.
>
> (Girard, 2018: 134, references omitted)

For these reasons, Girard pleads for the implementation of local ABS rules through the development of biocultural community protocols under the Nagoya Protocol (ibid).

Finally, developing sui generis legislation would also mean not questioning the utility and reason of the existing protection rights and equate to remaining in the dominant appropriation scheme (i.e. moving away from the inappropriability of seeds). Perhaps the latter proposal would go too far and would not be realistic in the current context.

This point emphasizes why facilitating access to seeds is so crucial, but even more, why it is necessary to effectively allow for all stakeholders (not only breeders and researchers, whether from public or private institutions) to use the global seed commons: farmers should not only be 'passive beneficiaries' of financial and non-monetary benefits, but active 'co-managers' in the design and implementation process.

Recommendation 4: Recognize a direct facilitated access to PGRFA for farmers; promote sui generis PVP systems; adapt existing IP regulation to recreate an effective farmers' exemption; and adapt existing certification and commercialization of seeds regulations.

Benefit-sharing and the Benefit-sharing Fund

For this fifth Treaty topic, there are two conceptual constraints related to benefit-sharing. The foremost issue obviously relates to the lack of funding impeding the realization of benefit-sharing activities through the BSF. The need to review the funding mechanism is already taken on board in the Treaty review process and constitutes a technical constraint. This falls outside the scope of these recommendations and is therefore only superficially covered. However, finding sufficient money to fund the system remains a crucial element as a matter of course. The conceptual constraint identified here is that the financial trigger is linked to accessing the material included in the MLS. The Treaty analysis in Chapter 4 showed that the SMTA money triggering clause is not effective because of the time scale in innovation processes and because the SMTA is not used by those breeders who have money and detain IPRs. Besides, there is almost no voluntary contribution to the BSF (with the exception of less than a handful of cases, by states or other stakeholders i.e. recently the European seed industry). Too little money is coming into the Benefit-sharing Fund. Finding funding that does not derive from the compulsory benefit-sharing obligation as foreseen in the SMTA might be a more viable option.

Along that line, the ongoing review process seeks to secure more stable and long-term funding means. When trying to implement the invariable principles of interdependence and anticommons dilemma in finding a solution to this conceptual constraint, one should keep in mind the need to protect the interdependence link between all PGRFA stakeholders and resources and to promote the widest possible use and involvement. One way would be to obtain money from other stakeholders such as the food industry[37] (i.e. through the implementation of Article 13.6 dealing with voluntary contributions to the MLS by the food processing industries) or consumers. Up to now, neither consumers associations nor the food industry have been much involved in the discussions between the various Treaty stakeholders (Frison et al., 2011). This is surprising if one considers that we are all consumers, whereas in developed countries, farmers represent only 3 per cent of the population, and that the food processing industry is economically strong and financially flourishing. Therefore, it is vital to raise awareness among consumers, to identify and encourage the food industry to contribute to the MLS and to design mechanisms for this purpose, as their future food supply depends on the continued availability of PGRFA. Strong incentives for the food industry to contribute to the Benefit-sharing Fund are required. An example could be to create a 'green tag' for products coming from these industries contributing to the Fund. With this label consumers would be able to decide to buy products that contribute to the conservation and sustainable use of PGRFA. But for this to happen, consumers should be conscious that their choices regarding food products provide them with considerable leverage to influence the food industry's economic and policy choices. Contracting Parties should therefore target consumers as well as farmers' organizations in their public awareness programmes (Frison et al., 2011: 273–274). However, to take this path, there needs to be strong and determined political will. Apart from some rare countries which have adopted alternative modes of voluntary payment (notably Norway), it seems that the momentum has not come yet for such political commitment (Petit et al., 2001).

The second conceptual constraint related to benefit-sharing deals with the position of farmers in the access/benefit-sharing relationship, that is to say their role and position within the Treaty community. Benefit-sharing provisions, including the BSF round of calls for funding benefit-sharing projects, put farmers in the position of beneficiaries, i.e. passive receivers. Notwithstanding the fact that the reduced funding of the BSF does not allow farmers to benefit from the Treaty, that the BSF procedure only allows to fund a limited number of selected projects following selective criteria (vs. benefiting all farmers), and that direct access to seeds is not formally recognized for farmers, this passive position contrasts with their ancestral role of main actors in the seed and food chain. It also departs from the community invariable principle which advocates for an active participation of all stakeholders in the commons. The Treaty MLS was designed for researchers, breeders and trainers, as bridging elements between the commercial stakeholders (seed industry) and the main users of seeds (farmers). Analysis of the Treaty implementation shows that the MLS is not a global seed commons (i.e. for all) but a researcher/breeder seed commons, where farmers are relegated to a passive position in the exchange of seeds. However, if one wants the Treaty to reach its objectives, farmers and other relevant stakeholders, will have to be included in the game as active participants to the management of the system. The redesign of the MLS should focus on farmers as primary stakeholders of the system.

Recommendation 5: The benefits of the Treaty should reach all beneficiaries and farmers should be repositioned as active stakeholders in the Treaty, MLS and BSF management.

Information and knowledge

For smallholder farmers, seeds (the material) and traditional knowledge (the related information) are indissociably linked. This type of information is crucial for the conservation and sustainable use of PGRFA as a seed without its associated information and traditional knowledge is of no use. This is recognized in Treaty Article 9.2 which stipulates that 'each Contracting Party should [. . .] take measures to protect and promote Farmers' Rights, including: a) protection of traditional knowledge relevant to plant genetic resources for food and agriculture'.[38] Therefore, information and (traditional) knowledge should be protected from misappropriation;[39] their access should be promoted and made available to all seed stakeholders, along with the material (and the adequate technology to handle it). Traditional knowledge is considered as 'information' relating to PGRFA. The Treaty provides for an instrument to address information: the Global Information System (GLIS), which aims at enhancing the documentation on PGRFA (that including that of crop wild relatives, on-farm and in situ material), as well as promoting its exchange (Plant Treaty Article 17).

The conceptual constraint related to this topic lies in the inadequate (or at least incomplete) tools developed by contracting parties[40] to facilitate access to information that is most relevant to farmers, in particular traditional knowledge. Ideally, the GLIS could be used as a clearinghouse mechanism sharing traditional knowledge while protecting it from misappropriation. However, to date, there is no efficient mechanism protecting traditional knowledge from misappropriation, which then hinders access. The tools that are proposed to contribute to the GLIS (*inter alia* the DivSeek Initiative[41] or the Global Open Genome Sequence Data Framework)[42] respond to experts' requirements, far from a majority of farmers' knowledge, needs and practices at local levels. It is understood that the end-beneficiary of this type of initiative is to be the farmer, but it does not recognize the fact that the majority of the seeds used by smallholder farmers does not originate from commercial breeders but from informal seed exchange systems between farmers. This confirms the fact that the Treaty system is designed for breeders and researchers and fails to integrate farmers' role and needs in that system, thereby failing to develop means to reach the Treaty overall goals of food security and sustainable agriculture. This reflects again the imbalance in the Treaty between breeders and farmers. Addressing issues related to traditional knowledge through the GLIS could be a way to provide the most relevant information and knowledge to farmers and to limit the cases of misappropriation, which have inflamed polemics about biopiracy (Isaac and Kerr, 2004; Ho, 2006; Shiva, 1997). The Digital Object Identifier (DOI) tool, which is currently being set up by the Treaty Secretariat,[43] could contribute to develop the necessary sharing and protection tools if it were expanded to traditional knowledge. For now, the DOI project focuses on the accessions deposited by parties in the MLS.

Therefore, measures should be enforced to re-establish a true public research, which aims are to serve the public interest (i.e. reach food security and sustainable agriculture) and the needs of smallholder farmers and which funding does not depend on private interests. This would mean *inter alia* refocus public research on traditional varieties (with intra-variety genetic heterogeneity) (Bonneuil et al., 2006). The GLIS should include information systems that allow sharing of traditional knowledge while protecting it from misappropriation. One way to limit misappropriation could be to transfer the burden of proof on the patent or the PVP owner. When applying for a patent or a PVP, the applicant should demonstrate that the innovation does not originate from a misappropriated material/knowledge.

Recommendation 6: Develop the GLIS keeping in mind the overall goals of the Treaty, by rendering available, visible and accessible information relevant for all stakeholders, in particular farmers; seek means to turn the MLS/global seed commons into a space where traditional knowledge would be protected from misappropriation.

Third Party Beneficiary

The seventh Treaty topic deals with the legal mechanisms to enforce Treaty rights. These rules and procedures intervene at two levels: 1) at the level of the MLS and its SMTA, where PGRFA users may act; and 2) at the level of the Treaty, where contracting parties are the major stakeholders at play. The focus will be placed on the most innovative part of the system: the Third Party Beneficiary (3PB). The analysis in Chapter 4 showed that the 3PB concept bridges the gap between the private contractual law relationship between parties to the SMTA and the public international law setting where contracting parties to the Treaty are anchored. It creates a triangular relationship between contracting parties, stakeholders and the MLS, in which every participant (whether directly involved in the SMTA contractual relationship or not) may trigger the 3PB when suspecting a breach of rights in an SMTA. Indeed, the 3PB is the virtual entity (materialized by FAO) representing the MLS, designed to assert its rights and to allow for enforcement of SMTAs.[44] It functions as a warrant for the respect of the MLS' collective rights and obligations, which can be triggered by 'any natural or legal person'; even those not party to the contractual agreement (SMTA) at the origin of the breach of right (Procedures for the Operation of the Third Party Beneficiary, Article 4.2; see also Resolution 11/2013, § 4).

The conceptual constraint related to the 3PB lies in the lack of transparency and advertising of the system at two levels: prior to a case and once a case is triggered. Indeed, to be fully effective, advertising and informing members of the Treaty community on the existence of the 3PB and its procedure in the widest and most transparent manner is crucial. The 3PB webpage briefly explains the procedures under its scope. However, the Plant Treaty website does not explicitly mention the first 3PB case that occurred in 2012–2013, nor does it publish a specific report on the case on the 3PB webpage. The information published is limited and can only be found in the documents of the Governing Body session (if one is aware that a case has occurred), not on the 3PB webpage. Besides, no action is taken by the Governing Body or the 3PB to advertise and inform the public on its role and procedures. There could be an 'Easy-3PB' online tool, similar to the 'Easy-SMTA' online tool, to facilitate triggering the 3PB procedure.

Furthermore, once the 3PB is triggered, information on the resolution of the case should be more transparent. Article 9 of the 3PB Procedures provides that the 3PB shall submit a report to the Governing Body at every Regular Sessions. Such report[45] should contain information on a number of items regarding its operations[46] in the biennium (that is to say for the part of year 2011 that was not covered by the previous report to the Governing Body, and for the biennium 2012–2013). At the last session of the Governing Body, a 'Report on the Operations of the Third Party Beneficiary' was submitted to the Governing Body, but its summary of the first case was very limited (IT/GB-6/15/10), and access to direct information on the case is not possible. In the 2013–2014 case, the situation seems to have been swiftly solved, *inter alia* because the CGIAR centres involved were embarrassed with the publicity that the ETC group had made on these cases. Reputation is a strong motive for complying with the rules. This is why transparency is so important.

Finally, contracting parties have refused to expand the scope of action of the 3PB to compliance issues (and have consequently developed a parallel compliance mechanism), limiting the 3PB scope of action to the breaches in SMTAs. While this is understandable from a political point of view during the Treaty negotiation, one could question this decision. Indeed, expanding the 3PB's action to situation of non-compliance outside the SMTA contractual relationship could enhance implementation of major Treaty obligations. Reputation is a good incentive for respecting the rules of the game. Enlarging the role of the 3PB would trigger this reputational spectrum. However, the Governing Body might not be ready yet for opening such a debate . . .

Recommendation 7: Advertise on the 3PB's role and procedures to the Treaty community and the public and deal with 3PB cases in a more transparent way; eventually, expand the 3PB's competency to situations of non-compliance with Treaty provisions outside the SMTA contractual relationship.

Participation and governance

In the Treaty analysis, information on the rules and procedures for the governance of the Treaty showed that there is little space for other actors than states and international organizations to govern the PGRFA issues at stake. This is consistent with international law. However, the arguments made in this chapter emphasize why it is indispensable to effectively allow all stakeholders to participate in the governance of the global seed commons. In particular, farmers should not only be passive beneficiaries of financial and non-monetary benefits, but active co-managers in the design and implementation process. The conceptual constraint related to this Treaty topic deals with involving all stakeholders in the governance of the Treaty, in particular de facto holders of seeds (i.e. farmers) along with breeders, researchers or the seed industry.[47] Taking such a multi-stakeholder approach to governing the Treaty (Nasiritousi et al., 2015;

Bäckstrand, 2006), inspired from the functioning of the FAO Committee on World Food Security, could contribute to designing an effective global seed commons through the current MLS review process. The Committee on World Food Security (CFS) defines itself as

> the foremost inclusive international and intergovernmental platform for all stakeholders to work together to ensure food security and nutrition for all. The Committee reports to the UN General Assembly through the Economic and Social Council (ECOSOC) and to FAO Conference.[48]

Chapter 6 has shown that many different stakeholders, with diverse (and sometimes opposing) interests are involved with the Treaty. However, the number and heterogeneity of these actors makes it difficult for the Treaty to be effectively implemented (Libecap, 1994). The lack of stakeholders' participation is problematic at different levels: the needs and specificities of all stakeholders are not reflected in the system as designed; it poses the question of lack of trust between stakeholders (Six et al., 2015; Henry and Dietz, 2011) and of legitimacy in governing the resource (i.e. are public institutions, breeders and researchers more legitimate to manage PGRFA than farmers, who have acted as stewards during millennia?). What role is left for informal dialogues[49] and informal networks within the formal system?[50]

NGOs and farmers' organizations have demonstrated that they are able to provide concrete, useful, important information on the conservation and sustainable use of seeds. As primary actors, smallholder farmers in particular (in number, they are the majority group of seed users and as food producer, a majority of the world population), should be able to participate in the international management of seeds. They should collaborate in the identification of what material/technology should be researched or developed.[51] Recognizing their needs and practices, their modes of functioning, their networks, would be a way to implement Ostrom's eighth design principle on nested enterprises and respect the embeddedness of seed systems. It would allow to respect the heterogeneous, complex and diverse characteristics of the plural seed systems within the global seed commons. It would also fit quite well with recent studies demonstrating the need to shift from a uniform, industrial agricultural mode of production to diversified agroecological systems (IPES-Food, 2016).

Smallholder farmers or farmers more generally as well as breeders and researchers should be formally invited to take part of the global seed commons. As main game player of the agricultural input market, the Big Five should also be part of the debate. The several multi-stakeholders dialogues which occurred during the negotiation of the Treaty, and recently during its implementation, showed that having all stakeholders sitting at the same table at the same time could favour a constructive dialogue between historically opposed groups, and eventually unblock difficult negotiations.

Finally, should other stakeholders at different levels of the food chain be involved: consumers and the food processing industry? Consumers certainly have a say in what they want to eat (healthy, diverse, local food?) and hence they could influence the type of research activities funded by public research organizations. Citizen initiatives are also active in promoting the recognition of a right to exchange and grow traditional seeds (which can be contrary to EU seed legislation, see Winge, 2012, 2015). As for the food processing industry, the Treaty highlights their role and responsibility in the food chain (Article 13.6) and identifies them as potential financial contributors to the MLS. To what degree should these stakeholders be involved in the Treaty governance? Schlager and Ostrom (1992: 251) distinguish 'between rights at an operational-level (access and withdrawal to a CPR) and rights at a collective-choice level' (management, exclusion and alienation) that is to say 'the difference between exercising a right and participating in the definition of future rights to be exercised' (see above in the paragraph on 'Ostrom's institutional analysis'). These issues should be further discussed in order to reinvent how to govern the Treaty and its instruments.

However, rather than conceptual constraints, these are technical challenges regarding the governing mechanism of the Treaty (i.e. adapt rules and procedures of the Governing Body to create a decision-making space for all Treaty stakeholders). Concretely modifying the general governance mechanism of the Treaty so as to better reflect the reality of the PGRFA field (in terms of heterogeneity of actors and their needs, of networks and of institutions) falls outside the direct scope of this thesis. Notwithstanding the fact that this would necessitate quite some courage and creativity for changing universally recognized international law of treaty rules, I believe that there is hope in this regard. Indeed, the Treaty forum has undoubtedly demonstrated its capacity to be creative and innovative in designing new concepts and instruments under international law. Moreover, there are examples of such evolution in international governing systems from which the Governing Body could draw inspiration for its renovation, notably the reform of the Committee on World Food Security (De Schutter, 2013).[52] It could even go further by establishing voting rights to each stakeholder groups, not only to contracting parties. Adapting the Treaty governance mechanisms towards an inclusive multi-stakeholder approach might contribute to resolve constraints identified in the Treaty analysis, *inter alia* by ensuring a better political power balance. In order to build a system which all stakeholders will abide by and will implement, a rebalancing of powers needs to take place in the Governing Body, through the recommended inclusive multi-stakeholder approach. The recognition of FRs and the farmers' participation in the decision-making process are preconditions for this to take place.

Recommendation 8: Allow all stakeholders to effectively participate in the global seed commons governance, taking into account the various levels of governance.

Conclusion

Implementing these eight recommendations are first steps towards constructing a political commons and towards addressing the difficult issue of PGRFA management with an 'ecology of law' perspective. Indeed, establishing the conditions for a real dialogue and a collective decision-making space for all stakeholders is the first step towards constructing the political global seed commons.

1 Formally recognizing food security and sustainable agriculture as direct objectives of the Treaty would be a strong political signal, shifting the purpose of the MLS from being a tool that upholds seed commodification to a tool enhancing seed commoning. This would reset sustainability at the heart of the global seed commons' project and recognize food production as the core of farmers' role.

2 Harmonizing the scope of the MLS with that of the Treaty to include all PGRFA and expand the boundaries of the Treaty to make it truly global would respond to the anticommons dilemma. A sustainable PGRFA management functions like a virtuous circle where the more seeds and related information are used and shared, the more they develop, expand and gain value. Therefore, a wide community of users for a wide spectrum of resources is crucial.

3 and 4) Formally recognizing farmers' rights at the international law level and committing to implement these rights at the national level is a sine qua non condition for an effective global seed commons to function and reach its objectives. This recommendation goes hand in hand with the fourth one recognizing a direct facilitated access to PGRFA for farmers, promoting sui generis PVP systems to recreate effective farmers' exemption, and softening the commercial seed certification and registration regulations. They both necessitate strong political will to formalize de facto rights into internationally recognized de jure rights.

5 Benefits of the Treaty should reach all beneficiaries and farmers should be repositioned as active stakeholders in the Treaty, the MLS and the BSF management. This implies that the BSF should be widened and increased. This fifth recommendation resonates with the role of each stakeholder in a community, and hence in a network of related communities. Farmers are those tilling the land, sowing the seeds, harvesting the fruits of their labour, producing our food. They cannot be relegated to passive beneficiaries in the Treaty system. Farmers' knowledge and practices are valuable. Humanity is benefiting from them. Implementing an 'ecology of law approach' (Capra and Mattei, 2015) to the MLS requires this move to happen.

6 Developing the Treaty's Global Information System, keeping in mind the overall goals of the Treaty, by rendering available, visible and accessible relevant information to all stakeholders, in particular farmers, participates to the rebalancing of powers that needs to take place. Similarly, seeking means to transform the global seed commons into a space where seeds and traditional knowledge would be protected from misappropriation is an important element for the MLS to function effectively. Enlarging the Annex I list of crops to all PGRFA will not happen as long as the rights

of de facto right holders are not protected from the excesses of the 'second enclosure movement' (Boyle, 2003b). However, this enlargement has to take place, to respect the invariable principle of sustainability, interdependence and anticommons dilemma.

7 In order to contribute to the rebalancing of power between stakeholders, the potential of the innovative Third Party Beneficiary instrument has to be unlocked. Advertising on the 3PB's role and procedures to the Treaty community and to the public and dealing with 3PB cases in a more transparent way are first steps in that direction. Envisaging to expand the scope of the 3PB to situations of non-compliance would be a second step.

8 Finally, allowing all stakeholders to effectively participate in the governance of the global seed commons is the last but not least recommendation that results from this work. As Mattei (2011b: 12; see also 2011a: 101–102) puts it 'each individual's survival depends on its relationship with others, with the community, with the environment'. Constructing a political global seed commons taking this into account imposes 'commoning' (Bollier, 2014), or, as Dardot and Laval (2014: 580) formulate it, 'act in common' ('agir en commun'). According to them, participating to an activity constitutes the basis of all political obligations, i.e. from 'co-activity results co-obligation' (ibid). The political obligation then 'derives its strength from practical engagement binding all those who have drawn up the rules of their activity together', (ibid)[53] thereby imposing a multi-stakeholder approach to any political common governing system. Rebalancing the economic and political powers within the Treaty forum is crucial to allow for a democratic governance to take place (Lucarelli, 2013). Only then can adequate political and governing decisions be made that will address the issue at stake in responding to all its aspects and to all stakeholders needs. Only then can the collective interest (i.e. sustainable agriculture and food security) be safeguarded and translated into public policies (Lucarelli, 2011a, 2011b, 2013).

Throughout these eight recommendations, there is one important cross-cutting aspect that appears in almost every topic: the lack of recognition of the role and rights of smallholder farmers. By lack of recognition, the author means the lack of translation of their role and de facto rights into concrete obligations, instruments and procedures in the Treaty implementation, i.e. de jure rights. This participates in the identified imbalance of rights between Treaty stakeholders; an imbalance which has to be redressed in order to effectively implement a global seed commons and eventually reach the Treaty's objectives.

To conclude, I am not going as far as affirming that inappropriability is the only possible normative political construct for 'the common' to happen (Dardot and Laval, 2014: 583). To suggest that the MLS should be a space for inappropriability of all seeds and making a normative statement as this being THE solution to the constraints identified in the PGRFA management system would presume from stakeholders' discussions and decisions. Other solutions can be envisaged and lead to an effective politically constructed global seed commons. Such a political decision can only be the result of a collective debate and choice between all stakeholders. This decision could therefore only be taken and the above recommendations implemented, once the balance of powers in the Treaty forum is re-equilibrated, thereby allowing for an equitable and fair discussion and decision-making process between all stakeholders.

Notes

1 Although authors rather see this role played by the CGIAR as a means to maintain an easy access to resources for developed countries and their breeding companies as a warrant to secure their economic development through the Green Revolution (Lenné and Wood 2011: 150; Mooney, 2011: 142).
2 Resolution 8/83 recognized that 'plant genetic resources are a heritage of mankind to be preserved, and to be freely available for use, for the benefit of present and future generations'.
3 Sections 2 and 3 of the present chapter have been published earlier this year in Frison (2018, chapter 17 of the book) Girard and Frison (2018) 'Commons, Plant Breeding and Agricultural Research', Routledge, pp. 272–89. With permission from the publisher.
4 There is some confusion on the term and definition used in the literature regarding the commons at a 'global' / universal / world level. Some authors talk about global commons when dealing exclusively with the high seas and deep seabed, Antarctica, outer space and the global atmosphere (Buck, 1998; Vogler, 2012; Ashley Roach, 2012). Regarding the Treaty, authors also talk about the 'global seed commons', but without associating it with the latter school of thought.
5 In game theory, this has been modelled under the prisoner's dilemma (Rapoport and Chammah, 1965).
6 US Supreme Court, *Diamond v. Chakrabarty*, 447 U.S. 303 (1980), where the court ruled (5–4) in favour of Chakrabarty, holding that: 'A live, human-made micro-organism is patentable subject matter under 35 U.S.C. § 101. [. . .] Respondent's micro-organism constitutes a "manufacture" or "composition of matter" within that statute'. For an analysis of the evolution of plant patenting in the USA and Europe see Batur (2014: 69 et seq). For an analysis of the *Chakrabarty* case and following relevant case law see Aoki (2004).
7 For a summary history of Ostrom's research projects see Coriat (2013: §§ 20–62); Van Laerhoven and Ostrom (2013).
8 For early thoughts on the bundle of rights concept see Ciriacy-Wantrup and Bishop (1975). The bundle of rights approach can also be found in Alchain and Demsetz' work, although leading to an opposite conclusion.
9 International Association for the Study of the Commons website, www.iasc-commons.org/
10 International Journal of the Commons website, www.thecommonsjournal.org/
11 See various web-based information: 'Economic Crisis in a Globalized World' – Wide-Angle (21 November 2008) at www.pbs.org/wnet/wideangle/uncategorized/how-global-is-the-crisis/3543/; 'Financial Crisis' – IMF, at www.imf.org/external/np/exr/key/finstab.htm; and 'Times of Crisis' – Reuters: Multimedia interactive charting the year of global change, at http://timesofcrisis.reuters.com/app/
12 Authors that are included in this trend do not necessarily all share the same ideas. What brings them together in this trend is the fact that they push further Ostrom's thinking, although exploring different directions.
13 European Water Movement, http://europeanwater.org/fr/

14 Personal translation: 'c'est la finalité des "communs" qui est mise en exergue, une finalité de *justice globale dans les droits d'accès et d'usage des ressources essentielles*'.

15 Parts of this reflexion will be further developed in the ANR research project 'Bioculturalis' (2019–2021), upon acceptance of funding, with Fabien Girard, Université de Grenoble.

16 Nuits debout, https://nuitdebout.fr

17 Tatro della Valle, www.teatrovalleoccupato.it/

18 Acqua Bene Comune, Napoli prima in Italia per la gestione pubblica dell'acqua, www.comune.napoli.it/flex/cm/pages/ServeBLOB.php/L/IT/IDPagina/20076

19 Movimiento M15, www.movimiento15m.org/

20 'Transition Network: A Movement of Communities Coming Together to Reimagine and Rebuild Our World', https://transitionnetwork.org/ and https://citiesintransition.eu/

21 End of 2007, less than 25 scientific publications on the Plant Treaty were collected, more than half of which are authored by non-academics. FAO documents and publications are not counted in this list.

22 These principles are not exclusively part of the theory of the commons and can be found in other theories and disciplines. Sustainability for example is clearly found in environmental international law and sustainable development principles.

23 In 1997, FAO released the results of a worldwide study aimed at assessing the degree of dependence of a country's main food crops on genetic diversity in areas of origin and primary diversity located elsewhere. It shows that all regions in the world are highly dependent upon resources originating for another region, North America being the highest dependent region, and Asia and the Pacific region being the least dependent region (Palacios, 1997). This study was requested by the CGRFA and complements the first report of the State of the World's Genetic Resources for Food and Agriculture. Palacios adds that 'crops such as cassava, maize, groundnut and bean originated in Latin America but have become food staples in many countries of sub-Saharan Africa, illustrating the interdependence of cropped species in the developing countries. Cassava is the main food crop for 200 million Africans in 31 countries and has a farmgate value of over US$ 7 billion. At the same time, Africa and its indigenous varieties of millet and sorghum have helped feed other parts of the world such as southern Asia (13%) and Latin America (8%)' (Palacios, 1997: 3). In 2015, another study confirms these findings, highlighting that our estimation for countries' interdependence is even higher and more diverse than foreseen (Khoury et al., 2015).

24 FAO (2010). There is not a single self-sufficient country for crop genetic resources. All countries are both donors and recipients of PGRFA. This means that breeding new varieties repeatedly necessitates genetic material from other countries (Frison and Halewood, 2005). The average degree of interdependence among countries for their most important crops is around 70 per cent (Esquinas-Alcazar, 2005). Most of the efforts that are necessary to manage plant genetic resources can therefore only be carried out through international cooperation (Fowler and Hodgkin, 2004).

25 Specifically, as Fowler and Hodgkin noted, 'materials held in genebanks eventually require regeneration, ideally in the same environment in which they were collected in order to avoid changes in the genetic composition of the sample, and even loss of some genes or alleles. Because most collections contain materials from many countries, cooperation is needed if high conservation standards are desired. In Europe, there is increasing collaboration. In some cases, different genebanks concentrate on maintaining different crops, and for a number of crops, common information resources have been developed' (Fowler and Hodgkin, 2004: 148).

26 Latin American and the Caribbean countries stated – during the negotiation of the Treaty, in the Bogota Declaration (18–22 March 1996) – that '[t]he trend toward globalization of the international economy and the inherent growing interdependency find clear expression in the issue of sharing of and access to the world's Genetic Resources for Food and Agriculture' (FAO, 1996a: 4). This view was shared by most negotiating countries.

27 This idea relates to the fifth invariable principle of 'community' and to questions related to stakeholder participation in governing a collective system.
28 Another argument supporting this conclusion is the one developed by Pistor and De Schutter on 'essential resources', which 'calls attention to distributional equity and sustainability'. According to them, governance of resources 'should be promoted with the proviso that nobody should be excluded from resources that are essential for satisfying basic needs; further, the exploitation of the resource today should not jeopardize the ability of future generations to satisfy their needs' (Pistor and De Schutter, 2015: 24).
29 Although now with CRISPR technology, the physical material is no longer needed to create an improved variety. As long as the genetic code is known, access to the physical seed variety becomes superfluous, thereby rendering the control on seed access almost impossible.
30 See note 27 in Chapter 3 for an explanation of the OSSI system.
31 This raises interesting questions on the operational-choice versus collective-choice level of actions developed by Schlager and Ostrom in their implementation of the bundle of rights in the CPRs (Schlager and Ostrom, 1992).
32 The Global Plan of Action for the Conservation and Sustainable Utilization of Plant Genetic Resources for Food and Agriculture was formally adopted by representatives of 150 countries during the Fourth International Technical Conference on Plant Genetic Resources, which was held in Leipzig, Germany, from 17 to 23 June 1996. The World Food Summit took place in Rome, Italy on 13–17, November 1996, where both the Rome Declaration on World Food Security and the Global Plan of Action were adopted.
33 Self-determination is understood as the choice left to farmers to decide what, how and for whom to produce food, freed from the neo-liberal economic power and bargain imposed on them. Focus is set on autonomy of all seed-related activities (Van der Ploeg, 2008; Demeulenaere, 2014; Schneider and Niederle, 2010; Stock et al., 2014; Stock and Forney, 2014).
34 At the Sixth Governing Body, during a side event session, Eng Siam Lim had launched the idea that 'a MLS for farmers' should be created.
35 Global Consultation on Farmers' Rights website, www.planttreaty.org/sites/default/files/008_GB7_NCP_FREC_en.pdf
36 Although the EU directive on biotechnological innovations provides in its Article 11 paragraph 1 allows farmers to use the harvest deriving from patented material (this does not include the right to exchange seeds for example). See also recitals 46–51 of Directive 98/44/EC and Article 14 of Council Regulation No. 2100/94 of 27 July 1994 on Community Plant Variety Rights. Article 14 recognizes that small farmers may freely use harvested material of some plant species (listed in its paragraph 2.a), while other farmers should pay an equitable remuneration (see Council Regulation, Article 14, paragraph 3).
37 Resolution 1/2006, point 11 '[i]nvites Contracting Parties, the private sector, *including the Food Processing* and other value-added Industries, non-governmental organizations, and all other interested parties, to make voluntary contributions to the Funding Strategy' (emphasis added).
38 Traditional knowledge is an important question to further seek.
39 Current negotiations on an international legal instrument to ensure the effective protection of TK, traditional cultural expressions and genetic resources are taking place within the WIPO Intergovernmental Committee on Intellectual Property and Genetic Resources, Traditional Knowledge and Folklore (WIPO-IGC). See www.wipo.int/tk/en/igc/
40 The Governing Body adopted a Vision paper which States that the 'development of a truly effective Global Information System as foreseen in the International Treaty involves, *inter alia*: strengthening existing systems and, where gaps remain, establishing new systems and initiatives; promoting inter-connectivity among systems; and providing overarching mechanisms to ensure ready access to the information and services provided'. See IT/

GB-6/15/Report, Appendix A, Resolution 3/2015, Annex 'Vision for the Global Information System on PGRFA'.

41 The objective of this community-driven initiative is to bridge the gap between the information requirements of genebank curators, plant breeders and more targeted upstream biological researchers, in order to support applied germplasm curation, forward-looking breeding programmes and strategic research. DivSeek, 'Harnessing the power of crop diversity to feed the future'. White Paper, available at http://static1.squarespace.com/static/537207e3e4b0d4555960edfd/t/53b08ea6e4b0efba71ed6fbc/1404079782586/White+Paper+DivSeek.pdf

42 The Global Open Genome Sequence Data Framework pledge for a 'universal access to genome information, needing nothing more than a web browser' in order to 'transform plant breeding [. . .] [and] spawn innovation around the world'. Warthmann and Chiarolla (2015: 2) propose to establish a 'public license for genomic information on crop germplasm' as the first mechanism to ensure that 'such data will be systematically treated as a public good for the benefits of mankind'. Available at https://sustainabledevelopment.un.org/content/documents/5934Thinking%20a%20global%20open%20genome%20sequence%20data%20framework%20for%20sustainable%20development.pdf.

43 Digital Object Identifiers (DOI) of the Global Information System (GLIS) of Article 17 of the International Treaty on Plant Genetic Resources for Food and Agriculture (ITPGRFA) webpage, www.fao.org/plant-treaty/areas-of-work/global-information-system/doi/it/

44 The Plant Treaty website states that's the 'Third Party Beneficiary is an entity designated by the Governing Body of the International Treaty and which acts on behalf of the Governing Body itself and the Multilateral System to ensure observance of the contractual terms and conditions of the SMTA by the individual providers and recipients'. www.planttreaty.org/content/what-third-party-beneficiary

45 By Resolution 5/2009 and Resolution 5/2011, the GB requested the Secretary to provide such report in accordance with Article 9 of the 3PB Procedures.

46 Article 9 of the 3PB procedures states that the report should contain information on: 'a) the number, and a summary, of cases where it received information regarding noncompliance with the terms and conditions of a Standard Material Transfer Agreement; b) the number, and a summary, of cases where it initiated dispute settlement; c) the number, and a summary, of disputes settled through amicable dispute settlement, mediation or arbitration; d) the number, and a summary, of pending disputes; e) any legal questions that appeared in the context of dispute settlement and that may require the attention of the Governing Body; f) the expenditure from the Third Party Beneficiary Operational Reserve; g) any estimate of the needs of the Third Party Beneficiary Operational Reserve in the forthcoming biennium; h) any other relevant non-confidential information'.

47 In some developed countries' delegation, representatives from the seed industry are often participating as experts to the delegation, thereby constituting an officially invisible presence of the sector to the negotiations, but officiously actively participating to the process. This reflects once again, the imbalance of powers between stakeholders.

48 United Nations' Committee on World Food Security (CFS) website, www.fao.org/cfs/cfs-home/en/

49 With the Crucible Group and Keystone Dialogue initiatives, History has shown the utility of involving stakeholders in discussing problems and imagining solutions that would contribute to collectively face major future challenges. Indeed, such informal dialogue constitutes a good place to tackle highly technical issues (and take these issues far from the political sphere for a while), by those very persons who deal with these aspects in their everyday work (contrary to negotiators in official GB meetings who are generally representatives of ministries and not direct users of PGRFA).

50 The tentative informal multi-stakeholder dialogue held in 2013–2014 aborted. See Chapter 3 in this book.

51 The CGIAR already builds significant partnerships with local organizations and universities in developing countries.
52 Using a multi-stakeholder, inclusive approach, the CFS develops and endorses policy recommendations and guidance on a wide range of food security and nutrition topics. These are developed starting from scientific and evidence-based reports produced by the High Level Panel of Experts on Food Security and Nutrition (HLPE) and/or through work supported technically by The Food and Agricultural Organization (FAO), The International Fund for Agricultural Development (IFAD), World Food Programme (WFP) and representatives of the CFS Advisory Group. CFS holds an annual plenary session every October in FAO, Rome.
53 Personal translation: 'tire toute sa force de l'engagement pratique liant tous ceux qui ont élaboré ensemble les règles de leur activité'.

7 Conclusion

Redesigning a global seed commons

Synthesis

Access to seeds for farmers (like access to land or to water) is an essential component for reaching food security and sustainable agriculture. However, there are several impediments to such access, including erosion of agrobiodiversity, legal and technological means enclosing PGRFA, or political hurdles. These impediments are amplified by risks and hazards resulting from climate change. These are immediate challenges, which Humanity has to address now in the collective interest.

The present book attempted to unravel some of the questions and difficulties related to these challenges by analysing in detail the International Treaty on Plant Genetic Resources for Food and Agriculture, which aims at conserving, sustainably using and facilitating access to PGRFA. Using an inductive research approach, where several disciplines, theories, concepts and methods are mixed, a thorough legal analysis of the Treaty was carried out and complemented by a more sociological stakeholders' analysis. This mixed method allowed to capture a 360° view of the political situation and to understand the issues at stake in the international setting regulating access to seeds.

The research results showed that, although the Treaty and its instruments (Multilateral System of access and benefit-sharing (MLS), Third Party Beneficiary (3PB), Benefit-sharing Fund (BSF), etc.) are very innovative from an international law perspective, their implementation revealed major dysfunctions. Their examination enabled to spot eight important conceptual constraints in the Treaty's structure, which hinder contracting parties in reaching the set objectives. The theory of the commons has been identified as a useful theoretical framework to address these constraints. Six commons' invariable principles were set forward to mitigate these constraints and eight recommendations were formulated in an attempt to improve the Treaty at the conceptual level. Table 7.1 below provides a summary of the conceptual constraints and recommendations.

By transforming the current intergovernmental multilateral legal instrument into an effective and collectively constructed political *Global Seed Commons*, the overall objective of this work is to contribute to designing an alternative path to the current seed regulatory and legal setting entangled in an out-of-date public/

Table 7.1 Summary table of recommendations

	Treaty topics	*Conceptual constraints*	*Recommendations*
1	**Sustainable agriculture & food security**	Overall goals of Treaty are not reached *inter alia* because they are not recognized as direct objectives	Formally recognize food security and sustainable agriculture as direct objectives of the Treaty
2	**Scope**	Difference between the scope of the Treaty and the scope of the MLS, leading to dysfunction	Harmonize the scope of the MLS with that of the Treaty to include all PGRFA Expand the Treaty boundaries to make it truly global
3	**Farmers' Rights**	The role of farmers in PGRFA management and their associated rights are not recognized at the international level in the same terms as IPRs	Formally recognize Farmers' Rights at the international law level Commit to implementing these rights at the national level
4	**Facilitated access**	Facilitated access is absent for the ultimate beneficiaries i.e. farmers	Recognize a direct facilitated access to PGRFA for farmers Promote sui generis PVP systems to recreate an effective farmers' exemption Adapt existing regulations on seed certification and commercialization
5	**Benefit-sharing/ Benefit-sharing Fund**	Farmers are considered as passive beneficiaries denying their de facto active role as main stakeholders in the food production chain Benefit-sharing is taking place in a very limited way	Reposition farmers as active stakeholders in the Treaty, MLS and BSF management Increase size of BS Fund Benefits of the Treaty should reach all beneficiaries
6	**Information/ knowledge**	Appropriation, Protection Availability mainly of one type of information of interest to breeders	Develop the GLIS keeping in mind the overall goals of the Treaty and the needs of smallholder farmers Seek means to turn the MLS into a space where traditional knowledge would be protected from misappropriation
7	**Third Party Beneficiary**	Preservation of MLS rights, but not of all stakeholders' rights. Lack of system to balance powers	Advertise the 3PB's role to the Treaty community and to the public to increase its efficiency and use Deal with 3PB cases in a more transparent way Expand 3PB's mandate to compliance
8	**Participation/ governance**	Governance of MLS remains at state level Lack of inclusion of all stakeholders at all levels Problem of trust	Allow all stakeholders to effectively participate in the global seed commons governance

private good dichotomy appropriation scheme. One cross-cutting aspect that appears all along the analysis is the lack of recognition of the role and rights of smallholder farmers. Recognition of Farmers' Rights – as a collective right of groups – at the international level would be an essential step to overcome the imbalance of rights pertaining to seeds and to reach the food security and sustainable agriculture overall goals of the Treaty.

Further developments

This book provides recommendations for the political construct of a global seed commons, which are hoped to be useful in the current review process of the Treaty. It does not provide all the answers, but rather opens many more questions. In the following last paragraphs, further research paths are proposed at the theoretical and technical levels.

At the theoretical level, several directions could be followed. First, as a continuation to the present use of the theory/ies of the commons, further work could be carried out in line with what has been called 'the new vogue of the commons' (Mattei, Dardot and Laval, etc.). Dardot and Laval (2010, 2014) question the notion of appropriation and promote the collective and political decision to design specific resources or services as not appropriable. Inappropriability is envisaged as a necessary new category, next to the public and the private ownership and management of resources or services, if the objective is to serve the collective interest and sustainability requirement. Could inappropriability be envisaged for governing PGRFA? Capra and Mattei (2015) call for a new vision of the role of Law, as an all-embracing science, an integral part of a whole, i.e. taking an ecological perspective of the Law. This is an attractive perspective for those observing the functioning of nature and humanity within nature as a 'holistic system'. Applying this concept to PGRFA management would require us to position ourselves differently; to rethink our approach to the farmer-seed (human-nature) relationship.

This questioning could be developed using feminist theories and methodologies, *inter alia* by exploring Haraway's concept of 'situated knowledges'(1988). Haraway calls for a new understanding of (scientific) objectivity, which would allow us to connect different points of view and recognize that all views are partial (i.e. knowledge come from different and specific positions, therefore we should be more critical about ours and others points of view). She states that '[s]ituated knoweldges are about communities, not about isolated individuals' (ibid: 590). In addressing the issue of power in the imbalance of rights pertaining to seeds, one could explore the feminist conceptualization of power, which envisages power as a resource to be (re)distributed (Okin, 1989), as domination (Young, 1990), and as empowerment (Miller, 1992; Starhawk, 1987). Another example can be taken with Keenan's work on property (2015). Her analysis of the concept of property focuses on 'the space in which the subject is embedded rather than on the propertied subject and rights of exclusion' (Langegger, 2016: 1). Keenan conceptualizes property as spatially contingent relations of belonging rather than as a bundle of rights. 'Within this theoretical frame, she then deconstructs an often-neglected pivot point of property: networks of possession. She theorizes property as a network of relations of belonging, instead of a spatially flat bundle of rights to a particular valued resource' (ibid). This theorization of property over a subject as networks, relations, connections and evolution resembles, at first sight, with the important characteristics of seeds and of the human-seed relationship.

A complementary research path could be grounded on the recent Latin American Decolonial movement, which Alvarez Villareal and Coolsaet have applied to environmental justice (Alvarez Villarreal and Coolsaet, forthcoming). This movement does not refer to colonialization, i.e. the political and historical moments that ended with the political independence of the last colonies in the 1960s, but to 'coloniality' i.e. the diversity of practices that derive from the matrix of power created by colonialism but that are still at work within contemporary or post-colonial societies (Quijano, 2007). Coloniality is sustained by three different but correlated dimensions: power, knowledge and being. If coloniality dehumanizes humanity and objectifies nature, then decoloniality refers to 'efforts at re-humanizing the world, to breaking hierarchies of difference that dehumanize subjects and communities and that destroy nature' (Maldonado-Torres, 2016: 10). At a theoretical level, decolonization requires recognizing that theory is always place-bound (Escobar, 2008), and demands from the intellectual to decentre her perspective from the Global North to the periphery, starting from the experience of the wretched of the earth (Dussel, 2011). This allows shedding light on practices, while at the same time provides us with alternatives to a Eurocentric model of society. Through this perspective, what can be grasped is that the Treaty (and law in general) needs to take into account the heterogeneity and customary practices of societies. For what is at stake is not only people's right to food/or modes of subsistence in the most material sense of the term, but also their knowledge, their traditional practices and therefore the 'reproduction' of their modes of life and subsistence as a community (i.e. the fight against the 'epistemicide' of de Sousa Santos, 2006).

Furthermore, the Treaty, and the present research findings, could be examined through the lens of the Global Public Goods[1] (GPG) theory developed in the early 2000s by Inge Kaul et al. (1999, 2003) and Kaul and Conceição (2006). The GPG theory attempts to provide answers to problems related to globalization. Kaul and Conceição (2006) argue that many contemporary international crises – such as food crises – have their roots in serious Global Public Goods undersupply. They identify three policy gaps to be filled for their theory to reach normative and effective impacts on international legal regimes: a 'jurisdictional gap', a 'participation gap', and an 'incentive gap'. The jurisdictional gap focuses on the discrepancy between a globalized world and national, separate units of policy-making (Kaul et al., 1999). The participation gap highlights that today, international cooperation is still mainly an intergovernmental process, whereas important new global actors, such as international non-governmental organizations or citizens' actions, have emerged. The incentive gap stresses the importance of promoting international cooperation in the implementation of international agreements. These gaps match quite well many of the Treaty constraints identified above. Further research could assess if and how mitigating these gaps would improve the effectiveness of the Global Seed Commons. A general questioning of the role of states in international law would need to be addressed with regard to the necessary transition towards agro-ecologically sustainable systems (De Schutter, 2015b: 9–10). Indeed, states ought to find a new role, responding to the challenges of our transitioning Anthropocene, different from the welfare state or the liberal state, and facilitating or even empowering citizens in their initiatives towards sustainable livelihoods.

Further studying the links, roles and responsibilities of the state, of citizens' movement and of other stakeholders could be useful to deepen the connection between regulatory systems of seed exchanges based on (intellectual) property rights and informal seed exchange systems based on sharing social and cultural norms. In the continuation of a preliminary study carried out with other researchers on the role of trust and social capital in the design and evolution of institutions for collective action (Six et al., 2015), one could explore the link between legal norms designed around strong (intellectual) property rights and soft social norms designed around trust and reciprocity. Nodal governance (Burris et al., 2005; Drahos, 2004) could be a pertinent theoretical framework used to analyse the social sharing aspect of seed exchange networks. Nodal governance is an elaboration of network theory (Newman, 2010) that explains how a variety of actors operating within social systems (Wasserman and Faust, 1994) interact along networks to govern their systems. This framework could significantly develop the understanding of the role of, *inter alia*, the interdependence, the community as well as the complexity/diversity/heterogeneity invariable principles of the Global Seed Commons.

Studying the Treaty through the lens of Human Rights could also complement the present work, in particular regarding the formal recognition of Farmers' Rights at the international level. Developments currently taking place within different fora in promoting and recognizing specific rights to seeds, land, water, food and all embracing peasants' rights could greatly enhance the Treaty's implementation. Using the concept of 'essential resource' (Pistor and De Schutter, 2015) as a complementary concept promoting the common management of PGRFA could be one way to enter this human rights approach. In addition, the negotiations on the UN Declaration on the Rights of Peasants and Other People Working in Rural Areas seem promising, in particular if correlated with the development of access and benefit-sharing policies centred around communities (i.e. such as in the Nagoya Protocol and the development of 'biocultural Protocols' promoting biocultural rights of groups such as local and indigenous people, see Girard, 2018). Developing and promoting such collective rights of communities could be a way to strengthen the recognition and enforcement of Human Rights, by bridging each recognition level of these rights together: the individual, the group, and the universal levels.

Several other theoretical frameworks could be useful to work on the Treaty. Behavioural studies could be an interesting field to research in order to unravel the delicate question of trust during Treaty negotiations and in collective management systems, especially within communities constituted by heterogeneous seed stakeholders. In international relations, studies could further build on the results of this book by exploring the difficult question of designing horizontally coherent international policies. That is to say, to develop policies with a holistic view of the general system in which the policy is designed (i.e. link it with neighbouring policies), taking into account the recent citizens movements on the transition (Hopkins, 2008; De Schutter, 2015a; de Nooy van Tol, 2016). Applied to PGRFA management, this would mean to relate the international agricultural policy to a (currently inexistent) international food policy, involving health-, environmental- and economic-related policies, in a systemic manner as well as to support the vivid emulation of citizens' movements on a diversity of food and agriculture projects. The list of future areas of work is infinite!

Further research at a more technical level could also supplement this work. While it is not the direct purpose of this book to propose ready-to-implement solutions to the Treaty implementation constraints identified throughout the analysis, the overall objective is to serve the discussions taking place in the ongoing Treaty review process which aims at mitigating the said constraints. Along that line, several suggestions are made to propose additional technical investigations.

First, in the IP field, further exploration on how to (re-)design a farmer's exemption in the MLS in relation to the existing legislation on plant variety protection and patents is greatly needed, in furtherance of Correas' (2015) or Van der Kooij's (2010) proposals. Could the MLS, coupled with sui generis plant variety protection laws, recreate an effective farmer's exemption? Additionally, a clearer understanding and vision of how to protect PGRFA-related traditional knowledge is required. Besides, further work is needed to mitigate the impediments of access to PGRFA due to national seed laws on certification and commercialization. Digging into the technicalities of intellectual property protection legislation and seed laws is therefore crucial.

In addition, in international relations studies and public international law, further exploration is needed to review the governance systems of the Treaty. Would an adaptation of the Governing Body rules allow for a type of multi-stakeholder governance similar to that of the FAO Committee on World Food Security? How could participatory democracy (Blondiaux, 2008; Girard and Le Goff, 2010) be mobilized to promote an effective multi-stakeholders governance in the Treaty?

Furthermore, highly technical issues have been raised regarding the administrative burden of PGRFA management and exchanges between stakeholders. A deeper examination of the SMTA technical rules regarding tracking and identification would be useful in order to facilitate the access to Annex I PGRFA. Financial issues are also key to the dysfunction of the Treaty. Different means of funding the Treaty ought to be envisaged and tested, etc. Not to mention biodiversity-related studies, where innovative strategies and technologies for conservation and sustainable use of PGRFA would benefit the implementation of the Treaty, and in particular focus on in situ and on-farm conservation and sustainable development strategies (Maxted et al., 2013; Dulloo, 2015). The list of other possible technical developments could be quite long.

My own next steps will be to verify the validity of the six invariable principles in smallholder farmer communities in China, Yunnan, through a case study analysis. By conducting surveys with farmers in nine villages of the Yuan Yang Terraces, I hope to demonstrate the utility of these principles. The aim is to present the results of this case study at the next session of the Governing Body, so that Treaty stakeholders may envisage these invariable principles as useful input for the ongoing review of the Treaty's multilateral system. Thanks to the award of two Belgian fellowships (F.W.O and F.N.R.S.), I am very excited to begin this new research project in collaboration with Mourad Hannachi, INRA-Grignon (France), and to work at the University of Antwerp, besides my long-lasting affiliation to the Université catholique de Louvain. I hope that this more sociologically oriented postdoc fieldwork will enable me to touch upon and see in a more concrete manner the holistic vison of the human-seed relationship I sensed out of the theoretical research exposed in this book.

Besides, thanks to funding acceptance of a French ANR research project, I will collaborate with Prof. Fabien Girard and other academics on the 'Bioculturalis' project, which aims to explore the Biocultural Community Protocols (BCPs). BCPs are developed as empowerment tools by communities with a view to promoting substantive (equitable benefit-sharing, taking into account ILCs' perspectives) and procedural (avoiding misunderstanding, allowing enough time and money) rights in access to genetic resources and associated traditional knowledge in between communities and between communities and bioprospectors. By fostering communities' collective action and transforming their participation role at the local, national and international level (i.e. all levels of society), BCPs could also scale-up the active involvement of communities in biodiversity conservation. The state, by supporting the development of these initiatives and promoting self-government, plays a role of enabling social innovation processes. By conducting case studies on BPCs in Peru, Colombia and Madagascar, this project intends to analyse the empowerment processes these farming communities have designed. We hope to show that these grassroots innovations develop and promote mechanisms allowing the communities' active participation in governing their resources, traditional knowledge and their related innovation as a solution to social and economic development needs and to food security threats.

Epilogue

As I review the final lines of this book mid-2018, more and more initiatives arise and literature is published promoting sharing, inclusivity, community, respect of the living seed and resilience. In the Global North, the Transition Network gathers an ever-growing number of citizens willing to question and reappropriate their modes of production and consumption, many initiatives dealing with food and agriculture. In addition, academics increasingly question the dominant narrative about IPRs in food and agriculture R&D and the breeding sector, decrying that

> the dominant global IPR appears to have contributed to consolidation in the seed industry while failing to genuinely engage with the potential of alternatives to support social goods such as food security, adaptability and resilience. The dominant IPR also constrains collaborative and cumulative plant breeding processes that are built upon the work of countless farmers past and present.
>
> (Vanloqueren and Baret, 2009; Kloppenburg, 2010; Stiglitz, 2014; Frison, 2016; Howard, 2016, IPES-Food, 2016, 2017b), with Halpert and Chappell (2017: 1)

In the Global South, voices rise louder and louder, calling for the respect of smallholder farmers' traditional practices and ways of living, respect for the living and collective responses to food and nutrition needs (African Centre for Biodiversity, 2018; Shiva, 2018) as the only resilient path towards human survival in the Anthropocene.

Are these wake-up calls the sign of a profound change in society? Or are they simply the return of the pendulum going back and forth on the public/private property scale of international resource governance? I hope for the former and plead for a collective and equitable change towards resilience: a change in our Eurocentric-dominant mentality over the property of things towards a global community of stewards of Earth, our mother home.

Note

1 This analysis has already been carried out but has not been published yet. Part of it has been presented with Charlotte de Callataÿ at the 3rd Thematic IASC Conference on Knowledge Commons, October 2016, Paris, with a paper entitled '*Exploring the normativity and effectiveness of Global Public Goods with two case studies: the Global Seed Commons and the Convention on the Law of the Non-navigational Uses of International Watercourses*'

Bibliography

Aalberts, T. E. (2012) *Constructing Sovereignty between Politics and Law*, London: Routledge.

Adair, J. R. (1997) 'The Bioprospecting Question: Should the United States Charge Biotechnology Companies for the Commercial Use of Public Wild Genetic Resources?', *Ecology Law Quarterly* 24(1).

AEGIS (2009) *Strategic Framework and Memorandum of Understanding*, available at www.ecpgr.cgiar.org/AEGIS/AEGIS_home.htm [Accessed on 13/10/2015].

African Centre for Biodiversity (2018) *Report from the National Seed Dialogue and Celebration*, available at https://acbio.org.za/en/celebrating-smallholder-farmers-and-seed-diversity-south-africa-report-national-seed-dialogue-and?utm_source=phplist121&utm_medium=email&utm_content=HTML&utm_campaign=Politics%20of%20Seed%20in%20South%20Africa [Accessed on 20/04/2018].

AGRA (2013) *Africa Agriculture Status Report: Focus on Staple Crops*, Nairobi: Alliance for a Green Revolution in Africa. Available at: http://reliefweb.int/sites/reliefweb.int/files/resources/agrafinalaugust20akim.pdf [Accessed on 14/10/2017].

Agrawal, A. (2001) 'Common Property Institutions and Sustainable Governance of Resources', *World Development* 29(10): 1649–1672.

——— (2002) 'Common Resources and Institutional Sustainability', in Ostrom, E., Dietz, T., Dolsak, N., Stern, P. C., Stonich, S. and Weber, E. U (eds.), *The Drama of the Commons*, Washington, DC: National Academy Press, 41–85.

Agrawal, A. and Ostrom, E. (2008) 'Decentralization and Community-based Forestry: Learning from Experience', in Webb, E. L. and Shivakoti, G. P. (eds.), *Decentralization, Forests and Rural Communities: Policy Outcomes in South and Southeast Asia*, New Delhi, India: Sage, 44–67.

Ahmed, A. U., Vargas Hill, R., Smith, L. C., Wiesmann, D. M. and Frankenberger, T. (2007) 'The World's Most Deprived Characteristics and Causes of Extreme Poverty and Hunger', in *2020 Discussion Paper 43*, Washington, DC: International Food Policy Research Institute.

Alchion, A. A. and Demsetz, H. (1973) 'The Property Right Paradigm', *Journal of Economic History* 33(1).

Allaire, G. (2013) 'Les communs comme infrastructure institutionnelle de l'économie marchande', *Revue de la régulation. Capitalisme, institutions, pouvoirs* 14.

Almekinders, C. J. and Louwaars, N. P. (2002) 'The Importance of the Farmers' Seed Systems in a Functional National Seed Sector', *Journal of New Seeds* 4(1–2): 15–33.

Alston, P. and Weiler, J. H. (1998) 'An "Ever Closer Union" in Need of a Human Rights Policy', *European Journal of International Law* 9(4): 658–723.

Altieri, M. A. (1999) 'The Ecological Role of Biodiversity in Agroecosystems', *Agriculture, Ecosystems & Environment* 74(1): 19–31.

——— (2002) 'Agroecology: The Science of Natural Resource Management for Poor Farmers in Marginal Environments', *Agriculture, Ecosystems & Environment* 93(1): 1–24.

Altieri, M. A. and Faminow, M. D. (1996) 'Agroecology: The Science of Sustainable Agriculture', *Canadian Journal of Agricultural Economics* 44(2): 199–201.

Altieri, M. A., et al. (2012a) 'The Scaling Up of Agroecology: Spreading the Hope for Food Sovereignty and Resiliency', Sociedad Cientifica LatinoAmericana de Agroecologia – SOCLA's *Rio+ 20 Position Paper*, A contribution to discussions at Rio+20 on issues at the interface of hunger, agriculture, environment and social justice, available at https://foodfirst.org/wp-content/uploads/2014/06/JA11-The-Scaling-Up-of-Agroecology-Altieri.pdf

Altieri, M. A., et al. (2012b) 'Agroecologically Efficient Agricultural Systems for Smallholder Farmers: Contributions to Food Sovereignty', *Agronomy for Sustainable Development* 32(1): 1–13.

Altieri, M. A. and Merrick, L. (1987) 'In Situ Conservation of Crop Genetic Resources through Maintenance of Traditional Farming Systems', *Economic Botany* 41(1): 86–96.

Altieri, M. A. and Nicholls, C. I. (2012) 'Agroecology Scaling Up for Food Sovereignty and Resiliency', *Sustainable Agriculture Reviews* (Springer): 1–29.

Alvarez Villarreal, L. and Coolsaet, B. (forthcoming) 'Decolonizing Environmental Justice', *Capitalism, Nature, Socialism* 29.

American National Research Council (1993) *Managing Global Genetic Resources: Agricultural Crop Issues and Policies*, Washington, DC: The National Academies Press.

Amin, A. and Howell, P. (2016) 'Thinking the Commons', in Amin, A. and Howell, P. (eds.), *Releasing the Commons: Rethinking the Futures of the Commons*, London; New York, NY: Routledge.

Andersen, R. (2008) *Governing Agrobiodiversity. Plant Genetics and Developing Countries*, Oxon; New York, NY: Routledge.

——— (2015) 'Some Considerations on the Relation between Farmers' Rights, Plant Breeders Rights and Legislation on Variety Release and Seed Distribution', *Input Paper for the Second Meeting Ad Hoc Technical Committee on Sustainable Use of Plant Genetic Resources for Food and Agriculture Rome*, Italy, 02/03/2015–03/03/2015. Available at www.planttreaty.org/sites/default/files/Appendix13.pdf [Accessed on 24/03/2015].

Andersen, R. and Winge, T. (2009) *The Plant Treaty and Farmers' Rights: Implementation Issues for South Asia*, South Asia Watch on Trade, Economics & Environment, Kathmandu, Nepal.

Anderson, J. R. (1998) 'Selected Policy Issues in International Agricultural Research: On Striving for International Public Goods in an Era of Donor Fatigue', *World Development* 26(6).

Andersson, K. P. and Ostrom, E. (2008) 'Analyzing Decentralized Resource Regimes from a Polycentric Perspective', *Policy Sciences* 41(1): 71–93.

Anvar, S. L. (2008) *Semences et Droit. L'emprise d'un modèle économique dominant sur une réglementation sectorielle*, PhD thesis, Paris: Université Paris I Panthéon-Sorbonne.

Aoki, K. (1998) 'Neocolonialism, Anticommons Property, and Biopiracy in the (Not-So-Brave) New World Order of International Intellectual Property Protection', *Indiana Journal of Global Legal Studies* 6(1): 11–58.

——— (2004) 'Weeds, Seeds & (and) Deeds: Recent Skirmishes in the Seed Wars', *Cardozo Journal of International and Comparative Law* 11: 247.——— (2008) *Seed Wars: Controversies and Cases on Plant Genetic Resources and Intellectual Property*, Durham, NC: Carolina Academic Press.

——— (2010) 'Seeds of Dispute: Intellectual-Property Rights and Agricultural Biodiversity', *Golden Gate University Environmental Law Journal* 3(1).

——— (2011) 'Food Forethought: Intergenerational Equity and Global Food Supply-Past, Present, and Future', *Wisconsin Law Review* 2.

Aoki, K. and Luvai, K. (2007) 'Reclaiming Common Heritage Treatment in the International Plant Genetic Resources Regime Complex', *Michigan State Law Review* 35.

Ashley Roach, J. (2012) 'The Central Arctic Ocean: Another Global Commons', *Global Policy* 3(1): 82–84.

Ayres, I. and Braithwaite, J. (1992) *Responsive Regulation: Transcending the Deregulation Debate*, Oxford: Oxford University Press.

Bäckstrand, K. (2006) 'Democratizing Global Environmental Governance? Stakeholder Democracy after the World Summit on Sustainable Development', *European Journal of International Relations* 12(4): 467–498.

Badstue, L. B., Bellon, M. R., Berthaud, J., Juárez, X., Rosas, I. M., Solano, A. M. and Ramírez, A. (2006) 'Examining the Role of Collective Action in an Informal Seed System: A Case Study from the Central Valleys of Oaxaca, Mexico', *Human Ecology* 34(2): 249–273.

Badstue, L. B., Bellon, M. R., Berthaud, J., Ramírez, A., Flores, D. and Juárez, X. (2007) 'The Dynamics of Farmers' Maize Seed Supply Practices in the Central Valleys of Oaxaca, Mexico', *World Development* 35(9): 1579–1593.

Bailey, S. and Mattei, U. (2013) 'Social Movements as Constituent Power: The Italian Struggle for the Commons', *Indiana Journal of Global Legal Studies* 20(2): 965–1013.

Baland, J.-M. and Platteau, J.-P. (1996) *Inequality and Collective Action in the Commons*, CRED, Namur, Belgium: University of Namur.

——— (1997) 'Wealth Inequality and Efficiency in the Commons, Part I: The Unregulated Case', *Oxford Economic Papers* 49(4): 451–482.

——— (1998a) 'Division of the Commons: A Partial Assessment of the New Institutional Economics of Land Rights', *American Journal of Agricultural Economics* 80(3): 644–650.

——— (1998b) 'Wealth Inequality and Efficiency in the Commons, Part II: The Regulated Case', *Oxford Economic Papers* 50(1): 1–22.

Baret, P., et al. (2015) 'Research and Organic Farming in Europe', in *A Report Commissioned by the Greens: European Free Alliance in the European Parliament*, UK: Louvain-la-Neuve, Earth & Life Institute, Université catholique de Louvain, Louvain-la-Neuve, Belgium and Organic Research Centre, UK.

Barton, J. H. (1999) 'Intellectual Property Management', *Biotechnology for Developing-Country Agriculture: Problems and Opportunities: A 2020 Vision for Food, Agriculture, and the Environment*, Focus 2, IFPRI, Washington, DC, USA.

Barzel, Y. (1997) *Economic Analysis of Property Rights*, Cambridge: Cambridge University Press.

Baslar, K. (1998) *The Concept of the Common Heritage of Mankind in International Law*, The Hague; Boston; Cambridge, MA: M. Nijhoff Publishers and Kluwer Law International.

Batta-Bjørnstad, S. I. (2004) *Breakthrough for "the South"? An Analysis of the Recognition of Farmers' Rights in the International Treaty on Plant Genetic Resources for Food and Agriculture*, Fridtjof Nansen Institute, FNI Report 13/2004, Lysaker, Norway.

Batur, F. (2014) *Agrobiodiversity Conservation and Plant Improvement: Adjustments in Intellectual Property Rights Reclaiming the Public Domain towards Sustainability and Equity*, PhD thesis, Faculté de droit et de criminologie, Louvain-la-Neuve, Université catholique de Louvain.

Batur, F. and Dedeurwaerdere, T. (2014) 'The Use of Agrobiodiversity for Plant Improvement and the Intellectual Property Paradigm: Institutional Fit and Legal Tools for Mass Selection, Conventional and Molecular Plant Breeding', *Life Sciences, Society and Policy* 10(1): 1.

Bavikatte, K. S. (2014) *Stewarding the Earth: Rethinking Property and the Emergence of Biocultural Rights*, New Delhi: Oxford University Press.

Bavikatte, K. S. and Robinson, D. F. (2011) 'Towards a People's History of the Law: Biocultural Jurisprudence and the Nagoya Protocol on Access and Benefit Sharing', *Law, Environment and Development Journal* 7(1): 35–51.

Bechtold, B. (2015) 'Introduction: Beyond Hardin and Ostrom New Heterodox Research on the Commons', *Review of Radical Political Economics*, reprinted in 2016, 48(1): 5–8.

Beddington, J., Asaduzzaman, M., Fernandez, A., Clark, M., Guillou, M., Jahn, M., Erda, L., Mamo, T., Bo, N. V., Nobre, C. A., Scholes, R., Sharma, R. and Wakhungu, J. (2012) *Achieving Food Security in the Face of Climate Change: Final Report from the Commission on Sustainable Agriculture and Climate Change*, Copenhagen, Denmark: CGIAR Research Program on Climate Change, Agriculture and Food Security (CCAFS).

Bellon, M. R., Hodson, D. and Hellin, J. (2011) 'Assessing the Vulnerability of Traditional Maize Seed Systems in Mexico to Climate Change', *Proceedings of the National Academy of Sciences of the United States of America* 108(33).

Benkler, Y. (1997) 'Overcoming Agoraphobia: Building the Commons of the Digitally Networked Environment', *Harvard Journal of Law & Technology* 11: 287.

——— (2002) 'Coase's Penguin, or, Linux and "The Nature of the Firm"', *The Yale Law Journal* 112(3): 369–446.

——— (2007) *The Wealth of Networks: How Social Production Transforms Markets and Freedom*, New Haven, CT; London: Yale University Press.

Bente, G., et al. (2008) 'Avatar-Mediated Networking: Increasing Social Presence and Interpersonal Trust in Net-Based Collaborations', *Human Communication Research* 34(2): 287–318.

Berg, T., et al. (1991) 'Technology and the Gene Struggle', *NORAGRIC Occasional Paper: Series C, Development and Environment*, Ås, Norway: Norwegian Centre for International Agricultural Development and Agricultural University of Norway.

Berge, E. and van Laerhoven, F. (2011) 'Governing the Commons for Two Decades: A Complex Story', *International Journal of the Commons* 5(2): 160–187.

Berkes, F., et al. (2008) *Navigating Social-Ecological Systems: Building Resilience for Complexity and Change*, Cambridge: Cambridge University Press.

Bertacchini, E. (2007) 'Biotechnologies, Seeds and Semicommons', *Working Paper: Department of Economics*, Torino: University of Turin.

——— (2008) Property Rights and Plant Genetic Resources for Food and Agriculture, PhD, Economia, Scienze Economische e finanziarie, Scienze Giuridiche, Torino, University of Turin, 118.

Bessa, A. (2015) 'Traditional Local Communities: What Lessons Can Be Learnt at the International Level from the Experiences of Brazil and Scotland?', *Review of European Comparative & International Environmental Law* 24(3): 330–340.

Bevilacqua, D. and Duncan, J. (2010) 'Towards a New Cosmopolitanism: Global Reflexive Interactive Democracy as a New Mechanism for Civil Society Participation in Agri-Food Governance', *Global Jurist: Advances* 10(1).

Biber-Klemm, S. and Cottier, T. (eds.) (2006) 'Rights to Plant Genetic Resources and Traditional Knowledge', *Basic Issues and Perspectives*, Berne: CABI.

Blakeney, M. (2011) 'Patents and Plant Breeding: Implications for Food Security', *Amsterdam Law Forum*, VU University Amsterdam, Amsterdam, The Netherlands.

Blomley, N. (2015) 'The Right Not to Be Excluded: Common Property and the Struggle to Stay Put', in Amin, A. and Howell, P. (eds.), *Releasing the Commons: Rethinking the Futures of the Commons*, London; New York, NY: Routledge.

Blomquist, W. and Ostrom, E. (1985) 'Institutional Capacity and the Resolution of a Commons Dilemma', *Review of Policy Research* 5(2): 383–394.

Blondiaux, L. (2008) *Le Nouvel Esprit De La Démocratie: Actualité De La Démocratie Participative*, Paris, France: Editions du Seuil.

Bocci, R., et al. (2011) 'Farm Seed Opportunities, Recommendations for On-Farm Conservation in Europe', in Maxted, N., Dulloo, M. E., Ford-Lloyd, B. V. et al. (eds.), *Agrobiodiversity Conservation: Securing the Diversity of Crop Wild Relatives*, Wallingford, UK: CABI.

Bollier, D. (2007) 'The Growth of the Commons Paradigm', in Hess, C. and Ostrom, E. (eds.), *Understanding Knowledge as a Commons: From Theory to Practice*, Cambridge, MA; London: MIT Press.

——— (2013) *Silent Theft: The Private Plunder of Our Common Wealth*, Oxon, UK: Routledge.

——— (2014) *Think Like a Commoner: A Short Introduction to the Life of the Commons*, Gabriola Island, Canada: New Society Publishers.

Bollier, D. and Helfrich, S. (2014) 'Introduction: The Commons as a Transformative Vision', in Bollier, D. and Helfrich, S. (eds.), *The Wealth of the Commons: A World beyond Market & State*, Amherst, MA: Levellers Press.

——— (2014) *The Wealth of the Commons: A World beyond Market & State*, Amherst, MA: Levellers Press.

——— (2015) *Patterns of Commoning*, Commons Strategy Group and Off the Common Press, Levellers Press, Amherst, United States.

Bonadio, E. (2007) 'Crop Breeding and Intellectual Property in the Global Village', *European Intellectual Property Review* 29(5).

Bonneuil, C., et al. (2006) 'Innover autrement? La recherche face à l'avènement d'un nouveau régime de production et de régulation des savoirs en génétique végétale', *Dossiers de l'environnement de l'INRA* 30: 29–51.

Bonneuil, C. and Thomas, F. (2009) *Gènes, pouvoirs et profits, Recherche publique et régimes de production des savoirs, de Mendel aux OGM*, Versailles: Quæ.

Bordwin, H. J. (1985) 'The Legal and Political Implications of the International Undertaking on Plant Genetic-Resources', *Ecology Law Quarterly* 12(4).

Bosselmann, K. (1996) 'Plants and Politics: The International Legal Regime Concerning Biotechnology and Biodiversity', *Colorado Journal of International Environmental Law and Policy* 7: 111.

Bourguignon, L. and Bourguignon, C. (2015) 'La mort des sols agricoles', *Études sur la mort* 148(2): 47–53.

Bowie, N. (1988) 'The Moral Obligations of Multinational Corporations', *Problems of International Justice* 97.

Boyle, J. (1996) *Shamans, Software, and Spleens: Law and the Construction of the Information Society*, Cambridge, MA: Harvard University Press.

——— (2003a) 'Foreword: The Opposite of Property?', *Law and Contemporary Problems* 66(Winter/Spring, 1 and 2).

——— (2003b) 'The Second Enclosure Movement and the Construction of the Public Domain', *Law and Contemporary Problems* 66(Winter/Spring 2003, 1 and 2).

——— (2008a) *Public Domain: Enclosing the Commons of the Mind*, New Haven, CT; London: Yale University Press.

——— (2008b) *Remix: Making Art and Commerce Thrive in the Hybrid Economy*, London: Bloomsbury.

Bragdon, S. (2004) 'International Law of Relevance to Plant Genetic Resources: A Practical Review for Scientists and Other Professionals Working with Plant Genetic Resources', *Issues in Genetic Resources* 10.

Brahy, N. (2008) *The Property Regime of Biodiversity and Traditional Knowledge*, Bruxelles: Larcier.

Braunschweig, T., Meienberg, F., Pionetti, C. and Shashikant, S. (2014) *Owning Seeds, Accessing Food: A Human Rights Impact Assessment of Upov 1991 Based on Case Studies in Kenya, Peru and the Philippines*, The Berne Declaration, Zurich, Switzerland.

Bravo, G. and De Moor, T. (2008) 'The Commons in Europe: From Past to Future', *International Journal of the Commons* 2(2).

Brockway, L. H. (1979) 'Science and Colonial Expansion: The Role of the British Royal Botanical Gardens', *American Ethnologist* 6(3): 449–465, Interdisciplinary Anthropology.

Brousseau, E., et al. (2012) *Global Environmental Commons: Analytical and Political Challenges in Building Governance Mechanisms*, Oxford: Oxford University Press.

Bruins, M. (2009) 'The Evolution and Contribution of Plant Breeding and Related Technologies in the Future', in *Proceedings of the Second World Seed Conference*, Rome, September 8–10 (FAO Headquarters).

Brush, S. B. (2001) 'Genetically Modified Organisms in Peasant Farming: Social Impact and Equity', *Indiana Journal of Global Legal Studies* 9(1): 135–162.

——— (2004) *Farmers Bounty: Locating Crop Diversity in the Contemporary World*, New Haven, CT: Yale University Press.

——— (2005) 'Protecting Traditional Agricultural Knowledge', *Washington University Journal of Law & Policy* 17: 59.

——— (2007) 'Farmers' Rights and Protection of Traditional Agricultural Knowledge', *World Development* 35(9): 1499–1514.

Buchanan, J. M. and Yoon, Y. J. (2000) 'Symmetric Tragedies: Commons and Anticommons', *Journal of Law and Economics* 43(1): 1–14.

Buck, S. J. (1998) *The Global Commons: An Introduction*, Washington DC: Island Press.

Burchi, F. and De Muro, P. (2016) 'From Food Availability to Nutritional Capabilities: Advancing Food Security Analysis', *Food Policy* 60: 10–19.

Burris, S., Drahos, P. and Shearing, C. (2005) 'Nodal Governance', *Australian Journal of Legal Philosophy* 30.

Butchart, S. H., et al. (2010) 'Global Biodiversity: Indicators of Recent Declines', *Science* 328(5982): 1164–1168.

Byerlee, D. and Dubin, H. J. (2010) 'Crop Improvement in the CGIAR as a Global Success Story of Open Access and International Collaboration', *International Journal of the Commons* 4(1): 1–19.

Cadieux, K. V. and R. Blumberg (2014) 'Food Security in Systemic Context', *Encyclopedia of Food and Agricultural Ethics*: 979–988.

Calabresi, G. and Melamed, D. (1972) 'Property Rules, Liability Rules, and Inalienability: One View of the Cathedral', *Harvard Law Review* 85(6).

Calvert, J. and Joly, P.-B. (2011) 'How Did the Gene become a Chemical Compound: The Ontology of the Gene and the Patenting of DNA', *Social Science Information* 50(2): 157–177.

Candolle, A. de (1883) *Origine Des Plantes Cultivées*, Paris: G. Baillière et Cie.

Caplat, J. (2014) *Changeons d'agriculture-Réussir la transition*, Paris: Actes Sud.

Capra, F. (1996) *The Web of Life: A New Synthesis of Mind and Matter*, London, UK: HarperCollins.

Capra, F. and Mattei, U. (2015) *The Ecology of Law: Toward a Legal System in Tune with Nature and Community*, Oakland, CA: Berrett-Koehler Publishers.

CBD (2008) *Biodiversity and Agriculture: Safeguarding Biodiversity and Securing Food for the World, International Day for Biological Diversity*, Montreal: Secretariat of the Convention on Biological Diversity.

CFS Committee on World Food Security (2009) *Reform of the Committee on World Food Security: Final Version.* Document CFS: 2009/2Rev. 2 available at ftp://ftp.fao.org/docrep/fao/meeting/018/k7197e.pdf

CGIAR (2012) *Agreement Establishing the Consortium of International Agricultural Research Centers as an International Organization*, CGIAR System, available at https://cgspace.cgiar.org/bitstream/handle/10947/2592/Agreement_establishing_the_Consortium_of_International_Agricultural_Research_Centers_as_an_International_Organization.pdf?sequence=4 [Accessed on 21/08/2012].

CGRFA (1997) *Progress Report on The Global System for the Conservation and Sustainable Utilization of Plant Genetic Resources for Food and Agriculture*, Commission on Genetic Resources for Food and Agriculture, Seventh Session, Rome, 15–23 May 1997, document CGRFA-7/97/3, FAO, Rome.

Chang, H.-J. (2009) 'Rethinking Public Policy in Agriculture: Lessons from History, Distant and Recent', *The Journal of Peasant Studies* 36(3): 512.

Charles, D. (2001a) *Lords of the Harvest: Biotech, Big Money, and the Future of Food*, Cambridge, MA: Perseus Publishing, 92–125.

——— (2001b) 'Seeds of Discontent', *Science* 294(5543).

——— (2001c) 'Seed Treaty Signed: U.S., Japan Abstain', *Science* 294(5545).

Chayes, A. and Chayes, A. H. (1998) *The New Sovereignty: Compliance with International Regulatory Agreements*, Cambridge, MA: Harvard University Press.

Chen, J. (2006) 'There's No Such Thing as Biopiracy . . . and It's a Good', *McGeorge Law Review* 37(1): 1–32.

Chiarolla, C. (2006) 'Commodifying Agriculture Biodiversity and Development-Related Issues', *Journal of World Intellectual Property* 9(1): 25–60.

——— (2008) 'Plant Patenting, Benefit Sharing and the Law Applicable to the Food and Agriculture Organisation Standard Material Transfer Agreement', *Journal of World Intellectual Property* 11(1): 1–28.

——— (2011) *Intellectual Property, Agriculture and Global Food Security: The Privatization of Crop Diversity*, Cheltenham, UK; Northampton, MA: Edward Elgar.

Chiarolla, C. and Jungcurt, S. (2011) 'Outstanding Issues on Access and Benefit Sharing under the Multilateral System of the International Treaty on Plant Genetic Resources for Food and Agriculture', *Background Study Paper: Studie im Auftrag der Erklärung von Bern, Development Fund.* Verfügbar unter www.Evb

Chiarolla, C. and Shand, H. (2013) *An Assessment of Private Ex Situ Collections: The Private Sector's Participation in the Multilateral System of the FAO International Treaty on Plant Genetic Resources for Food and Agriculture*, Berne Declaration and Norwegian Development Fund, Zurich, Switzerland.

Christensen, E. (1987) 'A Proposal to Preserve Genetic Diversity for Future Generations', *Stanford Law Review* 40(1): 281.

Christinck, A. and Walloe Tvedt, M. (2015) *The UPOV Convention, Farmers' Rights and Human Rights*, F. M. f. E. C. a. D. (BMZ), Deutsche Gesellschaft für Internationale Zusammenarbeit (GIZ) GmbH.

Ciriacy-Wantrup, S. V. and Bishop, R. C. (1975) 'Common Property as a Concept in Natural Resources Policy', *Natural Resources Journal* 15: 713.

Claeys, P. (2012) 'The Creation of New Rights by the Food Sovereignty Movement: The Challenge of Institutionalizing Subversion', *Sociology* 46(5).

——— (2014a) *Human Rights and the Food Sovereignty Movement. Reclaiming Control*, PhD thesis, Université catholique de Louvain, Louvain-la-Neuve, Belgium.

——— (2014b) 'Food Sovereignty and the Recognition of New Rights for Peasants at the UN: A Critical Overview of La Via Campesina's Rights Claims over the Last 20 Years', *Globalizations* 12(4): 452–465.

——— (2015) 'The Right to Food: Many Developments, More Challenges', *Canadian Food Studies/La Revue canadienne des études sur l'alimentation* 2(2): 60–67.

Claeys, P. and Lambek, N. (2014) 'Introduction: In Search of Better Options: Food Sovereignty, the Right to Food and Legal Tools for Transforming Food Systems', in Lambek, N., Claeys, P., Wong, A. and Brilmayer, L. (eds.), *Rethinking Food Systems: Structural Challenges, New Strategies and the Law*, Dordrecht: Springer.

Clapp, J. and Fuchs, D. A. (2009) *Corporate Power in Global Agrifood Governance*, Cambridge MA: MIT Press.

Clark, C. W. (1973) 'Profit Maximization and the Extinction of Animal Species', *The Journal of Political Economy* 81(4): 950–961.

——— (1974) 'The Economics of Overexploitation', *Science* 181: 630–634.

Cleveland, D. A., Soleri, D. and Smith, S. E. (1994) 'Do Folk Crop Varieties Have a Role in Sustainable Agriculture?', *BioScience* 44(11).

Coase, R. H. (1960) 'The Problem of Social Cost', *Journal of Law and Economics* 3: 1–44.

Çoban, A. (2004) 'Caught between State-Sovereign Rights and Property Rights: Regulating Biodiversity', *Review of International Political Economy* 11(4).

Cochran, G. and Harpending, H. (2009) *The 10,000 Year Explosion: How Civilization Accelerated Human Evolution*, New York: Basic Books.

Cohen, W. M. (2004) 'Patents and Appropriation: Concerns and Evidence', *The Journal of Technology Transfer* 30(1–2): 57–71.

Conway, G. and Toenniessen, G. (1999) 'Feeding the World in the Twenty-First Century', *Nature* 402: C55–C56.

Cook-Deegan, R. and Dedeurwaerdere, T. (2006) 'The Science Commons in Life Science Research: Structure, Function, and Value of Access to Genetic Diversity', *International Social Science Journal* 58(188): 299–317.

Coolsaet, B., et al. (2015) *Implementing the Nagoya Protocol: Comparing Access and Benefit-Sharing regimes in Europe*, Boston, MA: Brill.

Coomes, O. T., et al. (2015) 'Farmer Seed Networks Make Limited Contribution to Agriculture? Four Common Misconceptions', *Food Policy* 56: 41–50.

Cooper, D. (1993) 'The International Undertaking on Plant Genetic Resources', *Review of European Community & International Environmental Law* 2(2).

——— (2002) 'The International Treaty on Plant Genetic Resources for Food and Agriculture', *RECIEL* 11(1): 1–16.

Cooper, D., et al. (1994) A Multilateral System for Plant Genetic Resources: Imperatives, Achievements and Challenges, Rome, Italy: Bioversity International.

Coradin, L. and Sampaio, M. J. A. (2011) 'Country Case-Study: Brazil: Actions and Reactions to the International Treaty on Plant Genetic Resources for Food and Agriculture', in Frison, C., López, F. and Esquinas-Alcazar, J. T. (eds.), *Plant Genetic Resources and Food Security: Stakeholder Perspectives on the International Treaty on Plant Genetic Resources for Food and Agriculture*, Oxon: Earthscan by Routledge and Bioversity International, 303.

Cordonier Segger, M.-C., Perron-Welch, F. and Frison, C. (2013) *Legal Aspects of Implementing the Cartagena protocol on Biosafety*, Cambridge: Cambridge University Press.

Coriat, B. (2013) 'Le retour des communs. Sources et origines d'un programme de recherche', *Revue de la régulation. Capitalisme, institutions, pouvoirs* 14.

——— (2015) *Le retour des communs: & la crise de l'idéologie propriétaire*, Paris, France: Éditions Les Liens qui libèrent.

Cornu, M., et al. (2017) *Dictionnaire des biens communs*, Paris, France: Presses Universitaires de France.

Correa, C. M. (1994) 'Sovereign and Property Rights Over Plant Genetic Resources', *Commission on Genetic Resources 2*, FAO Background Study Paper No. 2.

——— (1999) 'Access to Plant Genetic Resources and Intellectual Property Rights', *Commission on Genetic Resources for Food and Agriculture*, FAO Study Paper No. 8.

——— (2006) 'Considerations on the Standard Material Transfer Agreement under the FAO Treaty on Plant Genetic Resources for Food and Agriculture', *The Journal of World Intellectual Property* 9(2): 137–165.

——— (2011) 'An Innovative Option for Benefit-Sharing Payment under the International Treaty on Plant Genetic Resources for Food and Agriculture: Implementing Article 6.11 Crop-Related Modality of the Standard Material Transfer Agreement', in Frison, C., López, F. and Esquinas-Alcazar, J. T. (eds.), *Plant Genetic Resources and Food Security: Stakeholder Perspectives on the International Treaty on Plant Genetic Resources for Food and Agriculture*, Oxon: Earthscan by Routledge and Bioversity International, 321.

——— (2013) 'Plant Genetic Resources under the Management and Control of the Contracting Parties and in the Public Domain', in Halewood, M., Noriega, I. L. and Louafi, S. (eds.), *Crop Genetic Resources as a Global Commons*, Oxon: Earthscan by Routledge and Bioversity International, 177.

——— (2015) *Plant Variety Protection in Developing Countries: A Tool for Designing a Sui Generis Plant Variety Protection System: An Alternative to UPOV 1991*, by: Association for Plant Breeding for the Benefit of Society (APBREBES) and Its Member Organizations: Berne Declaration, The Development Fund, SEARICE and Third World Network.

——— (2017) 'Implementing Farmers' Rights Relating to Seeds', In *Research Papers*. Geneva, Switzerland: South Centre, 75.

Correa, C. M. and Yusuf, A. (eds.) (1998) *Intellectual Property and International Trade: The TRIPS Agreement*, 3rd edn, Alphen aan den Rijn: Wolters Kluwer, 2016.

Corten, O. and P. Klein (2011) *The Vienna Conventions on the Law of Treaties: A Commentary*, Oxford: Oxford University Press.Cox, M., et al. (2010) 'A Review of Design Principles for Community-Based Natural Resource Management', *Ecology and Society* 15(4): 38.

Cragg, G. M., et al. (2012) 'The Impact of the United Nations Convention on Biological Diversity on Natural Products Research', *Natural Product Reports* 29(12): 1407–1423.

Credit Suisse (2016) *Global Wealth Databook 2016*, available at www.credit-suisse.com/corporate/en/articles/news-and-expertise/the-global-wealth-report-2016-201611.html [Accessed on 11/05/2017].

(The) Crucible Group (1994) *People, Plants and Patents: The Impact of Intellectual Property on Trade, Plant Biodiversity, and Rural Society*, IDRC-CRDI.

(The) Crucible II Group (2000) 'Policy Options for Genetic Resources: People, Plants and Patents Revisited', *Seedling Solutions*, vol. 1, Rome, Italy: IPGRI.

——— (2001) 'Options for National Laws Governing Control over Genetic Resources and Biological Innovations', *Seedling Solutions*, vol. 2, Rome, Italy: IPGRI.

Cubero, J. I. (2011) 'Plant Breeders: The Point of View of a Plant Breeder on the International Treaty on Plant Genetic Resources for Food and Agriculture', in Frison, C., López, F. and Esquinas-Alcazar, J. T. (eds.), *Plant Genetic Resources and Food Security: Stakeholder Perspectives on the International Treaty on Plant Genetic Resources for Food and Agriculture*, Oxon: Earthscan by Routledge and Bioversity International, 197–208.

Cukier, K. N. (2006) 'Navigating the Future(s) of Biotech Intellectual Property', *Nature Biotechnology* 24(3): 249–251.

Cullet, P. (1999) 'Revision of the TRIPS Agreement Concerning the Protection of Plant Varieties', *The Journal of World Intellectual Property* 2(4): 617–656.

——— (2001) 'Plant Variety Protection in Africa: Towards Compliance with the TRIPs Agreement', *Journal of African Law* 45(1): 97–122.

——— (2004) 'Intellectual Property Rights and Food Security in the South', *The Journal of World Intellectual Property* 7(3): 261–286.

——— (2005) 'Seeds Regulation, Food Security and Sustainable Development', *Economic and Political Weekly* 40(32): 3607–3613.

——— (2007) 'Human Rights and Intellectual Property Protection in the TRIPS Era', *Human Rights Quarterly* 29: 403–430.

Cullet, P. and Kolluru, R. (2003) 'Plant Variety Protection and Farmers' Rights-Towards a Broader Understanding', *Delhi Law Review* 2: 41–59.

Cullet, P. and Raja, J. (2004) 'Intellectual Property Rights and Biodiversity Management: The Case of India', *Global Environmental Politics* 4(1): 97–114.

Dani, A. (2014) 'Il concetto giuridico di "beni comuni" tra passato e presente', *Historia et ius* 6: 1–48.

Dardot, P. and Laval, C. (2010) 'Du public au commun', *Revue du MAUSS* 35(1): 111–122.

——— (2014) *Commun: essai sur la révolution au XXIe siècle*, Paris: la Découverte.

Darwin, C. (1859) *On the Origin of Species by Means of Natural Selection*, London: J. Murray.

Dawyndt, P., et al. (2006) *Exploring the Microbiological Commons: Contributions of Bioinformatics and Intellectual Property Rights in Sharing Biological Information*, Oxford: UNESCO.

de Boef, W. S., Berg, T. and Haverkort, B. (1996) 'Crop Genetic Resources', in Bunders, J., Haverkort, B. and Hiemstra, W. (eds.), *Biotechnology: Building on Farmers' Knowledge*, London: Palgrave Macmillan.

de Carvalho, N. P. (2005) 'From the Shaman's Hut to the Patent Office: In Search of Effective Protection for Traditional Knowledge', *Washington University Journal of Law & Policy* 17: 111–186.

Dedeurwaerdere, T. (2006) 'The Institutional Economics of Sharing Biological Information', *International Social Science Journal* 58(188): 351–368.

——— (2008) 'The Role of Law, Institutions, and Governance in Facilitating Access to the Scientific Research Commons: A Philosopher's Perspective', in Van Overwalle, G. (ed.), *Fair Access to Health Care through Collaborative Licensing: Patent Pools, Clearing Houses, Open Source Models and Liability Regimes in Human Genetics*, Cambridge: Cambridge University Press.

——— (2010) 'Institutionalizing Global Genetic Resource Commons: Towards Alternative Models for Facilitating Access in the Global Biodiversity Regime', *SSRN Working Paper*, available at https://papers.ssrn.com/sol3/papers.cfm?abstract_id=1611549

——— (2012) 'Design Principles of Successful Genetic-Resource Commons for Food and Agriculture', *International Journal of Ecological Economics and Statistics* 26(3): 16–30.

——— (2013) 'Institutionalizing Global Genetic Resource Commons for Food and Agriculture', in Halewood, M., Lopez Noriega, I. and Louafi, S. (eds.), *Crop Genetic Resources as a Global Commons*. Oxon: Earthscan by Routledge and Bioversity International, 368.

——— (2014) *Sustainability Science for Strong Sustainability* Cheltenham, UK; Northampton, MA: Edward Elgar Publishing.

Dedeurwaerdere, T., et al. (2009) 'The Use and Exchange of Microbial Genetic Resources for Food and Agriculture', *Commission on Genetic Resources for Food and Agriculture* (46): 67.

Deibel, E. (2018) 'Open Sesame: Open Source and Crops', in Girard, F. and Frison, C. (eds.), *The Commons, Plant Breeding and Agricultural Research: Challenges for Food Security and Agrobiodiversity*, Oxon: Routledge.

De Jonge, B. and Khortals, M. (2006) 'Vicissitudes of Benefit Sharing of Crop Genetic Resources: Downstream and Upstream', *Developing World Business* 6(3): 144–157.

De Jonge, B. and Louwaars, N. (2009) 'The Diversity of Principles Underlying the Concept of Benefit Sharing', in Winter, G and Kamau, E. C. (eds.), *Genetic Resources, Traditional Knowledge and the Law: Solutions for Access and Benefit Sharing*, London: Earthscan, 37–56.

De Jonge, B., et al. (2015) 'A Solution to the Controversy on Plant Variety Protection in Africa', *Nature Biotechnology* 33(5): 487–488.

Demeulenaere, E. (2014) 'A Political Ontology of Seeds: The Transformative Frictions of a Farmers' Movement in Europe', *Focaal* 69.

——— (2018) 'Free Our Seeds! 'Strategies of Farmers' Movements to Reappropriate Seeds', in Girard, F. and Frison, C. (eds.), *The Commons, Plant Breeding and Agricultural Research: Challenges for Food Security and Agrobiodiversity*, Oxon: Routledge.

Demeulenaere, E. and Bonneuil, C. (2011) 'Des semences en partage. Construction sociale et identitaire d'un "collectif paysan" autour de pratiques semencières alternatives', *Techniques & Cultures* 54(2): 202–221.

Demeulenaere, E. and Goulet, F. (2012) 'Du singulier au collectif. Agriculteurs et objets de la nature dans les réseaux d'agricultures "alternatives"', *Terrains & Travaux* (1): 121–138.

De Moor, T. (2008) 'The Silent Revolution: A New Perspective on the Emergence of Commons, Guilds, and Other Forms of Corporate Collective Action in Western Europe', *International Review of Social History* 53(S16).

——— (2009) 'Avoiding Tragedies: A Flemish Common and Its Commoners under the Pressure of Social and Economic Change during the Eighteenth Century1', *The Economic History Review* 62(1).

——— (2011) 'Dossier "Le champ des commons en question: perspectives croisées": From Common Pastures to Global Commons: A Historical Perspective on Interdisciplinary Approaches to Commons', *Natures Sciences Sociétés* 19(4): 422–431.

Demsetz, H. (1967) 'Toward a Theory of Property Rights', *The American Economic Review* 57(2).

——— (2002) 'Toward a Theory of Property Rights II: The Competition between Private and Collective Ownership', *Journal of Legal Studies* 31(2).

De Nooy van Tol, J. (2016) *Transition to Agro-Ecology for a Food Secure World*, Bloomington, USA: Author House.

De Schutter, O. (2008) 'Building Resilience: A Human Rights Framework for World Food and Nutrition Security', *Report of the Special Rapporteur on the Right to Food*, United Nations General Assembly, Louvain-la-Neuve, Belgium: Université catholique de Louvain.

——— (2009a) 'Governing World Food Security: A New Role for the Committee on World Food Security', *Right to Food and Nutrition Watch*, 'Who Controls the Governance of the World Food System', FAO, Rome, Italy.

——— (2009b) 'Seed Policies and the Right to Food: Enhancing Agrobiodiversity and Encouraging Innovation', *Background Document to the Report (A/64/170) Presented at the 64th Session of the UN General Assembly*, Université catholique de Louvain, Louvain-la-Neuve, Belgium.

——— (2009c) 'Crisis into Opportunity: Reinforcing Multilateralism', *United Nations Report of the Special Rapporteur on the Right to Food*, Université catholique de Louvain, Louvain-la-Neuve, Belgium.

——— (2010a) 'The Right to Food', UN Doc. A/65/281, 11 August 2010, Université catholique de Louvain, Louvain-la-Neuve, Belgium.

——— (2010b) 'Agroecology and the Right to Food', United Nations. December 2010, Université catholique de Louvain, Louvain-la-Neuve, Belgium.

——— (2010c) 'Addressing Concentration in Food Supply Chains – The Role of Competition Law in Tackling the Abuse of Buyer Power', United Nations Briefing Note of the Special Rapporteur on the Right to Food. Briefing Note 03, December 2010, available at www.srfood.org/en/briefing-note-addressing-concentration-in-food-supply-chains

——— (2013) 'The Reform of the Committee on World Food Security: The Quest for Coherence in Global Governance', in *CRIDHO Working Paper 2013/8*, Louvain-la-Neuve, Belgium: Université catholique de Louvain.

——— (2014) 'Final Report, the Transformative Potential of the Right to Food', UN Doc. A/HRC/25/57, 24 January 2014, Université catholique de Louvain, Louvain-la-Neuve, Belgium.

——— (2015a) 'Food Democracy South and North: From Food Sovereignty to Transition Initiatives', Countercurrents.org, available at www.milanfoodlaw.org/wp-content/uploads/2015/03/15-03-22-Food-Democracy-South-And-North1.pdf [Accessed on 28/11/2015].

——— (2015b) 'La Cage Et Le Labyrinthe: S'évader De La Religion De La Croissance', in *21ème Congrès des économistes belges de langue française*, Liège, Belgium: Université de Liège.

De Schutter, O. and Vanloqueren, G. (2011) 'The New Green Revolution: How Twenty-First-Century Science Can Feed the World', *Solutions Journal* 2(4): 33.

de Sousa Santos, B. (2006) 'Capítulo I. La Sociología de las Ausencias y la Sociología de las Emergencias: para una ecología de saberes', in *Renovar la teoría crítica y reinventar la emancipación social (encuentros en Buenos Aires)*. ISBN 987-1183-57-7.

Dhar, B. (2002) *Sui Generis Systems for Plant Variety Protection*, Geneva, Switzerland: Quaker United Nations Office.

Di Robilant, A. (2012) 'Common Ownership and Equality of Autonomy', *McGill Law Journal/Revue de droit de McGill* 58(2): 263–320.

Dirzo, R. and Raven, P. H. (2003) 'Global State of Biodiversity and Loss', *Annual Review of Environment and Resources* 28(1).

Djoghlaf, A. (2006) 'Message of Dr Ahmed Djoghlaf, the New Executive Secretary of the Convention on Biological Diversity, to the Environmental NGOs of Our Planet', Montréal, Canada: Secretariat of the Convention on Biological Diversity.

Dorward, A., et al. (2004) 'A Policy Agenda for Pro-Poor Agricultural Growth', *World Development* 32(1): 73–89.

Drahos, P. (1996) *A Philosophy of Intellectual Property*, Aldershot, UK: Ashgate.

——— (1999) 'Biotechnology Patents, Markets and Morality', *European Intellectual Property Review* 21(9): 441–449.

——— (2004) 'Securing the Future of Intellectual Property: Intellectual Property Owners and Their Nodally Co-Ordinated Enforcement Pyramid', *Case Western Reserve Journal of International Law* 36.

——— (2006) 'A Defence of the Intellectual Commons', *Consumer Policy Review* 16: 2–5.

Drahos, P. and Braithwaite, J. (2002) *Information Feudalism: Who Owns the Knowledge Economy?* London, UK: Earthscan.

Drucker, A. and Caracciolo, F. (2012) 'The Economic Value of Plant Genetic Resources for Food and Agriculture: Study 1', in Stannard, C. and Moeller, N. (eds.) (2013), *Identifying Benefit Flows: Studies on the Potential Monetary and Non-Monetary Benefits Arising from the International Treaty on Plant Genetic for Food and Agriculture*, Rome: FAO.

Dulloo, E. (2015) 'Conservation and Availability of Plant Genetic Diversity: Innovative Strategies and Technologies', IV International Symposium on Plant Genetic Resources, Acta Horticulturae, Brisbane, Australia.

Dusollier, S. (2007) 'Sharing Access to Intellectual Property through Private Ordering', *Chicago-Kent Law Review* 82(3): 1391–435.

——— (2008) 'Le domaine public, garant de l'intérêt général en propriété intellectuelle?', in Buydens, M. and Dusollier, S. (eds.), *L'interêt général et l'accès à l'information en propriété intellectuelle*, Bruxelles: Bruylant.

——— (2013) 'The Commons as a Reverse Intellectual Property: From Exclusivity to Inclusivity', in Howe, H. and Griffiths, J. (eds.), *Concepts of Property in Intellectual Property Law*, Cambridge: Cambridge University Press, 258–282.

Dussel, E. (2011) 'From Critical Theory to the Philosophy of Liberation: Some Themes for Dialogue', *TRANSMODERNITY: Journal of Peripheral Cultural Production of the Luso-Hispanic World* 1(2).

Dutfield, G. (2002) 'Sharing the Benefits of Biodiversity', *The Journal of World Intellectual Property* 5(6): 899–931.

——— (ed.) (2004) *Intellectual Property, Biogenetic Resources and Traditional Knowledge*, London: Earthscan.

——— (2011) 'Food, Biological Diversity and Intellectual Property: The Role of the International Union for the Protection of New Varieties of Plants (UPOV)', QUNO (9).

——— (2018) 'Farmers, Innovation and Intellectual Property: Current Trends and Their Consequences for Food Security the Commons', in Girard, F. and Frison, C. (eds.), *Plant Breeding and Agricultural Research*, Oxon, UK: Routledge, 47–64.

Dutfield, G. and Suthersanen, U. (2008) *Global Intellectual Property Law*, Cheltenham: Edward Elgar Publishing Limited.

Eaton, D. and van Tongeren, F. (2004) 'Patents Versus Plant Varietal Protection', *8th ICABR International Conference on Agricultural Biotechnology: International Trade and Domestic Production*, Ravello, Italy.

Ebhojie, K. E. (2013) 'Patently Waiting for Sui Generis Rights: Systemic Biopiracy and Nigerian Traditional Knowledge in Vernonia Amygdalina', University of Liverpool, *Working Paper*.

Edelman, M. (1999) *Peasants against Globalization: Rural Social Movements in Costa Rica*, Stanford, CA: Stanford University Press.

Egziabher, T. B., et al. (2011) 'The African Regional Group: Creating Fair Play between North and South', in Frison, C., López, F. and Esquinas Alcázar, J. (eds.), *Plant Genetic Resources and Food Security: Stakeholder Perspectives on the International Treaty on Plant Genetic Resources for Food and Agriculture*, Oxon: Earthscan by Routledge and Bioversity International.

Ellen, R. and Platten, S. (2011) 'The Social Life of Seeds: The Role of Networks of Relationships in the Dispersal and Cultural Selection of Plant Germplasm', *Journal of the Royal Anthropological Institute* 17(3).

Ellickson, R. C. (1998) 'Law and Economics Discovers Social Norms', *The Journal of Legal Studies* 27(S2): 537–552.

ENB (2017) 'Summary of the Seventh Session of the Governing Body of the International Treaty on Plant Genetic Resources for Food and Agriculture', *Earth Negotiations Bulletin* Vol. 9 No. 691 (6 November 2017) IISD Reporting Service, available at http://enb.iisd.org/biodiv/itpgrfa/gb7/ [Accessed on 27/11/2017].

Engels, J. M. M., Dempewolf, H. and Henson-Apollonio, V. (2011) 'Ethical Considerations in Agro-Biodiversity Research, Collecting, and Use', *Journal of Agricultural & Environmental Ethics* 24(2): 108.

ESA (2011) *Position: On the Standard Material Transfer Agreement* (13 December 2011), available at www.euroseeds.eu/system/files/publications/files/esa_11.0938.1.pdf [Accessed on 27/02/2012].

——— (2012) *Position: On the Benefits of the Breeding Exemption to Society* (February 2012), available at www.euroseeds.eu/system/files/publications/files/esa_11.0156.1_0.pdf [Accessed on 27/02/2012].

——— (2014) *Press Release: European Seed Sector Shows Support to FAO Treaty on Plant Genetic Resources for Food and Agriculture* (16 October 2014), available at www.euroseeds.eu/system/files/publications/files/esa_14.0625.pdf [Accessed on 28/02/2014].

——— (2015) *Submission of Information on Interrelations with UPOV and WIPO* (5 February 2015), and *Comments of ESA European Seed Association with regard to Notification GB6-028 – Art.9 – UPOV/WIPO* (27.01.2015), ESA's views on Resolution 8/2013 adopted at the fifth meeting of the Governing Body of the International Treaty on Plant Genetic Resources for Food and Agriculture, 24–28 September 2013, Muscat, Oman, available at www.google.com/url?sa=t&rct=j&q=&esrc=s&source=web&cd=3&ved=2ahUKEwiOu7HorYXdAhXFUd4KHfdbDbwQFjACegQICBAC&url=http%3A%2F%2Fwww.fao.org%2F3%2Fa-bb929e.pdf&usg=AOvVaw1geCKwaLAKqjkNcH9RArEg

Escobar, A. (2008) *Territories of Difference: Place, Movements, Life, Redes*, Durham, NC: Duke University Press.

Esquinas-Alcazar, J. T. (1993) 'The Global System on Plant Genetic Resources', *Review of European Community & International Environmental Law* 2(2): 151–157.

——— (2005) 'Protecting Crop Genetic Diversity for Food Security: Political, Ethical and Technical Challenges', *Nature Reviews Genetics* 6(12): 946–953.

Esquinas-Alcázar, J. T., Frison, C. and López, F. (2011) 'Introduction', in Frison, C., López, F. and Esquinas-Alcázar, J. T. (eds.), *Plant Genetic Resources and Food Security: Stakeholder Perspectives on the International Treaty on Plant Genetic Resources for Food and Agriculture*, Oxon: Earthscan by Routledge and Bioversity International.

Esquinas-Alcázar, J. and Hilmi, A. (2008) 'Las negociaciones del tratado internacional sobre los recursos fitogenéticos para la alimentación y la agricultura The Negotiation of the International Treaty on Plant Genetic Resources for Food and Agriculture', *Recursos Naturales y Ambiente* (CATIE) 53: 20–29.

ETC Group (2012) The Greed Revolution. Mega Foundations, Agribusiness Muscle in on Public Goods, January/February 2012, available at www.etcgroup.org/fr/node/5317 (Accessed April 2015).

——— (2015) 'Breaking Bad: Big Ag Mega-Mergers in Play Dow + DuPont in the Pocket? Next: Demonsanto?', ETC Group Communiqué – Issue # 115 (December), available at www.etcgroup.org/fr/node/5317 (Accessed April 2015).

Evans, P. B., Jacobson, H. K. and Putnam, R. D. (1993) *Double-Edged Diplomacy: International Bargaining and Domestic Politics*, Berkeley: University of California Press.

Evenson, R. E. (2005) 'Agricultural Research and Intellectual Property Rights', in Maskus, K. E. and Reichman, J. H. (eds.), *International Public Goods and Transfer of Technology under a Globalized Intellectual Property Regime*, Cambridge, UK; New York, NY: Cambridge University Press.

Evenson, R. E. and Gollin, D. (2003) 'Assessing the Impact of the Green Revolution, 1960 to 2000', *Science* 300(5620): 758–762.

Falcon, W. P. and Fowler, C. (2002) 'Carving Up the Commons: Emergence of a New International Regime for Germplasm Development and Transfer', *Food Policy* 27(3): 197–222.

FAO (1996a) *First Report on the State of the World's PGRFA* (SoW1-PGRFA), Rome: FAO United Nations.

——— (1996b) *Report of the Commission on Genetic Resources for Food and Agriculture: Third Extraordinary Session*, Rome: FAO United Nations.

——— (1998) *Report of the Fifth Extraordinary Session of the Commission on Genetic Resources for Food and Agriculture*, Rome: FAO United Nations.

——— (2001) 'The Impact of Intellectual Property Rights (IPRs) on Food and Agriculture in Developing Countries Background Document' (Conference 6) of the FAO Electronic

Forum on Biotechnology in Food and Agriculture ran (20 March–13 May 2001), available at www.fao.org/biotech/c6doc.htm [Accessed on 17/01/2012].

——— (2010) *The State of Food Insecurity in the World Addressing Food Insecurity in Protracted Crises*, Rome: FAO United Nations.

——— (2013) *Report of the Resumed Seventh Meeting of the Ad Hoc Advisory Committee on the Funding Strategy*, IT/GB-5/13/Inf.4 Add.1, Rome Italy: FAO.

——— (2014a) *The State of Food and Agriculture: Innovation in Family Farming*, Rome: FAO United Nations.

——— (2014b) *The State of Food Insecurity in the World: Strengthening the Enabling Environment for Food Security and Nutrition*, Rome: FAO United Nations.

——— (2015) *The State of Food and Agriculture: Social Protection and Agriculture: Breaking the Cycle of Rural Poverty*, Rome: FAO United Nations.

Fennell, L. A. (2010) 'Commons, Anticommons, Semicommons', in Ayotte, K. and Smith, H. E. (eds.), *Research Handbook on the Economics of the Property Law*, Edward Elgar, 2011, Surrey, UK: Camberley.

Fernandez Diaz-Silveira, M. (2011) 'The Latin American and Caribbean Regional Group: A Long and Successful Process for the Protection, Conservation and Enhancing of Plant Genetic Resources for Food and Agriculture', in Frison, C., López, F. and Esquinas-Alcazar, J. (eds.), *Plant Genetic Resources and Food Security: Stakeholder Perspectives on the International Treaty on Plant Genetic Resources for Food and Agriculture*, Oxon: Earthscan by Routledge and Bioversity International.

Footer, M. E. (2000) 'Intellectual Property and Agrobiodiversity: Towards Private Ownership of the Genetic Commons', *Yearbook of International Environmental Law* 10(1).

Foresight UK (2011) *The Future of Food and Farming: Challenges and Choices for Global Sustainability*, London: The Government Office for Science (GO-Science).

Fowler, C. (1994) *Unnatural Selection: Technology, Politics, and Plant Evolution*, Yverdon, Switzerland: Gordon and Breach Science Publishers.

Fowler, C. and Hodgkin, T. (2004) 'Plant Genetic Resources for Food and Agriculture: Assessing Global Availability', *Annual Review of Environment & Resource* 29(1): 143–179.

Fowler, C. and Mooney, P. R. (1990) *Shattering: Food, Politics, and the Loss of Genetic Diversity*, Tucson: University of Arizona Press.

Fraleigh, B. and Harvey, B. L. (2011) 'The North American Group: Globalization That Works', in Frison, C., López, F. and Esquinas-Alcazar, J. (eds.), *Plant Genetic Resources and Food Security: Stakeholder Perspectives on the International Treaty on Plant Genetic Resources for Food and Agriculture*, Oxon: Earthscan by Routledge and Bioversity International.

Frankel, O. H. (1970) *Genetic Conservation in Perspective*, Oxford; Edinburgh: Blackwell Scientific Publications.

Frankel, O. H. and Hawkes, J. G. (1975) *Crop Genetic Resources for Today and Tomorrow*, Cambridge, UK; New York, NY: Cambridge University Press.

Freeman, R. E. (2010) *Strategic Management: A Stakeholder Approach*, Cambridge: Cambridge University Press.

Friedman, A. L. and Miles, S. (2002) 'Developing Stakeholder Theory', *Journal of Management Studies* 39(1): 1–21.

Frison, C. (2006) 'The Principles of Sustainable Development in the Context of the International Treaty on Plant Genetic Resources in Food and Agriculture', *McGill International Journal of Sustainable Development Law & Policy* 2(2).

——— (2016) *Towards Redesigning the Plant Commons: A Critical Assessment of the Multilateral System of Access and Benefit-Sharing of the International Treaty on Plant Genetic Resources for Food and Agriculture*, PhD thesis, Leuven: UCLouvain/KU Leuven.

——— (2018) 'Planting the Commons: Towards Redesigning an Equitable Global Seed Commons', in Girard, F. and Frison, C. (eds.), *The Commons, Plant Breeding and Agricultural Research*, Oxon: Routledge.

Frison, C. and Claeys, P. (2014) 'Right to Food in International Law', in P. B. Thompson and D. M. Kaplan (eds.), *Encyclopedia of Food and Agricultural Ethics*, Dordrecht: Springer: 1617–1624.

Frison, C., Dedeurwaerdere, T., and M. Halewood (2010) 'Intellectual Property and Facilitated Access to Genetic Resources under the International Treaty on Plant Genetic Resources for Food and Agriculture', *European Intellectual Property Review* 32(1).

Frison C. and Halewood M. (2005) 'Annotated Bibliography Addressing the International Pedigrees and Flows of Plant Genetic Resources for Food and Agriculture', Information document submitted by the System-wide Genetic Resources Programme (SGRP) of the CGIAR to the eighth Conference of the Parties of the Convention on Biological Diversity (COP 8) and the Ad hoc Open-ended Working Group on Access and Benefit-sharing, 20–31 March 2006, Curitiba, Brazil.

Frison, C. and Joie, T. (2006a) 'Elaboration d'une Réglementation de Biosécturité par Certains Pays en Développement: Expériences dans la Mise en Œuvre du Protocole de Cartagena en Afrique de L'Ouest', *Law, Environment and Development Journal* 2: 164.

——— (2006b) 'Expériences sur l'élaboration de nouvelles lois de développement de la biosécurité et de la biotechnologie: perspectives de réformes légales en Afrique de l'Ouest', in McInerney, T. F. (ed.), *Searching for Success: Narrative Accounts of Legal and Institutional Reform in Developing Countries*, Rome: International Development Law Organization IDLO, 1, 151–164.

Frison, C., López, F. and Esquinas-Alcázar, J. T. (2011a) *Plant Genetic Resources and Food Security: Stakeholder Perspectives on the International Treaty on Plant Genetic Resources for Food and Agriculture*, Oxon: Earthscan by Routledge and Bioversity International.

——— (2011b) 'General Conclusions: Summary of Stakeholders' Views and Suggestions to Cope with the Challenges in the Implementation of the International Treaty on Plant Genetic Resources for Food and Agriculture', in Frison, C., López, F. and Esquinas-Alcázar, J. T. (eds.), *Plant Genetic Resources and Food Security: Stakeholder Perspectives on the International Treaty on Plant Genetic Resources for Food and Agriculture*, Oxon: Earthscan by Routledge and Bioversity International.

Frison, E. A., Cherfas, J., and Hodgkin, T. (2011) 'Agricultural Biodiversity Is Essential for a Sustainable Improvement in Food and Nutrition Security', *Sustainability* 3(1): 238–253.

Frison, E. and Hodgkin, T. (2016) 'Strategic Opportunities to Strengthen Community-Based Approaches to Seed Agrobiodiversity', in *The Future of Food: Seeds of Resilience – A Compendium of Perspectives on Agricultural Biodiversity from around the World*, The Global Alliance for the Future of Food, available at https://futureoffood.org/report/the-future-of-food-seeds-of-resilience/opportunities-report/

Galluzzi, G., Halewood, M., Noriega, I. L. and Vernooy, R. (2016) 'Twenty-Five Years of International Exchanges of Plant Genetic Resources Facilitated by the CGIAR Genebanks: A Case Study on Global Interdependence', *Biodiversity and Conservation* 25(8): 1421–1446.

Garforth, K., et al. (2005) 'Overview of the National and Regional Implementation of Access to Genetic Resources and Benefit-Sharing Measures', in *Centre for International Sustainable Development Law REPORT*, 103, 3rd ed, Montreal, Canada: CISDL.

Garforth, K. and Cabrera Medaglia, J. (2006) 'Legal Reform for the Development and Implementation of Measures on Access to Genetic Resources and Benefit-Sharing', in McInerney, T. F. (ed.), *Searching for Success: Narrative Accounts of Legal and Institutional Reform in Developing and Transition Countries*, Rome: International Development Law Organization IDLO, 1, 139–149.

Garforth, K. and Frison, C. (2007) 'Key Issues for the Relationship between the Convention on Biological Diversity and the International Treaty on Plant Genetic Resources for Food and Agriculture', *Occasional Paper 2*, Quaker International Affairs Programme 2, 20.

Garnett, T. (2014) 'Three Perspectives on Sustainable Food Security: Efficiency, Demand Restraint, Food System Transformation: What Role for Life Cycle Assessment?', *Journal of Cleaner Production* 73: 10–18.

Gepts, P. (2004) 'Crop Domestication as a Long-Term Selection Experiment', *Plant Breeding Reviews* 24(2): 1295–1307.

Gerbasi, F. (2011) 'Overview of the Regional Approaches: The Negotiating Process of the International Treaty on Plant Genetic Resources for Food and Agriculture', in Frison, C., López, F. and Esquinas-Alcázar, J. T. (eds.), *Plant Genetic Resources and Food Security: Stakeholder Perspectives on the International Treaty on Plant Genetic Resources for Food and Agriculture*, Oxon: Earthscan by Routledge and Bioversity International, 321.

Gerstetter, C., et al. (2007) 'The International Treaty on Plant Genetic Resources for Food and Agriculture within the Current Legal Regime Complex on Plant Genetic Resources', *Journal of World Intellectual Property* 10(3/4).

Gintis, H., et al. (eds.) (2005) *Moral Sentiments and Material Interests: The Foundations of Cooperation in Economic Life*, Cambridge, MA: MIT Press.

Girard, C. and Le Goff, A. (2010) *La Démocratie Délibérative: Anthologie de Textes Fondamentaux*, Paris, France: Editions Hermann.

Girard, F. (2015) '"Though the Treasure of Nature's Germens Tumble All Together": The EPO and Patents on Native Traits or the Bewitching Powers of Ideologies', *Prometheus* 33(1): 1–23.

——— (2018) 'Composing the Common World of the Local Bio-Commons in the Age of the Anthropocene', in Girard, F. and Frison, C. (eds.), *The Commons, Plant Breeding and Agricultural Research*, Abingdon, UK: Routledge, 117–144.

Girard, F. and Frison, C. (2018a) *The Commons, Plant Breeding and Agricultural Research: Challenges for Food Security and Agrobiodiversity*, Oxon: Routledge.

——— (2018b) 'Introduction: Commoning the Seeds: The Future of Agrobiodiversity and Food Security', in Girard, F. and Frison, C. (eds.), *The Commons, Plant Breeding and Agricultural Research*, Oxon: Routledge.

Girard, F. and Noiville, C. (2014) 'Propriété industrielle et biotechnologies végétales: la Nova Atlantis', *Revue internationale de droit économique* 28(1): 59–109.

Glasius, M. and Pleyers, G. (2013) 'The Global Moment of 2011: Democracy, Social Justice and Dignity', *Development and Change* 44(3): 547–567.

Glowka, L. (1997) 'Emerging Legislative Approaches to Implement Article 15 of the Convention on Biological Diversity', *Review of European, Comparative & International Environmental Law* 6(3): 249–262.

Goffaux, R., et al. (2011) 'Quels indicateurs pour suivre la diversité génétique des plantes cultivées? Le cas du blé tendre cultivé en France depuis un siècle', *Rapport FRB, Série Expertise et synthèse* 44.

Gonzalez, C. G. (2015) 'Bridging the North-South Divide: International Environmental Law in the Anthropocene', *Pace Environmental Law Review* 32.

Gotor, E. and Caracciolo, F. (2009) 'An Empirical Assessment of the Effects of the 1994 in Trust Agreements on IRRI Germplasm Acquisition and Distribution', *International Journal of the Commons* 4(1).

Graeub, B. E., Chappell, M. J., Wittman, H., Ledermann, S., Kerr, R. B. and Gemmill-Herren, B. (2016) 'The State of Family Farms in the World', *World Development* 87: 1–15.

Graff, D. G., Wright, B. D., Bennett, A. B. and Zilberman, D. (2004) 'Access to Intellectual Property Is a Major Obstacle to Developing Transgenic Horticultural Crops', *California Agriculture* 58(2): 120–126.

GRAIN (2007) 'The End of Farm-Saved Seed? Industry's Wish List for the Next Revision of UPOV', *GRAIN Report*, available at www.grain.org/article/entries/58-the-end-of-farm-saved-seed-industry-s-wish-list-for-the-next-revision-of-upov [Accessed on 08/09/2011].

Greiber, T., et al. (2015) 'Conservation with Justice: A Rights-Based Approach', *IUCN Environmental Law and Policy Paper* 71: 115.

Grosheide, W. and Brinkhof, J. (eds.) (2002) *Articles on Cultural Expressions and Indigenous Knowledge*, Antwerp: Intersentia.

Gudeman, S. (2008) *Economy's Tension: The Dialectics of Community and Market*, New York, NY; Oxford: Berghahn Books, 2012.

Guiramand, M., Moeller, N. I. and Marino, M. (2012) 'Plant Breeding and the Use of the Standard Material Transfer Agreement: Consultation with Plant Breeding Experts', Study 5 in Stannard, C. and Moeller, N. (eds.) (2013), *Identifying Benefit Flows: Studies on the Potential Monetary and Non-Monetary Benefits Arising from the International Treaty on Plant Genetic for Food and Agriculture*, Rome: FAO.

Gutwirth, S. and Stengers, I. (2016) 'Le droit à l'épreuve des commons', *Revue Juridique de l'Environnement* 306–343.

Haas, E. B. (1975) 'Is There a Hole in the Whole? Knowledge, Technology, Interdependence, and the Construction of International Regimes', *International Organization* 29(3): 827–876.

Hadiprayitno, I. (2014) 'Food Security', *Encyclopaedia of Food and Agricultural Ethics* 957–964.

Halewood, M. (2000) *The Crucible Group Experience: Global Forum on Agricultural Research* (GFAR/00/17–06), Dresden: GFAR.

——— (2010) 'Governing the Management and Use of Pooled Microbial Genetic Resources: Lessons from the Global Crop Commons', *International Journal of the Commons* 4(1): 404–436.

——— (2013) 'What Kind of Goods Are Plant Genetic Resources for Food and Agriculture? Towards the Identification and Development of a New Global Commons', *International Journal of the Commons* 7(2): 278–312.

——— (2014) 'International Efforts to Pool and Conserve Crop Genetic Resources in Times of Radical Legal Change', in *Intellectual Property Rights: Legal and Economic Challenges for Development*, Oxford, UK: Oxford University press, 288–322.

Halewood, M., et al. (2006) 'Farmers, Landraces and Property Rights: Challenges to Allocating Sui Generis Intellectual Property Rights to Communities over Their Varieties', in Biber-Klemm, S. and Cottier, T. (eds.), *Rights to Plant Genetic Resources and Traditional Knowledge: Basic Issues and Perspectives*, Wallingford, UK: CABI.

Halewood, M., et al. (2018) 'Plant Genetic Resources for Food and Agriculture: Opportunities and Challenges Emerging from the Science and Information Technology Revolution', *New Phytologist* 217(4): 1407–1419.

Halewood, M., Lopez Noriega, I. and Louafi, S. (2013) *Crop Genetic Resources as a Global Commons: Challenges in International Governance and Law*, Oxon: Routledge.

Halewood, M. and Nnadozie, K. (2008) 'Giving Priority to the Commons: The International Treaty on Plant Genetic Resources for Food and Agriculture', in Tansey, G.and Rajotte, T. (eds.), *The Future Control of Food: A Guide to International Negotiations and Rules on Intellectual Property, Biodiversity and Food Security*, London: Earthscan.

Halpert, M.-T. and Chappell, M. J. (2017) 'Prima facie Reasons to Question Enclosed Intellectual Property Regimes and Favor Open-Source Regimes for Germplasm', *F1000Research* 6: 284.

Hamilton, C. (2006) 'Biodiversity, Biopiracy and Benefits: What Allegations of Biopiracy Tell us about Intellectual Property', *Developing World Bioethics* 6(3).

Haraway, D. (1988) 'Situated Knowledges: The Science Question in Feminism and the Privilege of Partial Perspective', *Feminist Studies* 14(3): 575–599.

Hardin, G. (1968) 'The Tragedy of the Commons', *Science* 162(3859): 1243–1248.

Hardon, J. J., Vosman, B. and van Hintumn, Th. J. L. (1994) 'Identifying Genetic Resources and Their Origin: The Capabilities and Limitations of Modern Biochemical and Legal Systems', *Background Study Paper n°4* (7–11 November), Commission on Plant Genetic Resources, First Extraordinary Session, Rome, FAO, Rome.

Harlan, J. R. (1975) *Crops & Man*, Madison, WI: American Society of Agronomy.

——— (1995) *The Living Fields: Our Agricultural Heritage*, Cambridge, UK; New York, NY: Cambridge University Press.

Hardoon, D. (2015) 'Wealth: Having It All and Wanting More: Oxfam International', *Paper Presented at the World Economic Forum*, Oxford, UK: Oxfam GB.

Hardoon, D., Ayele, S. and Fuentes-Nieva, R. (2016) 'An Economy for the 1%', Oxford: Oxfam, available at http://policy-practice.oxfam.org.uk/publications/an-economy-for-the-1-how-privilege-and-power-in-theeconomy-drive-extreme-inequ-592643 [Accessed on 21/02/2017].

Hawtin, G. and Fowler, C. (2011) 'The Global Crop Diversity Trust: An Essential Element of the Treaty's Funding Strategy', in C. Frison, F. López and J. T. Esquinas-Alcázar (eds.), *Plant Genetic Resources and Food Security: Stakeholder Perspectives on the International Treaty on Plant Genetic Resources for Food and Agriculture*, Oxon: Earthscan by Routledge and Bioversity International.

Hayden, C. (2007) 'Taking as Giving: Bioscience, Exchange, and the Politics of Benefit-Sharing', *Social Studies of Science* 37(5): 729–758.

HCB (Haut Conseil des biotechnologies, France) (2014) *Biotechnologies végétales et propriété industrielle*, Paris: La Documentation Française.

Heald, P. J. and Chapman, S. (2012) 'Veggie Tales: Pernicious Myths about Patents, Innovation, and Crop Diversity in the Twentieth Century', *University of Illinois Law Review* 4: 1051.

Hecquet, C. (2015) 'Comment Faire Circuler Les Semences? Enjeux Et Perspectives Pour Les Alternatives' (unpublished).

Helfer, L. R. (2004) 'Regime Shifting: The TRIPs Agreement and New Dynamics of International Intellectual Property Lawmaking', *Yale Journal of International Law* 29(1): 1.

——— (2005) 'Using Intellectual Property Rights to Preserve the Global Genetic Commons: The ITPGRFA', in Maskus, K. and Reichman, J. (eds.), *International Public Goods and Transfer of Technology under a Globalized Intellectual Property Regime*, Cambridge: Cambridge University Press.

——— (2015) 'Mapping the Interface between Human Rights and Intellectual Property', in Geiger, C. (ed.), *Research Handbook on Human Rights and Intellectual Property*, Cheltenham: Edward Elgar Publishing Limited, 6–18.

Heller, M. A. (1998) 'The Tragedy of the Anticommons: Property in the Transition from Marx to Markets', *Harvard Law Review* 111(3): 621–688.

——— (2010) The Gridlock Economy: How Too Much Ownership Wrecks Markets, Stops Innovation, and Costs Lives, New York: Basic Books.

Heller, M. A. and Dagant, H. (2001) 'The Liberal Commons', *Yale Law Journal* 110(4): 549–623.

Heller, M. A. and Eisenberg, R. S. (1998) 'Can Patents Deter Innovation? The Anticommons in Biomedical Research', *Science* 280(5364): 698–701.

Henry, A. D. and Dietz, T. (2011) 'Information, Networks, and the Complexity of Trust in Commons Governance', *International Journal of the Commons* 5(2): 188–212.

Henry, C. and Stiglitz, J. E. (2010) 'Intellectual Property, Dissemination of Innovation and Sustainable Development', *Global Policy* 1(3): 237–251.

Herdt, R. W. (1999) *Enclosing the Global Plant Genetic Commons* (14 January). T. R. Foundation, Institute for International Studies, Stanford, CA: Stanford University.

Hess, C. and Ostrom, E. (2003) 'Ideas, Artifacts, and Facilities: Information as a Common-Pool Resource', *Law and Contemporary Problems* 66(1/2): 111–145.

——— (2006) 'A Framework for Analysing the Microbiological Commons', *International Social Science Journal* 58(188): 335–349.

——— (eds.) (2007) *Understanding Knowledge as a Commons: From Theory to Practice*, Cambridge, MA; London: MIT Press.

Ho, C. M. (2006) 'Biopiracy and beyond: A Consideration of Socio-Cultural Conflicts with Global Patent Policies', *University of Michigan Journal of Law Reform* 39(3): 443–542.

Holder, J. B. and Flessas, T. (2008) 'Emerging Commons', *Social & Legal Studies* 17(3): 299–310.

Holley, C. and Shearing, C. (2017) 'A Nodal Perspective of Governance: Advances in Nodal Governance Thinking', in Drahos, P. (ed.), *Regulatory Theory: Foundations and Applications*, Canberra: ANU Press: 163.

Honoré, A. M. (1961) 'Ownership', in Guest, A. G. (ed.), *Oxford Essays in Jurisprudence*, Oxford: Oxford University Press.

Hope, J. (ed.) (2008) *Biobazaar. The Open Source Revolution and Biotechnology*, Cambridge; London: Harvard University Press.

Hopkins, R. (2008) *The Transition Handbook*, Totnes: Green Books.

Howard, P. H. (2009) 'Visualizing Consolidation in the Global Seed Industry: 1996–2008', *Sustainability* 1(4): 1266–1287.

——— (2016) *Concentration and Power in the Food System: Who Controls What We Eat?*, New York: Bloomsbury Publishing.

Hubacek, K. and Mauerhofer, V. (2008) 'Future Generations: Economic, Legal and Institutional Aspects', *Futures* 40(5).

IAASTD (2008) 'Agriculture at a Crossroads', Global Summary for Decision Makers, available at www.unep.org/dewa/agassessment/reports/IAASTD/EN/Agriculture%20at%20a%20 Crossroads_Global%20Summary%20for%20Decision%20Makers%20(English).pdf

——— (2009) *Synthesis Report: A Synthesis of the Global and Sub-Global International Assessment of Agricultural Knowledge, Science and Technology for Development Reports*, McIntyre, B. D., Herren, H. R., Wakhungu, J. and Watson, R. T. (eds.), Washington, DC: Island Press.

——— (n.d.) 'Issues in Brief', *Business as Usual Is Not an Option: Trade and Markets*, available at www.globalagriculture.org/fileadmin/files/weltagrarbericht/IAASTDBerichte/Issues-BriefTrade.pdf [Accessed on 10/05/2016].

IFAD (2011) *The Rural Poverty Report 2011*, printed by Quintily, Rome, Italy: International Fund for Agricultural Development (IFAD).

IPES-Food (2016) 'Thematic Report 1', *From Uniformity to Diversity: A Paradigm Shift from Industrial Agriculture to Diversified Agroecological Systems* (June), Louvain-la-Neuve, UCLouvain.

——— (2017a) 'Thematic Report 2', *Unravelling the Food-Health Nexus: Addressing Practices, Political Economy, and Power Relations to Build Healthier Food Systems* (October), Louvain-la-Neuve, UCLouvain.

——— (2017b) 'Thematic Report 3', *Too Big to Feed: Exploring the Impacts of Mega-Mergers, Consolidation and Concentration of Power in the Agri-Food Sector* (October), Louvain-la-Neuve, UCLouvain.

IPGRI (1996) 'Access to Plant Genetic Resources and the Equitable Sharing of Benefits: A Contribution to the Debate on Systems for the Exchange of Germplasm', *Issues in Genetic Resources* 4, Rome, Italy.

Isaac, G. E. and Kerr, W. A. (2004) 'Bioprospecting or Biopiracy?', *The Journal of World Intellectual Property* 7(1): 35–52.

IUCN (2004) *Accessing Biodiversity and Sharing the Benefits: Lessons from Implementing the Convention on Biological Diversity*, Cambridge, UK: IUCN-The World Conservation Union.

Jaffee, S. and Srivastava, J. (1994) 'The Roles of the Private and Public Sectors in Enhancing the Performance of Seed Systems', *The World Bank Research Observer* 9(1): 97–117.

Janis, M. D. (2001) 'Sustainable Agriculture, Patent Rights, and Plant Innovation', *Indiana Journal of Global Legal Studies* 9(1): 117.

Jarvis, D. I., et al. (2011) 'An Heuristic Framework for Identifying Multiple Ways of Supporting the Conservation and Use of Traditional Crop Varieties within the Agricultural Production System', *Critical Reviews in Plant Sciences* 30(1–2): 125–176.

Jarvis, D. I., Hodgkin, T., Brown, A. H. D., Tuxill, J., López Noriega, I., Smale, M. and Sthapit, B. (2016) *Crop Genetic Diversity in the Field and on the Farm: Principles and Applications in Research Practices*, New Haven, CT: Yale University Press.

Jayne, T. S. and Rashid, S. (2013) 'Input Subsidy Programs in Sub-Saharan Africa: A Synthesis of Recent Evidence', *Agricultural Economics* 44(6): 547–562.

Johnston, S. (1993) 'Conservation Role of Botanic Gardens and Gene Banks', *Review of European Community & International Environmental Law* 2(2).

Johnston, S. (1997) 'The Convention on Biological Diversity: The Next Phase', *Review of European Community & International Environmental Law* 6(3): 219–230.

Jones, S. S. (2004) 'Breeding Resistance to Special Interests', *Organic Farming Research Foundation Information Bulletin* 14.

Jungcurt, S. (2007) 'Institutional Interplay in International Environmental Governance: Policy Interdependence and Strategic Interaction in the Regime Complex on Plant Genetic Resources for Food and Agriculture', in *Landwirtschaftlich – Gärtnerischen Fakultät*. Berlin: Humboldt Universät, 278.

Kahneman, D. and Tversky, A. (2000) *Choices, Values, and Frames*, Cambridge: Cambridge University Press.

Kalaugher, E., et al. (2002) 'A Summary and Analysis of Existing of International Plant Genetic Resources Networks', *Commission on Genetic Resources for Food and Agriculture* 16, FAO, Rome, Italy.

Kamau, E. C. and Winter, G. (eds.) (2009) *Genetic Resources, Traditional Knowledge and the Law: Solution for Access and Benefit Sharing*, London: Earthscan.

Karjala, D. S. (1994) 'Misappropriation as a Third Intellectual Property Paradigm', *Columbia Law Review* 94(8).

Kaul, I. and Conceição, P. (2006) *The New Public Finance: Responding to Global Challenges*, New York, NY: Oxford University Press.

Kaul, I., Conceição, P., Le Goulven, K., and Mendoza, R. U. (2003) *Providing Global Public Goods – Managing Globalization*, Oxford: Oxford University Press, 646.

Kaul, I., et al. (eds.) (1999) *Global Public Goods: International Cooperation in the 21st Century*, Oxford: Oxford University Press.

Kaul, I. and Mendoza, R. U. (2003) 'Advancing the Concept of Public Goods', in Kaul, I., Conceição, P., Le Goulven, K. and Mendoza, R. U. (eds.), *Providing Global Public Goods: Managing Globalization*, Oxford: Oxford University Press, 646.

Keenan, S. (2015) *Subversive Property: Law and the Production of Spaces of Belonging*, Abingdon, UK: Routledge.

Keohane, R. O. and Ostrom, E. (1995) *Local Commons and Global Interdependence: Heterogeneity and Cooperation in Two Domains*, Thousand Oaks: Sage.

Kestemont, L. (2015) *Methods for Traditional Legal Research*, Reader 'Methods of Legal Research', research Master in Law, KU Leuven-University of Tilburg.

Khoury, C. K., et al. (2015) 'Where Our Food Crops Come from: A New Estimation of Countries' Interdependence in Plant Genetic Resources', *CIAT Policy Brief*, International Center for Tropical Agriculture 25.

Kiesler, S., et al. (1984) 'Social Psychological Aspects of Computer-Mediated Communication', *American Psychologist* 39(10): 1123.

Ki-Moon, B. (2009) 'Official Remarks of the United Nations Secretary-General Ban Ki-Moon at the Plenary Session of the Fifth Summit of the Americas', in Fifth Summit of the Americas (Port of Spain, Trinidad & Tobago) (17–19 April 2009).

Kloppenburg, J. R. (2004) *First the Seed: The Political Economy of Plant Biotechnology, 1492–2000*, 2nd edn, Madison, WI: University of Wisconsin Press (1988).

——— (2010) 'Impeding Dispossession, Enabling Repossession: Biological Open Source and the Recovery of Seed Sovereignty', *Journal of Agrarian Change* 10(3): 367–388.

——— (2014) 'Re-Purposing the Master's Tools: The Open Source Seed Initiative and the Struggle for Seed Sovereignty', *Journal of Peasant Studies* 41(6): 1225–1246.

Klose, N. (1950) *America's Crop Heritage: The History of Foreign Plant Introduction by the Federal Government*, Iowa City, IA: Iowa State College Press,.

Kock, M. A. and ten Have, F. (2016) 'The "International Licensing Platform-Vegetables": A Prototype of a Patent Clearinghouse in the Life Science Industry', *Journal of Intellectual Property Law & Practice* 11(7): 496–515.

Koo, B., et al. (2004) 'Plants and Intellectual Property: An International Appraisal', Science 306(5700): 1295–1297.

Koonan, S. (2014) 'India's Sui Generis System of Plant Variety Protection', *Briefing Paper*, Geneva, Switzerland: Quaker United Nations Office, 4.

Kotschi, J. and Rapf, K. (2016) 'Liberating Seeds with an Open Source Seed (OSS) Licence', *Working Paper*, Guggenhausen: AGRECOL.

Laird, S. A. and Wynberg, R. (2005) 'The Commercial Use of Biodiversity: An Update on Current Trends in Demand for Access to Genetic Resources and Benefit-Sharing, and Industry Perspectives on ABS Policy and Implementation', *Working Document CBD* UNEP/CBD/WG-ABS/4/INF/5, Montreal, Canada.

——— (2008) 'Access and Benefit-Sharing in Practice: Trends in Partnerships across Sectors', *Technical Series No. 38*, CBD Secretariat, Montreal, Canada.

Lakshmi Poorna, R., et al. (2014) 'Preservation and Protection of Traditional Knowledge: Diverse Documentation Initiatives across the Globe', *Current Science* 107(8): 1240–1246.

Lambek, N., et al. (2014) *Rethinking Food Systems: Structural Challenges, New Strategies and the Law*, London: Springer Science & Business Media.

Lampe, R. and Moser, P. (2016) 'Patent Pools, Competition, and Innovation: Evidence from 20 US Industries under the New Deal', *Journal of Law, Economics, and Organization* 32(1): 1–36.

Langegger, S. (2016) Book Review. *Sarah Keenan, Subversive Property: Law and the Production of Spaces of Belonging*, London: Routledge, in *Antipode: A Radical Journal of Geography*.

Laplace, M. (2015) 'Le Salvador Bannit Le Roundup De Monsanto Et Connaît Des Récoltes Records', *Notre Terre: L'info Ecologique au Quotidien* (27 April 2015), available at www.notreterre.org/archive/2015-04/ (Accessed 27 April 2015).

Lappé, F. M., Collins, J. and Fowler, C. (1979) *Food First: Beyond the Myth of Scarcity*, New York, NY: Ballantine Books.

Lappé, F. M., Collins, J. and Rosset, P. (1998) *World Hunger: 12 Myths*, New York, NY: Grove Press.

Laroche Dupraza, C. and Postollec, A. (2013) 'Food Sovereignty and Agricultural Trade Policy Commitments: How Much Leeway do West African Nations Have?', *Food Policy* 38: 115–125.

Larson, B. A. and Bromley, D. W. (1990) 'Property Rights, Externalities, and Resource Degradation: Locating the Tragedy', *Journal of Development Economics* 33(2): 235–262.

Lawson, C. (2009) 'Intellectual Property and the Material Transfer Agreement under the International Treaty on Plant Genetic Resources for Food and Agriculture', *European Intellectual Property Review* 31(5): 244–254.

La Via Campesina and GRAIN (2015) *Seed Laws That Criminalize Farmers*, La Via Campesina – GRAIN.

LeBuanec, B. (2007) 'Evolution of the Seed Industry during the Past Three Decades', *Seed Testing: International ISTA News Bulletin* 134, Bologna, Italy.

Leiserowitz, A. A., et al. (2006) 'Sustainability Values, Attitudes, and Behaviors: A Review of Multinational and Global Trends', *Annual Review of Environment and Resources* 31(1): 413–444.

Lenné, J. M. and Wood, D. (2011) *Agrobiodiversity Management for Food Security: A Critical Review*, Wallingford: CABI Publishing: [distributor] Cab International.

Lessig, L. (1995) 'The Regulation of Social Meaning', *The University of Chicago Law Review* 62(3): 943–1045.

Libecap, G. D. (1994) 'The Conditions for Successful Collective Action', in Keohane, R. O. and Ostrom, E. (eds.), Local Commons and Global Interdependence: Heterogeneity and Cooperation in Two Domains, 161–190, London: Sage Publications.

Lim, E. S. (2011) 'The Asian Regional Group', in Frison, C., López, F. and Esquinas-Alcázar, J. T. (eds.), *Plant Genetic Resources and Food Security: Stakeholder Perspectives on the International Treaty on Plant Genetic Resources for Food and Agriculture*, Oxon: Earthscan by Routledge and Bioversity International.

Lim, E. S. and Halewood, M. (2008) 'A Short History of the Annex 1 List', in Tansey, G. and Rajotte, T. (eds.), *The Future Control of Food: A Guide to International Negotiations and Rules on Intellectual Property, Biodiversity and Food Security*, London: Earthscan, Annex 3, 249 (Earthscan, IDRC and QIAP).

Llewelyn, M. and Adcock, M. (2006) *European Plant Intellectual Property*, Oxford: Hart Publishing.

Locher, F. (2013) 'Cold War Pastures: Garrett Hardin and the "Tragedy of the Commons"', *Revue d'histoire moderne et contemporaine* 60(1): 7–36.

——— (2016) 'Third World Pastures: The Historical Roots of the Commons Paradigm (1965–1990)', *Quaderni storici* 51(1): 301–334.

Louwaars, N. P. (2008) *Seeds of Confusion: The Impact of Policies on Seed Systems*, Wageningen: Wageningen Universiteit.

Louwaars, N. P., et al. (2005) 'Impacts of Strengthened Intellectual Property Rights Regimes on the Plant Breeding Industry in Developing Countries: A Synthesis of Five Case Studies', *Report Commissioned by the World Bank*, Wageningen, The Netherlands.

Louwaars, N., Tripp, R. and Eaton D. (2006) 'Intellectual Property Rights in the Breeding Industry: Farmers' Interests', *Agricultural and Rural Development* (14).

Louwaars, N. P. and de Boef W. S. (2012) 'Integrated Seed Sector Development in Africa: A Conceptual Framework for Creating Coherence between Practices, Programs, and Policies', *Journal of Crop Improvement* 26(1): 39–59.

Lucarelli, A. (2011a) 'Note Minime Per Una Teoria Giuridica Dei Beni Comuni', *Espaço Jurídico* 12(2): 11–20.

——— (2011b) *Beni comuni. Dalla teoria all'azione politica*, Viareggio, Italia: Dissensi.

——— (2013) *La Democrazia Dei Beni Comuni*, Rome; Bari: Laterza, Italia.

Madison, M., Frischmann, B. and Strandburg, K. (2010) 'Constructing Commons in the Cultural Environment', *Cornell Law Review* 95(4): 657–709.

Maggioni, L., Engels, J. and Lipman, E. (2015) 'Plant Genetic Resources Conservation in Europe: The AEGIS Experience', ECPGR Secretariat, presentation at the 110° CONGRESSO DELLA SOCIETÀ BOTANICA ITALIANA, Pavia, 16 September 2015.

Magretta, J. (1997) 'Growth through Global Sustainability: An Interview with Monsanto's CEO, Robert B. Shapiro', *Havard Business Review, January–Ferbuary 1997 issue*, available at https://hbr.org/1997/01/growth-through-global-sustainability-an-interview-with-monsantos-ceo-robert-b-shapiro.

Maldonado-Torres, N. (2016) *Outline of Ten Theses on Coloniality and Decoloniality*, Paris, France: Fondation Frantz Fanon.

Manzella, D. (2013) 'The Design and Mechanics of the Multilateral System of Access and Benefit Sharing', in Halewood, M., Noriega, I. L. and Louafi, S. (eds.), *Crop Genetic Resources as a Global Commons*, Oxon: Earthscan by Routledge and Bioversity International, 150.

Marden, E. (1999) 'The Neem Tree Patent: International Conflict over the Commodification of Life', *Boston College International and Comparative Law Review* 22(2): 279.

Marella, M. R. (ed.) (2012) *Oltre il pubblico e il privato. Per un diritto dei beni comuni*, Verona: Ombre corte.

——— (2017) 'The Commons as a Legal Concept', *Law Critique* 28(1): 61–86.

Mason, P. (2015) *Post Capitalism: A Guide to Our Future*, London: Penguin Books.

Mattei, U. (2000) *Basic Principles of Property Law: A Comparative Legal and Economic Introduction*, Westport, CT: Greenwood Publishing Group.

——— (2011a) *Beni Comuni: Un Manifesto*, Rome; Bari: Laterza.

——— (2011b) 'The State, the Market, and Some Preliminary Questions about the Commons', Paper presented as part of the Project 'Human Rights of People Experiencing Poverty' at the University of Turin as part of the DGIII Social Cohesion of the Council of Europe [Accessed on 05/02/2012].

——— (2012) 'Proprietà (nuove forme di)', in *Enciclopedia del diritto. Annali, Vol. 5: Abuso di posizione dominante*, Vertici internazionali, Milan: Giuffré.

——— (2013) 'Protecting the Commons: Water, Culture, and Nature: The Commons Movement in the Italian Struggle against Neoliberal Governance', *South Atlantic Quarterly* 112(2): 366–376.

Mattei, U. and Nader, L. (2008) *Plunder: When the Rule of Law Is Illegal*, Oxford, UK: Blackwell Publishing.

Matthews, D. (2007) 'Role of International NGOs in the Intellectual Property Policy-Making and Norm-Setting Activities of Multilateral Institutions', *Chicago-Kent Law Review* 82(3): 1369–1390.

Maxted, N., et al. (2011) 'Options to Promote Food Security: On-Farm Management and In Situ Conservation of Plant Genetic Resources for Food and Agriculture', *Background Study Paper* 51. Rome, Italy: FAO.

Maxted, N., Ford-Lloyd, B. V. and Hawkes, J. G. (2013) *Plant Genetic Conservation: The in Situ Approach*, London: Kluwer Academic Publishing.

McCarthy, J. (2005) 'Commons as Counterhegemonic Projects', *Capitalism Nature Socialism* 16(1).

McDonald, B. and Matthew, R. A. (2009) 'Food Security in a Global Age: Addressing Challenges from Malnutrition, Food Safety and Environmental Change', Meeting Paper, PSA 2009, Toronto.

McGown, J. (2006) *Out of Africa: Mysteries of Access and Benefit Sharing* (Edited and introduced by B. Burrows), Washington, DC; Richmond: Edmonds Institute in Cooperation with African Centre for Biosafety.

McGuire, S. and Sperling, L. (2016) 'Seed Systems Smallholder Farmers Use', *Food Security* 8(1): 179–195.

Mekouar, A. (2002) 'A Global Instrument on Agrobiodiversity: The International Treaty on Plant Genetic Resources for Food and Agriculture', *FAO Legal Papers Online* 24, Rome, Italy: FAO.

Merges, R. P. (1994) 'Of Property Rules, Coase, and Intellectual Property', *Columbia Law Review* 94(8): 2655–2673.

——— (1996) 'Contracting into Liability Rules: Intellectual Property Rights and Collective Rights Organizations', *California Law Review* 84: 1293–1386.

——— (2001) 'Institutions for Intellectual Property Transactions: The Case of Patent Pools', Expanding the Boundaries of Intellectual Property', in Dreyfuss, R., Zimmerman, D. L. and First, H. (eds.), *Expanding the Boundaries of Intellectual Property Innovation Policy for the Knowledge Society*, Oxford: Oxford University Press.

Metaforum KU Leuven (2015) *Voedselproductie En Voedselzekerheid: De Onvolmaakte Waarheid*, Leuven, Belgium: KU Leuven.

Mgbeoji, I. (2001) 'Patents and Traditional Knowledge of the Uses of Plants: Is a Communal Patent Regime Part of the Solution to the Scourge of Bio Piracy?', *Indiana Journal of Global Legal Studies* 9(1): 163–186.

——— (2003) 'Beyond Rhetoric: State Sovereignty, Common Concern, and the Inapplicability of the Common Heritage Concept to Plant Genetic Resources', *Leiden Journal of International Law* 16(4): 821–837.

Miller, J. B. (1992) 'Women and Power', in Wartenberg, T. E. (ed.), *Rethinking Power*, New York: State University of New York Press.

Miller, N. (2014) 'Novel Open Source Seed Pledge Aims to Keep New Vegetable and Grain Varieties Free for All', *University of Wisconsin: Madison News* (15 April 2014), available at http://news.wisc.edu/22748 [Accessed on 02/10/2015].

Minssen, T. and Nordberg, A. (2015) 'The Impact of "Broccoli II" and "Tomatoes II" on European Patents in Conventional Breeding, GMOs, and Synthetic Biology: The Grand Finale of a Juicy Patents Tale?', *Biotechnology Law Report* 34(3): 81–98.

Moeller, N. I. (2014) 'Summary of User Opinions, Following Interviews with Members of the Seed Industry', *Research Study*, FAO – Plant Treaty Secretariat 7, Rome, Italy.

Moeller, N. I. and Stannard, C. (2013) *Identifying Benefit Flows. Studies on the Potential Monetary and Non-Monetary Benefits Arising from the International Treaty on Plant Genetic Resources for Food and Agriculture*, Rome, Italy: FAO – Treaty Secretariat.

——— (2014) 'Estimating Income to Be Expected from Possible Changes in the Provisions Governing the Functioning of the Multilateral System', *Background Study 1*, Treaty Secretariat, FAO, Rome, Italy.

Mol, A. P. (2001) *Globalization and Environmental Reform: The Ecological Modernization of the Global Economy*, Cambridge, MA: MIT Press.

Mooney, P. (2011) 'International Non-Governmental Organizations: The Hundred Year (or So) Seed War: Seeds, Sovereignty and Civil Society: A Historical Perspective on the Evolution of the Law of the Seed', in Frison, C., López, F. and Esquinas-Alcázar, J. T. (eds.), *Plant Genetic Resources and Food Security: Stakeholder Perspectives on the International Treaty on Plant Genetic Resources for Food and Agriculture*, Oxon: Earthscan by Routledge and Bioversity International, 135–148.

Moore, G. (2013) 'The Third Party Beneficiary', in Halewood, M., Noriega, I. L. and Louafi, S (eds.), *Crop Genetic Resources as a Global Commons: Challenges in International Law and Governance*, Oxon: Earthscan by Routledge and Bioversity International.

Moore, G. and Frison, E. (2011) 'International Research Centres: The Consultative Group on International Agricultural Research and the International Treaty on Plant Genetic Resources for Food and Agriculture', in Frison, C., López, F. and Esquinas-Alcázar, J. T. (eds.), *Plant Genetic Resources and Food Security: Stakeholder Perspectives on the International Treaty on Plant Genetic Resources for Food and Agriculture*, Oxon: Earthscan by Routledge and Bioversity International, 149–162.

Moore, G. and Tymowski, W. (2005) 'Explanatory Guide to the International Treaty on Plant Genetic Resources for Food and Agriculture', IUCN, Gland, Switzerland.

Moore, J. W. (2015) *Capitalism in the Web of Life: Ecology and the Accumulation of Capital*, London, UK: Verso Books.

Morgera, E. (2016) 'Under the Radar: Fair and Equitable Benefit-Sharing and the Human Rights of Indigenous Peoples and Local Communities Related to Natural Resources (20 December 2016)', *BENELEX Working Paper* N. 10 (Rev January 2017), available at SSRN https://ssrn.com/abstract=2887803 or http://dx.doi.org/10.2139/ssrn.2887803 [Accessed on 14/10/2017].

——— (2017) 'Reflections on 2016 UN Biodiversity Conference (Part II): Assessing the Mo'otz Kuxtal Guidelines on Benefit-Sharing from the Use of Traditional Knowledge', available at www.benelexblog.law.ed.ac.uk/2017/03/01/reflections-on-2016-un-biodiversity-conference-part-ii-assessing-the-mootz-kuxtal-guidelines-on-benefit-sharing-from-the-use-of-traditional-knowledge/ [Accessed on 14/10/2017].

Morgera, E., Buck, M and Tsioumani, E. (2012) 'The 2010 Nagoya Protocol on Access and Benefit-sharing', in *Perspective: Implications for International Law and Implementation Challenges*, Leiden, The Netherlands: Martinus Nijhoff Publishers.

Morgera, E., Tsioumani, E. and Buck, M. (2014) *Unraveling the Nagoya Protocol*, Boston, USA: Brill and Nijhoff.

Morley, A., Mcentee, J. and Marsden, T. (2014) 'Food Futures: Framing the Crisis', in Marsden, T. and Morley, A. (eds.), *Sustainable Food Systems: Building a New Paradigm*, Oxon: Routledge.

Morten Haugen, H. (2014) 'The Right to Food, Farmers' Rights and Intellectual Property Rights: Can Competing Law Be Reconciled?', in Lambek, N., Claeys, P., Wong, A. and Brilmayer, L. (eds.), *Rethinking Food Systems: Structural Challenges, New Strategies and the Law*, Dordrecht: Springer, 195–218.

——— (2015) 'Inappropriate Processes and Unbalanced Outcomes: Plant Variety Protection in Africa Goes beyond UPOV 1991 Requirements', *The Journal of World Intellectual Property* 18(5): 196–216.

Moser, P. (2016) 'Patents and Innovation in Economic History', SSRN. Leonard N. Stern School of Business – Department of Economics; National Bureau of Economic Research, New York, NY.

Mozafari Hashjin, J. (2011) 'The Near East Regional Group: Centring the Diversity for Unlocking the Genetic Potential', in Frison, C., López, F. and Esquinas-Alcázar, J. T. (eds.), *Plant Genetic Resources and Food Security: Stakeholder Perspectives on the International Treaty on*

Plant Genetic Resources for Food and Agriculture, Oxon: Earthscan by Routledge and Bioversity International.
Munyi, P. and Jonas, H. (2013) 'Implementing the Nagoya Protocol in Africa: Opportunities and Challenges for African Indigenous Peoples and Local Communities', in Morgera, E., Buck, M. and Tsioumani, E. (eds.), *The 2010 Nagoya Protocol on Access and Benefit-Sharing in Perspective: Implications for International Law and Implementation Challenges*, Leiden; Boston, USA: Martinus Nijhoff.
Mwila, G. (2013) 'From Negotiations to Implementation: Global Review of Achievements, Bottlenecks and Opportunities for the Treaty in General and for the Multilateral System in Particula', in Halewood, M., Noriega, I. L. and Louafi, S. (eds.), *Crop Genetic Resources as a Global Commons: Challenges in International Law and Governance*, Oxon: Earthscan by Routledge and Bioversity International, 226–242.
Nagan, W. P. and Hammer, C. (2004) 'The Changing Character of Sovereignty in International Law and International Relations', *Columbia Journal of Transnational Law* 43(1): 141.
Nahrath, S. (2015) 'Bien Commun', in Bourg, D. and Papaux, A. (eds.), *Dictionnaire de la pensée écologique*, Paris: PUF.
Narasimhan, S. and Robinson, D. (2008) *Towards a Balanced "Sui Generis" Plant Variety Regime: Guidelines to Establish a National PVP Law and an Understanding of TRIPS-Plus Aspects of Plant Rights*, New York, NY: UNDP, United Nations Development Program.
Nasiritousi, N., et al. (2015) 'Normative Arguments for Non-State Actor Participation in International Policymaking Processes: Functionalism, Neocorporatism or Democratic Pluralism?', *European Journal of International Relations* 22(4): 920–943.
Newman, M. (2010) *Networks* (2018 second edition), Oxford, UK: Oxford University Press.
Niada, L. (2006) 'Hunger and International Law: The Far-Reaching Scope of the Human Right to Food', *Connecticut Journal of International Law* 22(1): 131–203.
Oberthür, S. and Stokke, O. S. (2011) *Managing Institutional Complexity: Regime Interplay and Global Environmental Change*, Cambridge, MA: MIT Press.
Oguamanam, C. (2007) 'Agro-Biodiversity and Food Security: Biotechnology and Traditional Agricultural Practices at the Periphery of International Intellectual Property Regime Complex', *Michigan State Law Review* 1: 215–256.
——— (2015) 'Africa's Food Security in a Broken Global Food System: What Role for Plant Breeders' Rights?', *Queen Mary Journal of Intellectual Property* 5(4): 409–429.
Okin, S. M. (1989) *Justice, Gender and the Family*, New York, NY: Basic Books.
Oldham, P. and Hall, S. (2012) 'Intellectual Property, Informatics and Plant Genetic Resources: Study 4', in Stannard, C. and Moeller, N. (eds.) (2013), *Identifying Benefit Flows: Studies on the Potential Monetary and Non-Monetary Benefits Arising from the International Treaty on Plant Genetic for Food and Agriculture*, Rome, Italy: FAO.
Olson, M. (1965) *The Logic of Collective Action: Public Goods and the Theory of Groups*, Cambridge, MA: Harvard University Press.
Onwuekwe, C. B. (2004) 'The Commons Concept and Intellectual Property Rights Regime: Whither Plant Genetic Resources and Traditional Knowledge?', *Pierce Law Review* 2(1): 65–90.
Oosterveer, P. (2005) 'Global Food Governance', in *Mansholt Graduate School of Social Sciences*, Wageningen, The Netherlands: Wageningen Universiteit, 226.
The Open Source Seed Initiative (2018). Available at https://osseeds.org/
Orsi, F. (2013) 'Elinor Ostrom et les faisceaux de droits: l'ouverture d'un nouvel espace pour penser la propriété commune', *Revue de la régulation. Capitalisme, institutions, pouvoirs* 14, Paris, France.
Ost, F., et al. (2015) 'Propriété et biens communs. La propriété et ses limites', Congrès de l'Association Suisse de Philosophie du Droit et de Philosophie Sociale, 26 septembre 2015, Université de Genève, Geneva, Switzerland.

Ostrom, E. (1987) 'Institutional Arrangements for Resolving the Commons Dilemma: Some Contending Approaches', The Question of the Commons: The Culture and Ecology of Communal Resources, 250–265, 46th National Conference of the American Society for Public Administration, 23–27 March 1985, Indianapolis, Indiana, USA.

——— (1990) *Governing the Commons: The Evolution of Institutions for Collective Action*, Cambridge; New York, NY: Cambridge University Press.

——— (1998) 'A Behavioral Approach to the Rational Choice Theory of Collective Action: Presidential Address, American Political Science Association, 1997', *American Political Science Review* 92(1): 1–22.

——— (1999) 'Revisiting the Commons: Local Lessons, Global Challenges', *Sciencemag* 284(278).

——— (ed.) (2005) *Understanding Institutional Diversity*, Princeton, NJ: Princeton University Press.

——— (2007) 'A Diagnostic Approach for Going beyond Panaceas', *Proceedings of the National Academy of Sciences* 104(39): 15181–15187.

——— (2008) 'Institutions and the Environment', *Economic Affairs* 28(3): 24–31.

——— (2008a) 'The Challenge of Common-Pool Resources', *Environment: Science and Policy for Sustainable Development* 50(4): 8–21.

——— (2008b) 'Doing Institutional Analysis: Digging Deeper Than Markets and Hierarchies', *Handbook of New Institutional Economics* Springer: 819–848.

——— (2009a) 'Design Principles of Robust Property-Rights Institutions: What Have We Learned?', in Ingram, K. G., Hong, Y.-H. (eds.), *Property Rights and Land Policies*, Cambridge, MA: Lincoln Institute of Land Policy.

——— (2009b) 'A General Framework for Analyzing Sustainability of Social-Ecological Systems', *Science* 325(5939): 419–422.

——— (2010a) 'The Institutional Analysis and Development Framework and the Commons', *Cornell Law Review* 95: 807–816.

——— (2010b) 'Beyond Markets and States: Polycentric Governance of Complex Economic Systems', *Transnational Corporations Review* 2(2): 1–12.

——— (2011) 'Reflections on Some Unsettled Problems of Irrigation', *American Economic Review* 101(1): 49–63.

——— (2014) 'Collective Action and the Evolution of Social Norms', *Journal of Natural Resources Policy Research* 6(4): 235–252.

Ostrom, E. and Basurto, X. (2011) 'Crafting Analytical Tools to Study Institutional Change', *Journal of Institutional Economics* 7(3): 317–343.

——— (2013) 'Façonner des outils d'analyse pour étudier le changement institutionnel', *Revue de la régulation. Capitalisme, institutions, pouvoirs* 14.

Ostrom, E., et al. (1999) 'Revisiting the Commons: Local Lessons, Global Challenges', *Science* 284(5412): 278–282.

Ostrom, E. and Delville, P. L. (2009) *Pour des systèmes irrigués autogérés et durables: façonner les institutions, Groupe de recherche et d'échanges technologiques Inter-Reseaux Developpement Rural, Synthèse en français de Elinor Ostrom, 1992, "Crafting institutions for self-governing irrigation systems"*, San Francisco: ICS Press, Institute for Contemporary Studies.

Ostrom, E. and Hess, C. (2008) 'Private and Common Property Rights', *Encyclopedia of Law and Economics*, Northampton: Edward Elgar.

Ostrom, E., et al. (2007) 'Going beyond Panaceas', *Proceedings of the National Academy of Sciences* 104(39): 15176–15178.

Ostrom, V. and Ostrom, E. (1977) 'Public Goods and Public Choices', in Savas, E. S. (ed.), *Alternatives for Delivering Public Services: Toward Improved Performance*, Boulder, CO: Westview, 7–49.

Oxfam (2014) 'Working for the Few: Political Capture and Economic Inequality', *178 Oxfam Briefing Paper* (20 January 2014), available at www.oxfam.org/sites/www.oxfam.org/files/bp-working-for-few-political-capture-economic-inequality-200114-en.pdf [Accessed on 03/10/2015].

——— (2016) 'An Economy for the 1%: How Privilege and Power in the Economy Drive Extreme Inequality and How This Can Be Stopped', *210 Oxfam Briefing Paper* (18 January 2016), available at www.oxfam.org/sites/www.oxfam.org/files/file_attachments/bp210-economy-one-percent-tax-havens-180116-en_0.pdf [Accessed on 19/04/2016].

Palacios, X. F. (1997) 'Contribution to the Estimation of Countries' Interdependence in the Area of Plant Genetic Resources', Commission on Genetic Resources for Food and Agriculture, Rome, FAO, Italy.

Pascual, U. and Perrings, C. (2007) 'Developing Incentives and Economic Mechanisms for in Situ Biodiversity Conservation in Agricultural Landscapes', *Agriculture, Ecosystems & Environment* 121(3): 256–268.

——— (2012) 'Developing Mechanisms for in Situ Biodiversity Conservation in Agricultural Landscapes', in Ninan, K. N. (ed.), *Conserving and Valuing Ecosystem Services and Biodiversity: Economic, Institutional and Social Challenges*, London: Earthscan, 151–174.

Pautasso, M., et al. (2013) 'Seed Exchange Networks for Agrobiodiversity Conservation: A Review', *Agronomy for Sustainable Development* 33(1): 151–175.

Pelegrina, W. R. and Salazar, R. (2011) 'Farmers' Communities: A Reflection on the Treaty from Small Farmers' Perspectives', in Frison, C., López, F. and Esquinas-Alcázar, J. T. (eds.), *Plant Genetic Resources and Food Security: Stakeholder Perspectives on the International Treaty on Plant Genetic Resources for Food and Agriculture*, Oxon: Earthscan by Routledge and Bioversity International.

Perronne, R., et al. (2016) 'L'évolution de la filière blé tendre en France entre 1980 et 2006: quelle influence sur la diversité cultivée', *Notes et Etudes Socio-Economiques* (41): 83–113.

Peterson, N. D. and Isenhour, C. (2014) 'Introduction: Moving beyond the "Rational Actor" in Environmental Governance and Conservation', *Conservation and Society* 12(3): 229.

Petit, M., et al. (2001) *Why Governments Can't Make Policy: The Case of Plant Genetic Resources in the International Arena*, Lima: Centro Internacional de la Papa.

Piotrowski, J. (2015) 'Divseek Project Aims to Uncover Crops' Hidden Genetic Data', *sci-DevNet* (12 January 2015).

Pistor, K. and De Schutter, O. (2015) *Governing Access to Essential Resources*, New York: Columbia University Press.

Pistorius, R. (1997) *Scientists, Plants and Politics: A History of the Plant Genetic Resources Movement*, Rome, Italy: IPGRI.

Pleyers, G. (2014) 'Les jeunes alter-activistes: alter-mondialisme, indignés et transition écologique', *Formes contemporaines de l'engagement des jeunes*, Paris: Syllepse.

——— (2017) *Mouvements sociaux: quand le sujet devient acteur*, Paris, France: Éditions de la Maison des sciences de l'homme.

Pleyers, G. and Capitaine, B. (2016) 'Jeunes alteractivistes: d'autres manières de faire de la politique. Perspectives internationales,' *Agora Débats/jeunesses* 73, Paris: INJEP, Les Presses de Sciences Po.

Plucknett, D. L., et al. (1983) 'Crop Germplasm Conservation and Developing Countries', *Science* 220(4593): 163–169.

Posey, D. A. (1999) *Cultural and Spiritual Values of Biodiversity*, Intermediate Technology Publications.

Posey, D. A. and Dutfield, G. (1998) 'Plants, Patents and Traditional Knowledge: Ethical Concerns of Indigenous and Traditional Peoples', in Van Overwalle, G. (ed.), *Patent Law, Ethics and Biotechnology*, Brussels: Bruylant, 112–126.

Poteete, A. R., et al. (2010) *Working Together: Collective Action, the Commons, and Multiple Methods in Practice*, Princeton, NJ: Princeton University Press.

Prieto, I., et al. (2015) 'Complementary Effects of Species and Genetic Diversity on Productivity and Stability of Sown Grasslands', *Nature Plants* 1.

Prifti, V. (2015) *The Breeder's Exception to Patent Rights: Analysis of Compliance with Article 30 of the TRIPS Agreement*, London UK: Springer.

Prip, C. and Fauchald, O. K. (2016) 'Securing Crop Genetic Diversity: Reconciling EU Seed Legislation and Biodiversity Treaties', *Review of European, Comparative & International Environmental Law* 25(3): 363–377.

Putnam, R. D. (1988) 'Diplomacy and Domestic Politics: The Logic of Two-Level Games', *International Organization* 42(3): 427–460.

Quijano, A. (2007) 'Coloniality and Modernity/Rationality', *Cultural Studies* 21(2–3): 168–178.

Quinn, B. J. (2009) 'Failure of Private Ordering and the Financial Crisis of 2008', *The New York University Journal of Law & Business* 5: 549.

Quinn, C. H., et al. (2007) 'Design Principles and Common Pool Resource Management: An Institutional Approach to Evaluating Community Management in Semi-Arid Tanzania', *Journal of Environmental Management* 84(1): 100–113.

Radin, M. J. (1996) *Contested Commodities*, Cambridge, MA: Harvard University Press.

Rai, A. K. (2005) 'Proprietary Rights and Collective Action: The Case of Biotechnology Research with Low Commercial Value', in Maskus, K. and Reichman, J. (eds.), *International Public Goods and Transfer of Technology under Globalized Intellectual Property Regime*, Cambridge: Cambridge University Press.

Ramanna, A. (2003) 'India's Plant Variety and Farmers' Rights Legislation: Potential Impact on Stakeholder Access to Genetic Resources', *Discussion Paper*, International Food Policy Research Institute (IFPRI), Washington, DC.

Rao, K., et al. (2006) 'Manual of Seed Handling in Genebanks', *Handbooks for Genebanks*, Rome: Bioversity International.

Rapoport, A. and Chammah, A. M. (1965) *Prisoner's Dilemma: A Study in Conflict and Cooperation*, Ann Arbor: University of Michigan Press.

Raustiala, K. and Victor, D. G. (2004) 'The Regime Complex for Plant Genetic Resources', *International Organization* 58(2): 277–309.

Ravenhill, J. (1990) 'The North-South Balance of Power', *International Affairs* 66(4): 731–748.

Reed, C. A. (1977) *Origins of Agriculture*, The Hague: Mouton.

Reed, M. S., et al. (2009) 'Who's In and Why? A Typology of Stakeholder Analysis Methods for Natural Resource Management', *Journal of Environmental Management* 90(5): 1933–1949.

Reichman, J. H. and Uhlir, P. F. (2003) 'A Contractually Reconstructed Research Commons for Scientific Data in a Highly Protectionist Intellectual Property Environment', *Law and Contemporary Problems* 66(315): 315–462.

Reider, R. (2015) 'The Obstacles to Food Security Worldwide', *New Zealand Herald*, available at www.nzherald.co.nz/element-magazine/news/article.cfm?c_id=1503340&objectid=11489292 [Accessed on 02/12/2015].

Rifkin, J. (2000) *The Age of Access: How the Shift from Ownership to Access Is Transforming Modern Life*, London: Penguin Books.

Rindos, D. (1984) *The Origins of Agriculture: An Evolutionary Perspective*, Orlando: Academic Press.

Rival, L. (2018) 'An Anthropological Lens on Property and Access Gudeman's Dialectics of Community and Market', in Girard, F. and Frison, C. (eds.). *The Commons, Plant Breeding and Agricultural Research: Challenges for Food Security and Agrobiodiversity*, Abingdon, UK: Routledge.

Roa-Rodríguez, C. and Van Dooren, T. (2008) 'Shifting Common Spaces of Plant Genetic Resources in the International Regulation of Property', *The Journal of World Intellectual Property* 11(3): 176–202.

Rochfeld, J. (2009) 'Entre propriété et accès: la résurgence du commun', *Frontières* 1: 69–87.

Rodale Institute (2011) *The Rodale Institute's 30-Year Farming Systems Trial Report*, Kutztown, PA: Rodale Institute.

Rodotà, S. (2012) *Il Diritto Di Avere Diritti*, Rome, Bari: Laterza.

——— (2013) 'Constituting the Commons in the Context of State, Law and Politics', Keynote at the Economics and the Commons Conference, Heinrich Böll Foundation, Commons Strategies Group, Charles Léopold Mayer Foundation and Remix the Commons, Berlin, Germany, 22–24 May 2013.

Rojahn, J. (2010) *Fair Shares or Biopiracy? Developing Ethical Criteria for the Fair and Equitable Sharing of Benefits from Crop Genetic Resources*, Tübingen, Germany: Universität Tübingen.

Rose, C. (1986) 'The Comedy of the Commons: Custom, Commerce, and Inherently Public Property', *The University of Chicago Law Review* 53(3): 711–781.

Rosell, M. (1997) 'Access to Genetic Resources: A Critical Approach to Decision 391 Common Regime on Access to Genetic Resources' of the Cartagena Agreement', *Review of European, Comparative & International Environmental Law* 6(3): 274–283.

Rosendal, G. K. (2001) 'Impacts of Overlapping International Regimes: The Case of Biodiversity', *Global Governance* 7(1): 95.

Rothstein, R. L. (1984) 'Is the North South Dialogue Worth Saving?', *Third World Quarterly* 6(1): 155–181.

Rousseau, J.-J. (1755). *Discours Sur L'origine Et Les Fondements De L'inégalité Parmi Les Hommes*, Second partie, Amsterdam: Marc Michel Rey. Republié en 2012 sur Presses Électroniques de France.

Ruiz-Muller, M. (2006) 'The Farmers' Rights Project: Background Study 3: Farmers' Rights in Peru: A Case Study', *FNI Report* 5, Fridtjof Nansen Institute, Lysaker, Norway.

Ruiz-Muller, M. and Vernooy, R. (ed.) (2012) *The Custodians of Biodiversity: Sharing Access and Benefits to Genetic Resources*, Oxon; New York, NY: Earthscan.

Runge, C. F. and Defrancesco, E. (2006) 'Exclusion, Inclusion, and Enclosure: Historical Commons and Modern Intellectual Property', *World Development* 34(10): 1713–1727.

Saab, A. (2015) 'International Law and the Pyramid of Assumptions: Do We Need More Food to Tackle Hunger in the Face of Climate Change?', *Sustainable Consumption* 8: 50.

Saez, C. (2016) 'Plant Variety Protection to Meet Food Security Plant Treaty, But Where Are Farmers' Rights?', *Intellectual Property Watch*, webmagazine, available at www.ip-watch.org/2016/04/28/plant-variety-protection-to-meet-food-security-plant-treaty-but-where-are-farmers-rights/

Safrin, S. (2004) 'Hyperownership in a Time of Biotechnological Promise: The International Conflict to Control the Building Blocks of Life', *American Journal of International Law* 98(4): 641–685.

Sagasti, F. and Timmer, V. (2008) 'An Approach to the CGIAR as a Provider of International Public Goods', *System-Wide Review of the CGIAR System: Lima, and Vancouver*: 79.

Sage, C. (2014) 'Food Security, Food Sovereignty and the Special Rapporteur Shaping Food Policy Discourse through Realising the Right to Food', *Dialogues in Human Geography* 4(2): 195–199.

Sanchez, P. A., Denning, G. L. and Nziguheba, G. (2009) 'The African Green Revolution Moves Forward', *Food Security* 1(1): 37–44.

Sánchez Carpio, D. and Becheva, S. (2014) *Agro-Ecology: Building a New Food System for Europe*, Brussels, Belgium: Friends of the Earth–Europe, available at www.google.com/url?sa=t&r

ct=j&q=&esrc=s&source=web&cd=1&ved=2ahUKEwjO3sm81YPdAhUtyoUKHSU7BV0QFjAAegQIABAC&url=https%3A%2F%2Fwww.foeeurope.org%2Fsites%2Fdefault%2Ffiles%2Fnews%2Ffoee_agroecology_local_economies_280314.pdf&usg=AOvVaw1L4RX95eV8dplpExqEQRL6.

Santilli, J. (2012) *Agrobiodiversity and the Law: Regulating Genetic Resources, Food, Security and Cultural Diversity*, Oxon; New York, NY: Routledge, 2016.

Sauer, C. O. (1952) *Agricultural Origins and Dispersals*, New York, NY: American Geographical Society.

Saunders, F. (2014) 'The Promise of Common Pool Resource Theory and the Reality of Commons Projects', *International Journal of the Commons* 8(2): 636–656.

Sauvé, R. and Watts, J. (2003) 'An Analysis of Ipgri's Influence on the International Treaty on Plant Genetic Resources for Food and Agriculture', *Agricultural Systems* 78(2): 307–327.

Scherr, S. J. and McNeely, J. A. (2008) 'Biodiversity Conservation and Agricultural Sustainability: Towards a New Paradigm of "Ecoagriculture" Landscapes', *Philosophical Transactions: Biological Sciences* 363(1491): 477–494.

Schimmelpfenning, D. and Heisey, P. (2010) 'Us Public Agricultural Research: Changes in Funding Sources and Shifts in Emphasis, 1980–2005', in *Economic Information Bulletin*, Washington, DC: USDA, 45.

Schlager, E. and Ostrom, E. (1992) 'Property-Rights Regimes and Natural Resources: A Conceptual Analysis', *Land Economics* 67(3): 249–262.

Schmietow, B. (2012) 'Conceptualizing the Commons: Genetic Resource Management, Property Claims, and Global Innovation Ethic', *St Antony's International Review* 8(1): 70–87.

Schneider, S. and Niederle, P. A. (2010) 'Resistance Strategies and Diversification of Rural Livelihoods: The Construction of Autonomy among Brazilian Family Farmers', *The Journal of Peasant Studies* 37(2): 379–405.

SDSN (2013) 'TG7 Issue Brief: Transformative Changes of Agriculture and Food Systems' by the Thematic Group 7 Sustainable Agriculture and Food Systems', United Nations, available at http://unsdsn.org/resources/publications/sdsn-tg7-issue-brief-transformative-changes-of-agriculture-and-food-systems/ [Accessed on 18/01/2015].

Sedjo, R. A. (1992) 'Property Rights, Genetic Resources, and Biotechnological Change', *Journal of Law and Economics* 35(1): 199–213.

SEED (Hecquet, C. and Stassart, P. M.) (2017) 'Le système semencier en question', *Mille Lieux* 6.

Semences Paysannes (2015) 'Message from the Semences Paysannes (Farmers' Seeds) Networks to Member Governments of the Governing Body of the ITPGRFA' (5–9 October 2015), available at www.foodsovereignty.org/message-from-the-farmers-seeds-networks-to-member-governments-of-the-governing-body-of-the-international-treaty-on-plant-genetic-resources-itpgrfa-meeting-in-rome-5–9-october-2015–2/ [Accessed on 25/01/2016].

Sen, A. K. (1970) *Collective Choice and Social Welfare*, Amsterdam, The Netherlands: Elsevier.

——— (1981) *Poverty and Famines: An Essay on Entitlement and Deprivation*, Oxford: Oxford University Press.

——— (2003) *Ethique et économie*, Paris: PUF.

Serageldin, I. (1999) 'Biotechnology and Food Security in the 21st Century', *Science* 285(5426): 387–389.

SGRP (2003) 'Guidelines and Statements on Genetic Resources, Biotechnology and Intellectual Property Rights – Version II', *Booklet of CGIAR Centre Policy Instruments*, IPGRI, Rome, Italy, 38.

Shands, H. L. (1995) 'The U.S. National Plant Germplasm System', *Canadian Journal of Plant Science* 75(1): 9–15.

Shands, H. L., Castineiras, L. and Hintum, T. van (2008) 'Collective Action for the Rehabilitation of Global Public Goods in the CGIAR Genetic Resources System: Phase 2 (GPG2)', *Final Report*, CGIAR, Montpellier, France.

Shashikant, S. (2015) 'Draft ARIPO Plant Variety Protocol Undermines Farmers' Rights to Save, Exchange and Sell Seeds', *Seed Freedom Press Release*, available at http://seedfreedom.info/draft-aripo-plant-variety-protocol-undermines-farmers-right-to-save-exchange-sell-seeds/ [Accessed on 25/01/2016].

Shashikant, S. and Meienberg, F. (2015) 'International Contradictions on Farmers' Rights: The Interrelations between the International Treaty, Its Article 9 on Farmers' Rights, and Relevant Instruments of UPOV and WIPO', The Bern Declaration and Third World Network, Zurich, Switzerland.

Shaw, J. D. (2007) *World Food Security: A History since 1945*, Basingstoke, UK; New York: Palgrave Macmillan.

Shiva, V. (1997) *Biopiracy: The Plunder of Knowledge and Nature*, Boston: South End Press.

——— (2001) *Protect or Plunder? Understanding Intellectual Property Rights*, London: Zed Books.

——— (2018) 'Victory in the Supreme Court: Supreme Court of India Upholds Article 8j of Our Patent Law Which Says Seeds, Plants, Animals Are Not Inventions, Hence Not Patentable', available at www.ifoam.bio/en/news/2018/05/08/indian-supreme-court-says-seeds-plants-and-animals-are-not-patentable [Accessed on 10/05/2018].

Simanis, E. and Hart, S. (2001) *The Monsanto Company: Quest for Sustainability*, Washington, DC: World Resources Institute, Management Institute for Environment and Business.

Singh, A. P., Manchikanti, P., et al. (2011) 'Sui Generis IPR Laws vis-à-vis Farmers' Rights in Some Asian Countries: Implications under the WTO', *Journal of Intellectual Property Rights* 16: 107–116.

Singleton, B. E. (2017) 'What's Missing from Ostrom? Combining Design Principles with the Theory of Sociocultural Viability', *Environmental Politics* 26(6): 994–1014.

Six, B., et al. (2015) 'Trust and Social Capital in the Design and Evolution of Institutions for Collective Action', *International Journal of the Commons* 9(1): 151–176.

Smith, B. D. (1998) *The Emergence of Agriculture*, New York, NY: New York Scientific American Library.

Smolders, W. (n.d.) 'Commercial Practice in the Use of Plant Genetic Resources for Food and Agriculture: Background Study Paper N°27', *Commission on Genetic Resources for Food and Agriculture*, FAO, Rome, Italy.

SOFI (2014) *The State of Food Insecurity in the World*, Rome, Italy: FAO.

——— (2017) *The State of Food Security and Nutrition in the World*, Rome, Italy: FAO.

Spanò, M. and Quarta, A. (eds.) (2016) *Beni comuni 2.0. Contro-egemonia e nuove istituzioni*, Milan: Mimesis.

Srinivas, K. R. (2008) 'Traditional Knowledge and Intellectual Property Rights: A Note on Issues, Some Solutions and Some Suggestions', *Asian Journal of WTO and International Health Law and Policy* 3: 81.

Srinivasan, C. S. (2012) 'Assessing the Potential for Monetary Payments from the Exchange of Plant Genetic Resources under the Multilateral System of the International Treaty on Plant Genetic Resources for Food and Agriculture. Case Study 3', in Moeller, N. I. and Stannard, C. (eds.), *Identifying Benefit Flows: Studies on the Potential Monetary and Non-Monetary Benefits Arising from the International Treaty on Plant Genetic Resources for Food and Agriculture*, Rome, Italy: FAO-Treaty Secretariat.

Stamets, P. (2005) *Mycelium Running: How Mushrooms Can Help Save the World*, Berkeley: Ten Speed Press.

Stannard, C. (2013) 'The Multilateral System of Access and Benefit Sharing: Could It Have Been Constructed Another Way?', in Halewood, M., Noriega, I. L. and Louafi, S. (eds.), *Crop Genetic Resources as a Global Commons*, Oxon: Earthscan by Routledge and Bioversity International, 243.

Stannard, C., Caracciolo, F. and Hillery, P. (2012) 'Modelling Payments to the Benefit-Sharing Fund, Resulting from the Standard Material Transfer Agreement: Study 2', in Moeller, N. I. and Stannard, C. (eds.), *Identifying Benefit Flows: Studies on the Potential Monetary and Non-Monetary Benefits Arising from the International Treaty on Plant Genetic Resources for Food and Agriculture*, Rome, Italy: FAO-Treaty Secretariat, 2013.

Stannard, C. and Moeller, N. (2013) *Identifying Benefit Flows: Studies on the Potential Monetary and Non-Monetary Benefits Arising from the International Treaty on Plant Genetic for Food and Agriculture*, Rome, Italy: FAO.

Starhawk (1987) *Truth or Dare: Encounters with Power, Authority, and Mystery*, San Francisco, CA: Harper & Row.

Starik, M. (1995) 'Should Trees Have Managerial Standing? Toward Stakeholder Status for Non-Human Nature', *Journal of Business Ethics* 14(3): 207–217.

Starr, J. and Hardy, K. C. (1993) 'Not by Seeds Alone: The Biodiversity Treaty and the Role for Native Agriculture', *Stanford Environmental Law Journal* 12(1): 86–87.

Sterckx, S. and Cockbain, J. (2012) *Exclusions from Patentability: How Far Has the European Patent Office Eroded Boundaries?* Cambridge; New York, NY: Cambridge University Press.

Stern, P. C. (2011) 'Design Principles for Global Commons: Natural Resources and Emerging Technologies', *International Journal of the Commons* 5(2): 213–232..

Sthapit, B., et al. (2008) 'The Value of Plant Genetic Diversity to Resource-Poor Farmers in Nepal and Vietnam', *International Journal of Agricultural Sustainability* 6(2): 148–166.

Stiglitz, J. E. (2008) 'Economic Foundations of Intellectual Property Rights', *Duke Law Journal* 57(6): 1693–1724.

——— (2010) 'Lessons from the Global Financial Crisis of 2008', *Seoul Journal of Economics* 23(3): 321–339.

——— (2014) 'Intellectual Property Rights, the Pool of Knowledge, and Innovation', *NBER Working Paper No. w20014*, Cambridge, MA: National Bureau of Economic Research.

Stock, P. V. and Forney, J. (2014) 'Farmer Autonomy and the Farming Self', *Journal of Rural Studies* 36: 160–171.

Stock, P. V., et al. (2014) 'Neoliberal Natures on the Farm: Farmer Autonomy and Cooperation in Comparative Perspective', *Journal of Rural Studies* 36: 411–422.

Stoll, P.-T. (2009) 'Access to GRs and Benefit Sharing: Underlying Concepts and the Idea of Justice', in Winter, G. and Kamau, E. C. (eds.), *Genetic Resources, Traditional Knowledge and the Law: Solutions for Access and Benefit Sharing*, London: Earthscan, 3–18.

Sumner, A. and Tiwari, M. (2011) 'Global Poverty Reduction to 2015 and Beyond', *Global Policy* 2(2): 138–151.

Tangley, L. (1987) 'Beyond the Green Revolution', *BioScience* 37(3): 176–180.

Tansey, G. (2002) 'Patenting Our Food Future: Intellectual Property Rights and the Global Food System', *Social Policy & Administration* 36(6): 575–592.

Temudo, M. P. (2011) 'Planting Knowledge, Harvesting Agro-Biodiversity: A Case Study of Southern Guinea-Bissau Rice Farming', *Human Ecology* 39(3): 309–321.

Ten Kate, K. and Lasen Diaz, C. (1997) 'The Undertaking Revisited: A Commentary on the Revision of the International Undertaking on Plant Genetic Resources for Food and Agriculture', *RECIEL* 6(3).

Ten Kate, K. and Laird, S. A. (2000) 'Biodiversity and Business: Coming to Terms with the "Grand Bargain"', *International Affairs* 76(2): 241–264.

Ten Kate, K. and Laird, S. A. (2002). The Commercial Use of Biodiversity: Access to Genetic Resources and Benefit-Sharing, London, UK: Earthscan.

Third World Network (Hammond, E.) (2016) 'DivSeek Founder Offers Patent Rights on Climate Change Genes to Syngenta and DuPont in Exchange for US $400,000', *Paper Series on DivSeek # 3* (25 May 2016), available at www.twn.my/title2/intellectual_property/info.service/2016/ip160504/DivSeek%20Paper%203_25May2016.pdf [Accessed on 25/06/2016].

——— (2017) 'Sequence Information: A Pressing Concern for the Seed Treaty' (30 October 2017), available at www.twn.my/title2/biotk/2017/btk171006.htm [Accessed on 25/11/2017].

Thomas, F. (2017) 'TIRPAA (Traité international de la FAO sur les ressources phytogénétiques pour l'alimentation et l'agriculture)', in Cornu, M., Orsi F. and Rochfeld, J. (eds.), *Dictionnaire des biens communs*, Paris: Presses Universitaires de France.

Thrupp, L. A. (2000) 'Linking Agricultural Biodiversity and Food Security: The Valuable Role of Agrobiodiversity for Sustainable Agriculture', *International Affairs* 76(2): 283–297.

Tilman, V. (2016) *Propriété intellectuelle, soutenabilité et (biens) communs: approche philosophique et étude de cas sur l'appropriation de la biodiversité agricole*, PhD thesis, University of Namur, Namur, Belgium.

Toledo, A. and Burlingame, B. (2006) 'Biodiversity and Nutrition: A Common Path toward Global Food Security and Sustainable Development', *Journal of Food Composition and Analysis* 19(6): 477–483.

Treaty Secretariat (2012a) 'Non-Food/Non-Feed Uses of Plant Genetic Resources for Food and Agriculture, Third Meeting of the Ad Hoc Advisory Technical Committee on the Standard Material Transfer Agreement and the Multilateral System', New Delhi, India, 26–28 June 2012, document IT/AC-SMTA-MLS 3/12/2.

——— (2012b) 'Report, Third Meeting of the Ad Hoc Advisory Technical Committee on the Standard Material Transfer Agreement and the Multilateral System', New Delhi, India, 26–28 June 2012, document IT/AC-SMTA-MLS 3/12/2.

——— (2013) *The Funding Strategy of the International Treaty*, Rome, Italy: FAO.

——— (2014) 'Facilitator's Summary: Informal Stakeholder Workshop on Multilateral System of the ITPGRFA', document IT/OWG-EFMLS-2/14/Inf.4.1, Second Meeting of the Ad-Hoc Open-Ended Working Group to Enhance the Functioning of the Multilateral System, Geneva, Switzerland, 9–11 December 2014.

Tripp, R., et al. (2007) 'Plant Variety Protection in Developing Countries. A Report From the Field' *Food Policy* 32: 354–371.

Tsing, A. L. (2015) *The Mushroom at the End of the World: On the Possibility of Life in Capitalist Ruins*, Princeton, NJ: Princeton University Press.

Tsioumani, E. (2005) 'International Treaty on Plant Genetic Resources for Food and Agriculture: Legal and Policy Questions from Adoption to Implementation', *Yearbook of International Environmental Law* 15(1): 119.

——— (2018) 'Beyond Access and Benefit-Sharing: Lessons from the Emergence and Application of the Principle of Fair and Equitable Benefit-Sharing in Agrobiodiversity Governance', in Girard, F. and Frison, C. (eds.), *The Commons, Plant Breeding and Agricultural Research*, Abingdon, UK: Routledge, 67–86.

Tsioumani, E., et al. (2016) 'Following the Open Source Trail Outside the Digital World: Open Source Applications in Agricultural Research and Development', *TripleC: Journal for a Global Sustainable Information Society* 14(1): 145–162.

UN (2004) 'Report by the Special Rapporteur on the Right to Food, Submitted in Accordance with Commission on Human Rights Resolution 2003/25', E/CN.4/2004/10, Ziegler, J., Geneva, Switzerland.

——— (2015a) *Millennium Development Goals Report 2015*, United Nations.

——— (2015b) 'A New Global Partnership: Eradicate Poverty and Transform Economies through Sustainable Development: The Report of the High-Level Panel of Eminent Persons on the Post-2015 Development Agenda', available at www.post2015hlp.org/wp-content/uploads/2013/05/UN-Report.pdf [Accessed on 25/11/2015].

UNDP (2000) *Human Development Report 2000: Human Development and Human Rights*, New York, NY: Oxford University Press.

——— (2003) *A World of Development Experience*, New York, NY: UNDP, United Nations Development Programme.

UNEP-GEF (n.d.) 'Building Biosafety Capacity in Developing Countries: Experiences of the UNEP-GEF Project on Development of National Biosafety Frameworks', *UNEP-GEF Study*, available at www.unep.org/Biosafety/files/UNEPGEFstudyVersion170605.pdf [Accessed on 23/01/2011].

Unknown (2015) 'Growing Access to Phenotype Data', *Nature Genetics* 47: 99.

UPOV (2009) 'Explanatory Notes on Exceptions to the Breeder's Right under the 1991 Act of the UPOV Convention', adopted by the UPOV Council at Its Forty-Third Ordinary Session on 22 October 2009, document UPOV/EXN/EXC/1, Geneva, Switzerland.

——— ((2013) 'Guidance for the Preparation of Laws Based on the 1991 Act of the UPOV Convention', adopted by the Council at Its Forty-Seventh Ordinary Session on 24 October 2013, Geneva, Switzerland, available at http://upov.int/edocs/infdocs/en/upov_inf_6_3.pdf

van den Hurk, A. (2011) 'The Seed Industry: Plant Breeding and the International Treaty on Plant Genetic Resources for Food and Agriculture', in Frison, C., López, F. and Esquinas-Alcázar, J. T. (eds.), *Plant Genetic Resources and Food Security: Stakeholder Perspectives on the International Treaty on Plant Genetic Resources for Food and Agriculture*, Oxon: Earthscan by Routledge and Bioversity International.

van der Kooij, P. (2010) 'Towards a Breeder's Exemption in Patent law?', *European Intellectual Property Review* 32(11): 545–552.

van der Ploeg, J. (2008) *The New Peasantries: Struggles for Autonomy and Sustainability in an Era of Empire and Globalization*, London: Earthscan.

van Ginkel, R. (1989) 'Plunderers into Planters: Zeeland Oystermen and the Enclosure of the Marine Commons', in Boissevain, J. and Verrips, J. (eds.), *Dutch Dilemmas: Anthropologists Look at the Netherlands*, Assen: Van Gorcum, 89–105.

van Laerhoven, F. and Ostrom, E. (2013) 'Traditions et évolutions dans l'étude des communs', *Revue de la régulation. Capitalisme, institutions, pouvoirs* 14.

van Lenteren, J. C., et al. (2011) 'Will the Convention on Biological Diversity Put an End to Biological Control?' *Revista Brasileira De Entomologia* 55(1): 1–5.

Vanloqueren, G. and Baret, P. V. (2009) 'How Agricultural Research Systems Shape a Technological Regime That Develops Genetic Engineering But Locks Out Agroecological Innovations', *Research Policy* 38(6): 971–983.

van Overwalle, G. (1998) 'Patent Protection for Plants: A Comparison of American and European Approaches', *Idea* 39(2): 143–194.

——— (2002) 'Protection of Traditional Knowledge: A Critical Synthesis', in Grosheide, W. and Brinkhof, J. (eds.), *Articles on Cultural Expressions and Indigenous Knowledge*, Antwerp: Intersentia, 251–268.

——— (2005) 'Protecting and Sharing Biodiversity and Traditional Knowledge: Holder and User Tools', *Ecological Economics* 53(4): 585–607.

——— (2008) 'L'intérêt général, le domaine public, les commons et le droit des brevets d'invention', in Buydens, M. and Dusollier, S. (eds.), *L'intérêt général et l''accès à l'information en propriété intellectuelle*, Bruxelles: Bruylant, 149–176.

——— (2009) 'A Man of Flowers: A Reflection on Plant Patents, the Right to Food and Competition Law', in Drexl, J., Hilty, R., Boy, L., Godt, C. and Remiche, B. (eds.), *Technology and Competition: Technologie Et Concurrence: Contributions in Honour of Hanns Ullrich: Mélanges En L'honneur De Hanns Ullrich*, Brussels: Larcier, 311–329.

——— (2010) 'Human Rights' Limitations in Patent law', in Groseheide, W. (ed.), *Intellectual Property and Human Rights: A Paradox*, Cheltenham, UK: Edward Elgar, 236–271.

——— (2013) 'Exclusive Ownership versus Open Commons: The Case of Gene Patents', *The WIPO Journal* 4(2): 139–156.

——— (2018) 'Creating Universal and Sustainable Access to Plants and Seeds: The Role of Clearinghouses, Open Source Licenses, and Inclusive Patents', in Girard, F. and Frison, C. (eds.), *The Commons, Plant Breeding and Agricultural Research*, Abingdon, UK: Routledge, 114–132.

van Zimmeren, E. (2009) 'Clearinghouse Mechanisms in Genetic Diagnostics: Conceptual Framework', in Van Overwalle, G. (ed.), *Gene Patents and Collaborative Licensing Models: Patent Pools, Clearinghouses, Open Source Models and Liability Regimes*, Cambridge: Cambridge University Press.

——— (2011) *Towards a New Patent Paradigm in the Biomedical Sector? Facilitating Access, Open Innovation and Social Responsibility in Patent Law in the Us, Europe and Japan*, PhD thesis, KU Leuven, Leuven, Belgium.

Vavilov, N. I. and Dorofeev, V. F. (1992) *Origin and Geography of Cultivated Plants*, Cambridge, UK; New York, NY: Cambridge University Press.

Verhaegen, E. (2015) 'La forge conceptuelle. Le "commun" comme réinterprétation de la propriété', *Recherches sociologiques et anthropologiques* 46(46–2): 111–131.

Veteto, J. R. and Skarbø, K. (2009) 'Sowing the Seeds: Anthropological Contributions to Agrobiodiversity Studies', *Culture & Agriculture* 31(2): 73–87.

Visser, B. and Borring, J. (2011) 'The European Regional Group: Europe's Role and Positions during the Negotiations and Early Implementation of the International Treaty on Plant Genetic Resources for Food and Agriculture', in Frison, C., López, F. and Esquinas-Alcázar, J. T. (eds.), *Plant Genetic Resources and Food Security: Stakeholder Perspectives on the International Treaty on Plant Genetic Resources for Food and Agriculture*, Oxon: Earthscan by Routledge and Bioversity International, 321.

Visser, B., Eaton, D., Louwaars, N. and Engels, J. (2003) 'Transaction Costs of Germplasm Exchange under Bilateral Agreements', Proceedings of the GFAR-2000 Conference, Dresden.

Visser, B. and Smolders, H. (2003) 'An Analysis of Effectiveness in Plant Genetic Resources Networks', *Commission on Genetic Resources for Food and Agriculture*, Rome, Italy: FAO.

Vivero Pol, J. L. and Schuftan, C. (2016) 'No Right to Food and Nutrition in the SDGs: Mistake or Success?', *BMJ Global Health* 1(1).

Vogler, J. (2012) 'Global Commons Revisited', *Global Policy* 3(1).

Vollan, B. and Ostrom, E. (2010) 'Cooperation and the Commons', *Science* 330(6006): 923–924.

Von Lewinski, S. (2008) *Indigenous Heritage and Intellectual Property: Genetic Resources, Traditional Knowledge, and Folklore*, Alphen aan den Rijn, The Netherlands: Kluwer Law International.

Wade, R. (1988) *Village republics. Economic Conditions for Collective Action in South India.* Cambridge, UK: Cambridge University Press.

Waldron, J. (1988) *The Right to Private Property*, New York, NY; Oxford: Oxfordshire, Clarendon Press.

——— (1994) 'The Advantages and Difficulties of the Humean Theory of Property', *Social Philosophy and Policy* 11(2): 85–123.

Warthmann, N. (2014) 'Plant Genetic Resources and Genomics: Mainstreaming Agricultural Research Through', *Background Study 5*, FAO, Rome, Italy.

Warthmann, N. and Chiarolla, C. (2015) 'Thinking a Global Open Genome Sequence Data Framework for Sustainable Development', *Global Sustainable Development Report 2015 Brief*, of the 2015 session of the United Nations High Level Political Forum (HLPF) on Sustainable Development, New York, USA.

Wasserman, S. and Faust, K. (1994) *Social Network Analysis: Methods and Applications*, Cambridge: Cambridge University Press.

Wattnem, T. (2016) 'Seed Laws, Certification and Standardization: Outlawing Informal Seed Systems in the Global South', *The Journal of Peasant Studies* 43(4): 850–867.

Weinstein, O. (2013) 'Comment comprendre les "communs": Elinor Ostrom, la propriété et la nouvelle économie institutionnelle', *Revue de la régulation. Capitalisme, institutions, pouvoirs* 14.

Weissbrodt, D. and Schoff, K. (2003) 'Human Rights Approach to Intellectual Property Protection: The Genesis and Application of Sub-Commission Resolution 2000/7', *Minnesota Intellectual Property Review* 5: 1–46.

Wendland, W. B. (2002) 'Intellectual Property and the Protection of Cultural Expressions: The Work of the World Intellectual Property Organization (WIPO)', in Grosheide, W. and Brinkhof, J. (eds.), *Articles on Cultural Expressions and Indigenous Knowledge*, Antwerp: Intersentia, 101–138.

Weston, B. H. and Bollier, D. (2013) *Green Governance: Ecological Survival, Human Rights, and the Law of the Commons*, New York, NY: Cambridge University Press.

Wheeler, T. and von Braun, J. (2013) 'Climate Change Impacts on Global Food Security', *Science* 341(6145): 508–513.

Wilkes, G. H. (1988) 'Plant Genetic Resources over Ten Thousand Years: From a Handful of Seed to the Crop Specific Mega Genebank', in Kloppenburg, J. (ed.), *Seeds and Sovereignty: The Use and Control of Plant Genetic Resources*, Chapel Hill, NC: Duke University Press.

Wilkes, H. (1987) 'Plant Genetic Resources: Why Privatize a Public Good?', *BioScience* 37(3): 215–218.

Willetts, P. (2011) *Non-Governmental Organizations in World Politics: The Construction of Global Governance*, Oxon; New York, NY: Routledge.

Wilson, D. S., et al. (2013) 'Generalizing the Core Design Principles for the Efficacy of Groups', *Journal of Economic Behavior & Organization* 90: S21–S32.

Windfuhr, M. and Jonsen, J. (2005) *Food Sovereignty: Towards Democracy in Localized Food Systems*, Heidelberg, Germany: FIAN-International.

Winge, T. (2012) *A Guide to EU Legislation on the Marketing of Seed and Plant Propagating Material in the Context of Agricultural Biodiversity*, Lysaker, Norway: Fridtjof Nansen Institute.

——— (2015) 'Seed Legislation in Europe and Crop Genetic Diversity', in *Sustainable Agriculture Reviews*, Amsterdam, The Netherlands: Springer, 1–64.

Wittman, H., Chappell, M. J., Abson, D. J., Kerr, R. B., Blesh, J., Hanspach, J., Perfecto, I. and Fischer, J. (2017) 'A Social: Ecological Perspective on Harmonizing Food Security and Biodiversity Conservation', *Regional Environmental Change* 17(5): 1291–1301.

World Food Summit Plan of Action (13 November 1996) FAO, Rome, Italy, available at www.fao.org/docrep/003/w3613e/w3613e00.HTM

Young, I. M. (1990) *Justice and the Politics of Difference*, Princeton, NJ: Princeton University Press.

Ziegler, J., Mahon, C., Golay, C. and Way, S.-A. (2011) *The Fight for the Right to Food: Lessons Learned*, Basingstoke, UK; New York: Palgrave Macmillan.

Zürn, M. (2012) 'Global Governance as Multi-Level Governance', in Levi-Faur, D. (ed.), *The Oxford Handbook of Global Governance*, Oxford: Oxford University Press, 730–744.

Index

Note: Page numbers in italics indicate figures and in bold indicate tables on the corresponding pages.

www.ingramcontent.com/pod-product-compliance
Ingram Content Group UK Ltd.
Pitfield, Milton Keynes, MK11 3LW, UK
UKHW010955180926
475701UK00004B/382
* 9 7 8 0 3 6 7 5 8 2 9 7 5 *